HEAT TRANSFER IN FIRES:
thermophysics
social aspects
economic impact

ADVANCES IN THERMAL ENGINEERING

Editors:
JAMES P. HARTNETT
THOMAS F. IRVINE, Jr.

I   Blackshear   • Heat Transfer in Fires:
                        thermophysics
                        social aspects
                        economic impact

II   Afgan and Beer • Heat Transfer in Flames

III   deVries   • Heat and Mass Transfer in the Biosphere:
                        plant growth and productivity
                        soil effects, ecology and pollution

 A publication of the International Centre for Heat and Mass Transfer

Founding Members:

American Institute of Chemical Engineers
American Society of Mechanical Engineers
Canadian Society for Chemical Engineering
Canadian Society for Mechanical Engineering
Institution of Chemical Engineers, London
Institution of Mechanical Engineers, London
National Committee for Heat and Mass Transfer
  of the Academy of Sciences of the U.S.S.R.
Societé Française des Thermiciens
Verein Deutscher Ingenieure
Yugoslav Society of Heat Engineers

Sponsoring Members:

Associazione Termotecnica Italiana
Egyptian Society of Engineers
Indian National Committee for Heat and Mass Transfer
Institution of Engineers of Australia
Israel Institute of Chemical Engineers
Koninklijk Instituut van Ingenieurs, Netherlands
Society of Chemical Engineers of Japan

HEAT TRANSFER IN FIRES:
thermophysics
social aspects
economic impact

PERRY L. BLACKSHEAR, Editor

University of Minnesota

SCRIPTA BOOK COMPANY
a division of hemisphere publishing corporation
1974   Washington, D.C.

A HALSTED PRESS BOOK

JOHN WILEY & SONS
New York   Toronto   London   Sydney

Library of Congress Cataloging in Publication Data

Main entry under title:

Heat Transfer in Fires:
thermophysics
social aspects
economic impact

(Advances in thermal engineering, 1)
Lectures presented at an international school organized by the International Centre for Heat and Mass Transfer.
1. Fire—Congresses. I. Blackshear, Perry L., ed.
II. International Centre for Heat and Mass Transfer.
TP265.H4   1974   621.4'023   74-8824
ISBN 0-470-07780-8

Heat Transfer in Fires:
thermophysics
social aspects
economic impact

Scripta Book Company
a division of hemisphere publishing corporation
1025 Vermont Avenue, N.W.
Washington, D.C. 20005

Printed in the United States of America

CONTENTS

LIST OF CONTRIBUTORS

P. Blackshear, *University of Minnesota, USA*
E. A. Brun, *Membre de l' Academie des Sciences de l' Institut de France, Meudon, France*
R. C. Corlett, *University of Washington, USA*
E. R. G. Eckert, *University of Minnesota, USA*
R. Goulard, *Purdue University, USA*
P. G. Seeger, *Forschungsstelle für Brandschutz Technik, University of Karlsruhe, Germany*
F. R. Steward, *The University of New Brunswick, Canada*
P. H. Thomas, *Fire Research Station, Borehamwood Herts, UK*
F. A. Williams, *The University of California at San Diego, USA*

PREFACE

This volume in five parts represents an attempt to collect under one cover those aspects of fluid mechanics and reaction kinetics, (Part III) and radiation heat transfer theory (Parts IV & V), that hold promise for the development of a scientific basis for fire technology, correlating parameters that serve in predicting geometric effects in large fires (Part II) and the history and impact of fire on society and society's options, (Part I).

The reasons society should care about fire research are clearly innumerated in part I. Fire has played a key role in species selection in general, in the advance of early agriculture in particular and will continue to have an important impact on society so long as life exists. The present cost in the U.S.A. (12,000 deaths, 300,000 serious injuries, 11×10^9 dollars total cost in losses and management) is scandalously high and can and should be reduced.

Some of these reductions can be effected by implementing social change (an alienated public produces arsonists) and supporting education (institutions with good fire prevention programs rarely have fires). Some of the reductions in loss can be effected through appropriate restriction on use and through education.

In many instances, especially when new materials, new arrangements of materials, or new combinations of materials are introduced, fire risks cannot be adequately predicted with our present state of knowledge so that appropriate use restrictions cannot be formulated. Because the pace of change and the variety of novel materials, and arrangements and uses of them is so great the testing of full scale models of all potential hazards is impractical. There is hope that by an understanding of the fundamental processes involved techniques for predicting the behavior of new aggregates can be developed and the losses due to fire can be reduced. This book gives background information on present efforts to obtain this understanding.

The authors of the separate chapters have sometimes employed (and defined) different nomenclature (in keeping with the international flavor of this volume). Special thanks are due them for their efforts in bringing together the relevant aspects of the present state of the art in fire research.

The editor would like to express his gratitude to the International Centre for Heat and Mass Transfer for organizing the school that led to this publication and for convening the school at such an idyllic site, Trogir, Yugoslavia.

Finally, a special word of thanks to Professor James Hartnett whose gift for social invention continues to make the world of heat and mass transfer an agreeable world.

Perry L. Blackshear, Jr.
Professor, Mechanical Engineering
University of Minnesota
Minneapolis, Minnesota
June, 1974

HEAT TRANSFER IN FIRES:
thermophysics
social aspects
economic impact

PART I

SOCIAL AND ECONOMIC ASPECTS OF FIRE

Chapter 1

THE FIRE PROBLEM IN THE UNITED STATES

A Report on the President's Commission on Fire Prevention and Control

E. R. G. ECKERT
University of Minnesota, USA

I have been asked to give a report on the work of the Commission on Fire Prevention and Control, of which I was a member. The Commission had the charge to study the fire problem in the United States and to make recommendations on actions which the federal government should take to reduce the hazards and losses by fire. In my discussion, I plan to concentrate on those phases of the Committee study which are connected with research assuming that the attendants of this summer school are mainly interested in these subjects.

THE FIRE PROBLEM

In order to grasp the size of the fire problem, it is necessary to refer to some statistics. The following data are for the United States. 12,000 deaths through fire occur in the average each year. This makes fire the second largest killer surpassed by deaths through vehicle accidents only. An impressive comparison is also presented in Fig. 1, in which the deaths in the U.S. by fires during the years 1961 through 1972 are compared with the deaths of U.S. military personnel during the Vietnam War. Approximately 300,000 persons are injured annually by fire. The injuries of every sixth person are such that they require hospitilization for periods from six weeks to two years in special burn centers with numerous operations and skin grafting. Many of the injured suffer permanent deformations on the body and on the face and undergo severe psychologic disturbances.

NOTE: All figures quoted in text are at the end of the chapter.

Many are not able to resume fully their previous occupations. It is especially sad that large numbers of children are in this group.

The loss of property by fire is also heavy, but can be relegated to second place compared with loss of life and injuries. The value of the direct losses of property is reported as 27 billion dollars per year and the total cost to society, including maintenance of fire departments, insurance, loss of productive time, is estimated as 11.4 billion dollars. The magnitude of the fire problem revealed by these figures caused the 90th Congress of the United States to establish in 1968 the Public Law 90-259. It consists of two parts. Title I assigns the responsibility for a fire research and safety program to the National Bureau of Standards, Title II created a National Commission on Fire Prevention and Control. In the following year, President Nixon appointed 20 members to this commission selected from the fire services, from testing laboratories, manufacturers associations, insurance companies, government organizations, burn centers and others. Two congressmen and two senators acted as advisory members. The Commission held 10 hearings to study the fire problem and finished its work in June 1973 with a report AMERICA BURNING submitted to the President in which it discusses the fire problem and makes recommendations for actions by the federal government.

The Commission concluded from the statistics mentioned above that first priority in their efforts should be the reduction of loss of life. This decision leads to Fig. 2, from which one can see that deaths occur overwhelmingly in residential buildings. It was found additionally that one or two-family houses were mainly involved in such fires. This indicates that a special effort of reducing fires should be directed toward such buildings. One has, however, also to consider that in recent years there was a definite trend within the population of the United States to move into apartment buildings and highrise buildings. One can expect that this trend will continue and that more and more people will live in these buildings. There were no fires in highrise buildings in the United States which involved heavy losses of life. However, such fires occurred in other countries - for example, in Seoul, Korea and in Sao Paulo, Brazil. From the statements above, one concludes that losses of life can most effectively be reduced by concentrating on residential buildings, one-family houses, apartment houses and highrise buildings.

Statistics also show that approximately 80% of multiple death fires occur between 11:00 P.M. and 6:00 A.M. and that more people are killed by smoke, toxic gases - which are developed through incomplete combustion, and through lack of oxygen, than through the flames themselves. Many persons die in their sleep. This means that one has to study fires in their early stages at a time when they are confined to the rooms of the buildings, whereas much research in past years, has concentrated on open "unconfined" fires. One also has to know much more about the development of smoke and toxic gases in fires of various kinds. Efforts to make materials fire retardant, for instance, can actually increase the life hazard since the treated material may produce heavier smoke and more toxic gases than the untreated product.

HAZARDS THROUGH MATERIAL

The main causes of fires are: heating and cooking devices, smoking and matches, and faulty electrical wiring. This can be seen in Table 1, which has been prepared by the National Fire Protection Association.

Table 1.   Estimated U.S. Building Fire Causes*

	Percent of fires	Percent of dollar losses
Heating and cooking.........	16	8
Smoking and matches.......	12	4
Electrical....................	16	12
Rubbish, ignition source unknown...................	3	1
Flammable liquid fires and explosion...................	7	3
Open flames and sparks.....	7	4
Lightning....................	2	2
Children and matches.......	7	3
Exposures...................	2	2
Incendiary, suspicious.......	7	10
Spontaneous ignition........	2	1
Miscellaneous known causes...................	2	6
Unknown...................	17	44
Total...................	100	100

*NFPA estimates.

Prevention of fires obviously depends to a considerable degree on the care which people exercise in their homes. The importance of education on fire hazards has, therefore, been stressed in the report of the Fire Commission. The hazards, however, depend also on the inflammability of the materials with which we surround ourselves. Many new synthetic products which have been recently introduced are inflammable and the consumer can not be expected to know their characteristics. An example of the destruction which results from insufficient knowledge or appreciation of the inflammability is the fire which occurred in the new west wing of the British Overseas Airways Corporation Building at Kennedy International Airport. It caught fire on August 26, 1970. Swiftly flames moved from one seat to the next along the 110 meter length of the wing. Gases from the incomplete combustion of the plastic material and the foam rubber of the seats gathered in clouds along the ceiling. When flames approached the clouds, the gases ignited explosively and

knocked out the huge glass windows of the terminal. As
the ceiling melted, combustible liquid dropped to the
floor further spreading the fire. Fortunately, this fire
occurred before the terminal was opened to the public.
However, damages totaled 2 million dollars. The United
States, as many other countries, has codes and standards
for classification of materials with respect to fire
resistance issued by organizations as the American
Society for Testing and Materials and the National Fire
Protection Association. Underwriters' Laboratories,
Factory Mutual Research Corporation, and other organizations test materials to see that they comply with such
standards. A difficulty, however, is caused by the fact
that the fire resistance of building materials depends
on their surroundings and on the way in which they are
incorporated in a structure. As a consequence, standard
tests give only limited information. They also have to
be designed with some arbitrariness as long as the dynamics of ignition and combustion are not basically
understood. This is vividly demonstrated by Fig. 3,
taken from a paper by H. W. Emmons (Heat Transfer in
Fire, Journal of Heat Transfer, vol. 95 (1973) pp. 145-
151). The figure contains the results of tests on flam-
ability of 24 wall materials obtained by 6 different
national standard tests. The average rating of the materials is used as abscissa and the individual ratings
obtained through the various tests are plotted as ordinate. The scatter from a 45 degree line is evidence of
our ignorance. It can, for instance, be observed that
material 18 is rated best in Germany and worst in
Denmark. The scatter is not much worse if one selects
randomly numbers between 0 and 24 and plots them in the
same way.

HAZARDS THROUGH DESIGN

It has been noted that the behavior of materials in
a fire depends on its surroundings. Tests have been
carried out to study this effect. Figure 4, shows for
instance, a drawing of the setup used in corridor tests
at the National Bureau of Standards. Wood cribs are
ignited in a room connected with the corridor and the
spread of the fire is studied by the instrumentation indicated in the figure and by observation during the short
time when the fire moves along the corridor toward the
window in the east wall. Various materials can be used
as floor, walls, or ceilings and the test provides a
means to judge the behavior of these materials. I had
the opportunity to watch such a test when the floor was
covered with a synthetic carpet, whereas the walls and
the ceilings were made of gypsum boards. The fire moved
within a few minutes down the corridor towards the east
end, which at these tests was completely open. Figure 5
presents three sketches indicating the movement of smoke

and fire down the corridor. The figure shows that at
first there was a clear separation between the smoke
moving along the upper part toward the exit and the air
coming from the open end of the corridor and moving in
the lower part of the corridor toward the closed end.
The thickness of the smoke layer gradually increased.
During this period a person could still have crawled
along the bottom of the corridor. This was no longer
possible towards the end of the burn test when vortices
began to mix the air with the smoke as indicated in the
third sketch.

Such tests provide valuable information, but much
more should be done to study the movement of smoke and
fire within a room, into adjoining rooms, and to the
other floors of a building. Useful information can be
obtained by experiments on small scale models, especially
when these experiments are supported by a number of large
scale tests and when the important similarity parameters
are established in this way. Results obtained from small
scale models can then be transferred to actual buildings.
Analytical methods can be developed with the goal to predict the movement and accumulation of smoke and fire in
new designs. In this connection it is useful to keep in
mind that the early stages of a fire are of special importance for the preservation of life, and that the main
killer is smoke which moves into occupied rooms well
ahead of the flames. At this stage, the spread of smoke
is primarily a natural convection problem. Questions,
how the operation of a warm air heating system influences
the movement of smoke in a home or whether an opening for
smoke through the roof of a building should be provided
or whether such an opening only increases the temperature
by fanning the fire, can then be answered.

A prediction of the spread of smoke and flames is
especially important for highrise buildings where often
insufficient consideration has been given to the fire
hazard. In the afternoon of August 5, 1970, for instance,
fire broke out on the 33rd floor of such a building in
Manhattan. The air conditioning system spread smoke
throughout the building. Smoke and hot gases shot upward
through the gaps between floor slabs and exterior walls.
An elevator was automatically summoned to the 33rd floor
because the products of combustion activated the call
button. The elevator jammed there, and two people died.

The development of new construction methods and new
designs is such that building codes can not keep up with
them, and new approaches to the fire safety have to be
considered which devise an organized plan to be initiated
in the case of fire. Such a systems approach was, for
instance, taken in the design of San Francisco's Trans-
america Building. In addition to a full sprinkler

system, smoke detection devices, and a central alarm system, the designers provided the building with emergency refuge areas, two-way voice communications with public areas, and an underground communications and command control center. Windows pivot so that burning rooms can be vented. In the event of a power failure, diesel pumps maintain water pressure and a diesel powered generator lights exitways and powers the elevators. There is also an emergency water supply should the city mains be disrupted. The building inspectors agreed to substitute such a systems approach for the compliance with various of the normal building codes.

Fire safety of the building construction is, of course, only part of the story, because the occupants fill the rooms with carpets, curtains, draperies, wall furnishings, upholstered furniture, beds, clothing, and similar objects. Many of these furnishings are flammable and only education of the population, labeling of the fire resistance and prohibiting of highly flammable objects can help in this situation.

To understand the spreading of flames in the latter stages of a fire requires research and investigations into the basic processes which determine the rate of spreading through diffusion of radicals created in the combustion processes, through heat conduction, convection, and radiation. The interaction of these processes in a confined space are still poorly understood and much research on ignition, pyrolysis, and combustion will have to be carried out before we are able to conclusively predict the path and rate of spreading of fire. Especially treacherous is the process of flash-over where objects distant from the fire are ignited through radiation. In the latter stages of a fire the construction elements of the house - like wood, insulating material, roofing, become involved whereas the early spread involves only carpets, draperies, furnishings, and coatings of walls and ceilings.

DETECTION AND EXTINGUISHMENT

It has become obvious that early detection and warning in case of a fire is of paramount importance. Devices responsive to heat, smoke, or ionization are on the market for this purpose. Some of them are sufficiently low priced so that they can be used in single-family homes. Statistics show that devices responsive to smoke are more effective than those responsive to heat. The Ohio State Building Code prescribes, therefore, smoke detectors for modular dwelling units. It would be ideal if the detection device would also report the fire automatically to the fire services. Efforts to provide inexpensive components for the transfer of messages by ultra-high frequency waves may offer a solution.

Sprinklers have proved effective for automatic extinguishment of fires, especially in larger buildings, and are widely used. Still, the processes occurring are not well understood so that optimum droplet size and the amount of water required to extinguish a certain type fire can not be predicted from basic knowledge. Improvements are possible through additives (gels) which make the water adhere better to the surface of combustible material and can in this way prevent or delay ignition. The idea and the name for this product "ablating water" is a spin-off from the space effort, where ablating materials have been developed to protect vehicles reentering through the atmosphere to earth from extensive heating. Other extinguishing agents like bubble-foam and powders are available. All of these products and the means of their delivery can certainly be improved through research.

Additatives that reduce friction losses in hoses have proven their effectiveness but are not widely used. A drag reduction in a certain Reynolds number range is caused by polymer addition. In this way the diameter of the hoses transporting the water can be reduced.

FIRE SERVICE

The fire service is the most hazardous occupation and deserves every protection which can be devised. The customary clothing, the turn-out coat, can certainly be improved, utilizing experience gathered through the space effort. Tests carried out at Boston have shown that some of these coats even fail the flammability test for drapery fabrics used in places of public assembly. Most fire fighters' helmets conduct heat readily to their inside. Breathing apparatus is heavy and of limited duration. Typically it weighs 15 kg and is designed for 30 minutes use. It is also bulky and makes the movement of the fire fighter in buildings difficult. Lights, periscopes or closed circuit television might be provided on top of the aerial ladders. Sensors to locate trapped victims through smoke could be developed.

It is important that each unit of the fire service develops and continuously updates a plan for fire protection and a strategy of fire fighting. This should include prevention by periodic inspection of buildings, emergency medical services, as well as response to fire alarms. Questions like the optimum number and locations of fire stations, the number of fire fighters on duty during the various hours of the day, the optimum size of vehicles and others should be clarified. New ways of fighting fires have continuously to be studied. The helicopter, for instance, has proven its value in various fires.

CONCLUSIONS

The Fire Commission pointed out that increased research in all areas mentioned in this paper is required for an improvement of fire safety. Strangely, the faculty of universities show, with some notable exceptions, little interest in fire research. One of the most important recommendations of the Commission is the creation of a National Fire Academy to educate and to improve communication within the fire service which is, in the United States as in other countries, organized on a local basis. The Academy should also have the task to bring the results of research to the fire services, to building inspectors, architects, and to the general public in a form which can be readily understood and utilized.

It should be possible by the actions mentioned in this paper to decrease drastically losses of life and property through fires.

DEATHS – U.S. FIRES vs. VIETNAM WAR

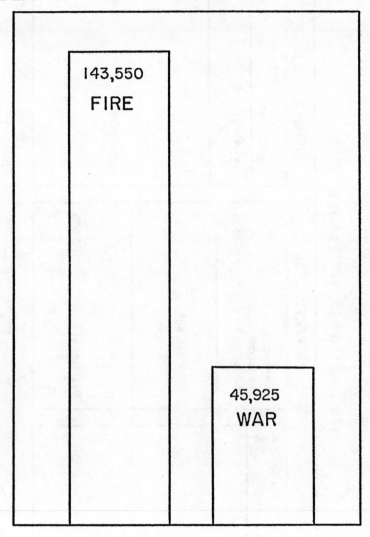

FIGURE 1

NOTE: Figure captions for this chapter are listed on page 16.

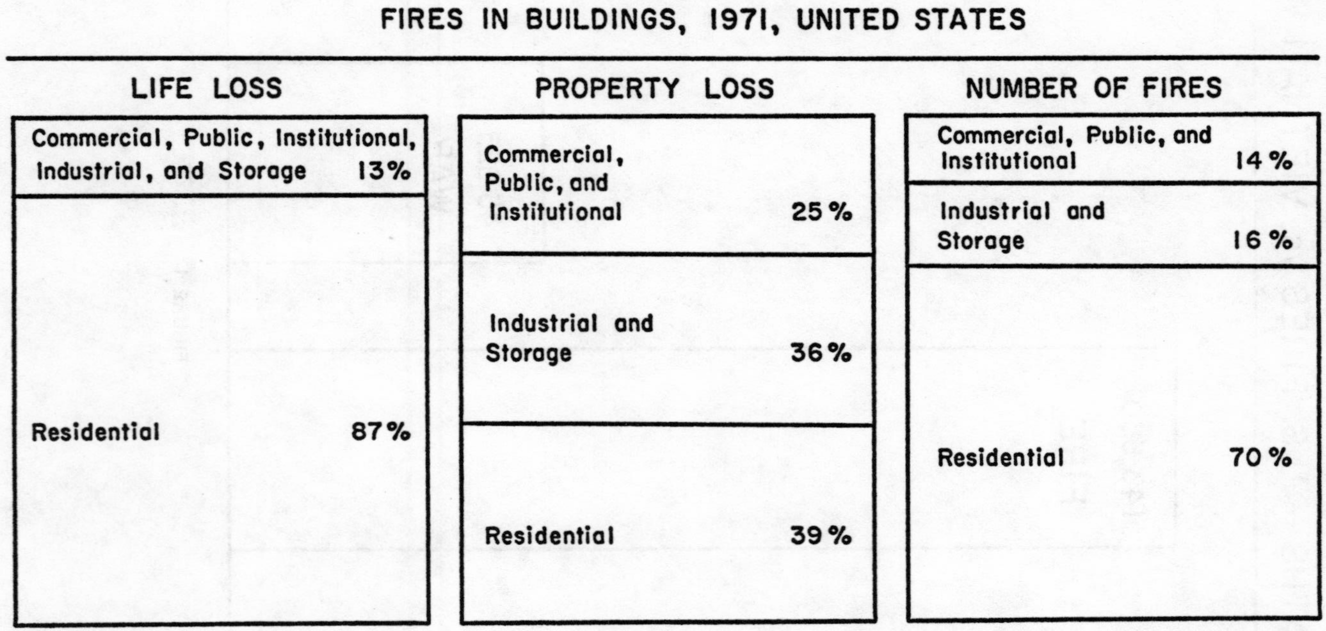

FIRES IN BUILDINGS, 1971, UNITED STATES

LIFE LOSS

Commercial, Public, Institutional, Industrial, and Storage 13%

Residential 87%

PROPERTY LOSS

Commercial, Public, and Institutional 25%

Industrial and Storage 36%

Residential 39%

NUMBER OF FIRES

Commercial, Public, and Institutional 14%

Industrial and Storage 16%

Residential 70%

FIGURE 2

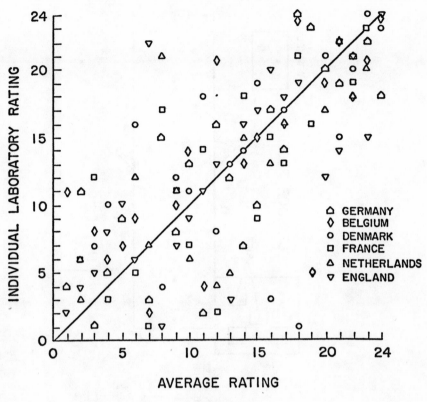

FIGURE 3

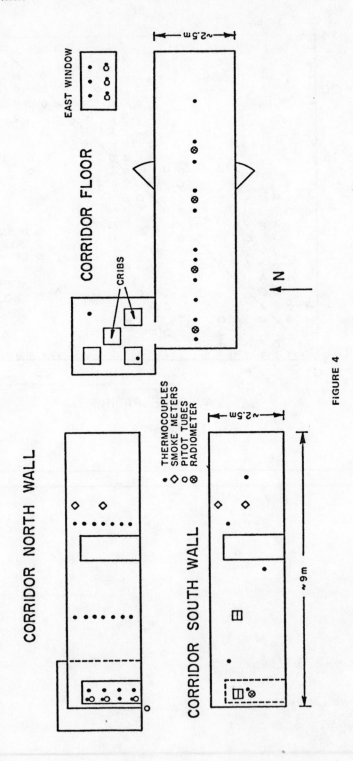

FIGURE 4

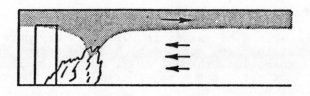

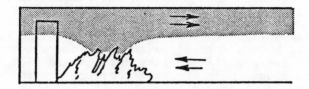

FIGURE 5

FIGURE CAPTIONS

Fig. 1   Comparison of deaths in U.S. military personnel
          resulting from actions by hostile forces in
          Vietnam, 1961 through 1972, and deaths from
          U.S. fires for the same period [from the report
          AMERICA BURNING].

Fig. 2   Fires in buildings within the United States
          during 1971 [from the report AMERICA BURNING].

Fig. 3   Flammability ratings of 24 wall covering mater-
          ials by 6 different national standard tests.
          The average flammability ratings are plotted on
          the abscissa in increasing order, the indivi-
          dual laboratory ratings are plotted on a verti-
          cal above the average rating [from H. W. Emmons
          "Heat Transfer in Fire", ASME preprint 72-WA/
          HT-63].

Fig. 4   Corridor test facility at the National Bureau
          of Standards [from F.C.W Fung, M. R. Suchomel
          and P. L. Oglesby "The NBS Program on Corridor
          Fires"] Fire Journal (1973), pp 41-48.

Fig. 5   Sketch of the smoke and flame movement in a
          corridor sketched at three consecutive times.

Chapter 2

THE FOREST FIRE PROBLEM[1]

E. A. BRUN
Membre de l'Academie des Sciences de l'Institut de France, Meudon, France

1. CONTROLLED FOREST FIRES

1.1 The Paleolithic and Neolithic man set forests on fire to clear land. At the time of the Roman conquest, Gaul and Germany were still completely forested. Since then, this original landscape has been gradually transformed with the aid of fire so as to develop land suitable for pasture and agriculture. This is done even today and even in developed countries. For example, in the United Kingdom Scottish heaths are periodically set on fire in order to clear land for sheep farming or grouse hunting.

Does this mean that land firing can sometimes be useful?

The problem occurs in the case of African savannas south of the Sahara and the prairies and woods of western and southern United States, where one wishes to prevent, through fire, brush encroachment of pasture land, particularly in areas where the land is not sufficiently valuable to employ mechanical or chemical clearing methods.

1.2 As far as the United States are concerned, many authors state that periodic firing of brush prairie, e.g., every two or three or five years, improves grazing conditions. In this case, firing vegetative waste such as pine needles, which inhibits grass growth, and may, over the long range, promote propagation of wild fires.

[1]Presented at the International Heat- and Mass-Transfer Center's summer conference on heat- and mass-transfer in fires, August 1973.

NOTE: All figures quoted in text are at the end of the chapter.

In the thirteen Southern States, more than a million hectares are fire-treated every year. A favorable day for such a fire treatment is one with low winds, generally at the beginning of spring, right after the first rains. Under these conditions, the surface temperature of the soil under the flame never exceeds 150° C, and the temperature of the soil one centimeter below the surface is only raised by a few degrees.

Such a fire delays the start of spring growth by one to four weeks. But the grass growth picks up right after that, its annual total yield is normal, and then increases in the next years. This is because, in spring the surface temperature of the burned-off tract is higher than that of the unburned one (Fig. 1),[2] since the newly-created surface absorbs more radiation than dry grass, and photosynthesis is thereby accelerated. During the night the burned surface is emitting more than the unburned one. Its temperature is thus reduced, which in turn decreases respiration.

Figure 2 shows a result of one of the many experiments involving meadows close to woods, where pine seeds must be suppressed to prevent loss of pasture. Photographs on the left were taken before the burn off, those on the right one year after (about 80 percent of the pine seeds have been burned off).

1.3 Bush fires are often started in South Africa, as indicated in this native song:

A drunkard drinks on a tree,
If he falls on the bare ground, he gets a bump,
If he fall on the grass, he falls softly,
Thanks to the grass and to the fire which raised it.

Nevertheless, the advisability of periodic controlled burnoff even for purposes of maintaining the pasture land is still under discussion. It seems sure that the gramineae of the Veld, a region of high plateaux, are maintained by such fires which prevent old but undecomposed vegetable matter from accumulating. But even in this case firing should be used with care, at time intervals of not less than two years, and after the first rains. Then meadows again become green right after the burning.

In West Africa, where forest dwellers are much more numerous than farmers, fires are not wanted as the natural cover of woods is regenerating the humus layer and should therefore be conserved.

However dangerous controlled fires might be for ecology, other dangers seem worse to many people, e.g., the growth of the elephant population in some South African savannas, where they eat the trees, and the building of large highways in the Amazonian forest, which is one of the world's largest producers of oxygen.

1.4 Of course, large controlled fires in Mediterranean region are out of question, especially during the dry season. I think, however, that some closely supervised little fires could often spare us from the wild fires that do occur. But this is my own opinion.

[2]Cited after Roger C. Anderson (Curtis Prairie).

2.   MEDITERRANEAN FOREST FIRES

2.1 The future of our Mediterranean forests is closely related to the problem of wild fires.

Since 1956, an average of 25,000 hectares of forests have been consumed by fire in the Provence-Riviera and Corsica region per year. But year-to-year differences are large, so that in 1970, a particularly disastrous year, 73 000 hectares were destroyed. Although these two regions represent only four out of the 95 "departements" (counties) of France, they account for two thirds of the burned forest areas [1].

In Italy, 2,320 wild fires consumed 23 500 hectares of forests in 1965. There too, the Mediterranean region was the most affected (more than 5,000 hectares in Liguria, and more than 4,000 hectares in Sardinia [2].

The Mediterranean slopes denuded by wild fires are subject to really extensive erosion. Moreover, the summer weather in the so-denuded areas becomes drier because of extermination of the brush, with disastrous effects on the surrounding cultivated land.

To these disastrous ecological consequences of forest fires we need to add the economic losses they cause. There is loss of ligneous material (the 1965 losses in Italy were estimated at 900 millions lire). Also, the physical appearance of an area that depends on tourism is ruined, and the social cost is thus large.

Given all this, what can be done ?

2.2 First, it is evident that starting of fires must be avoided. Here, we must distinguish between unintentional and deliberately started fires.

Nearly all unintentional fires are due to a lack of either caution or discipline on the part of individuals. In France, it is forbidden to throw a burning cigarette butt from a moving vehicle. But infractions being still numerous, the road sides should be kept cleared of brush. Unfortunately, this is expensive.

Schwab said that one million matches can be made from a tree, but one match is enough to destroy a million trees. People who smoke in forests and, what is even worse, those who light small fires for a picnic, should keep these words in mind. The danger here comes from the dry grass, the pine needles, and the dead wood covering the ground especially since loggers now leave pine branches on the site where they stay for years, whereas in the old days these branches were used for firewood. The cleaning of the underbrush is also expensive.

Firing of the field stubble after the harvest, and burning of household refuse cause some disasters. Among other unintentional causes of fires are also those due to natural phenomena, mainly lightning.

A few deliberately started fires are due to pyromaniacs or to playing children. But most of such fires were in the past due to

- arson directed against landowners who made their land off limits to hunters, and

- arson for swindling purposes, that is, fires started by ranchers who wanted new meadows for their cattle. This last reason disappeared in France, since using land for pasture in the ten years that follow a fire is now forbidden by law.

2.3 If the fire, whether intentional or not, propagates — and we shall examine the conditions for this propagation below — its presence must be detected as soon as possible in order to start a defense.

At least in Provence, there are fire lookouts, fire towers on hills and, in the most threatened regions, fire-scouting helicopters that alert patrol cars, which then move quickly to the affected area.

For continuous watches, automatic detectors are obviously more practical. I will briefly describe an infrared detector now being developed [3]. Radiation from the fire is detected by a thermistor bolometer located at the focus of a Cassegrain telescope (Fig. 3). Under these conditions, the spectral sensitivity is unlimited, and the apparatus may be used in the visible as well as the infrared. But experiments showed that in order to detect specifically the natural radiation from fires, filters should be placed before the receptor, to filter out all but the 8 to 12 μm band, in which emission from hot solid particles is high.

The apparatus, set on a rotating platform, is scanning the entire horizon; the angle of sight is also adjustable (Fig. 4). The first experiments gave satisfactory results, in that starting brushwood fires about twenty kilometers away were discovered.

2.4 After the fire has been detected, the damage must be limited.

To contain it on the ground, access roads must be built beforehand to allow men and equipment to penetrate the woods. So far, this has been done for 75 % of French forests. The necessary network of roads is about 1 km for 100 hectares of forests. This is a very large network, as the lanes must then be kept in good condition and forbidden to the public.

If the region is not too dry, so that the water can be used to douse the fire, this network of roads must be supplemented with either tanks or lakes along the access lanes. Otherwise, the only direct action possible involves shoveling, chemical spraying, or setting up counterfires.

If the access to the fire is either difficult or dangerous, which is true in about 25 ii of the cases, the only way left is to pout water from a water-carrier seaplane (the Canadian-built CL125 aircraft can carry 4.5 metric tons). The cost is ii 100 per metric ton of water dropped from aircraft (the cost per a metric ton of water delivered from the ground is about five times lower [4]).

The fight against fires requires the development of a real strategy in the distribution and use of the available means. Participants in this fight must first be trained. I will not speak about this training since we are not concerned with turning out firemen.

Finally, French statistics show that if the source of a fire is quickly detected and firefighting measures are implemented the fire should not destroy an area larger than 5 ha in 95 % of cases (Corsica is not included in these statistics).

2.5 The spread of fires can be limited, independently of fighting them, by preventive means.

First, fire breaks must be created (we will talk again about this later on). These fire breaks, clear of woods, should produce a pattern breaking up the forest into tracts of 4 000 to 5 000 hectares, generally rectangular, with the short sides parallel to the prevailing wind. The best fire breaks are cultivated areas, preferably irrigated in the forest regions.

Second, a logical reforestation program should be developed. The resinous trees should not be suppressed, since they grow surely and quickly and hold down the soil. But a larger fraction of the newly-planted trees must be given over to hardwood trees like hollies, pubescent oaks and even beeches, which in the past constituted the bulk of the Mediterranean forests. In terms of what has just been said, we can map regions of high, middle and low vulnerability, depending upon the tree concentration in the forests, the risk of fire, and the efficiency of the preventive measures.

2.6 In conclusion to this part, I would like to emphasize that one must inform and educate the public, and if people do not wish to be reasonable and cooperate, they must be prosecuted and punished. The forest areas not yet mutilated by fire must be protected, and sites already destroyed must be rebuilt.

3. FIRE PROPAGATION

3.1 The mechanism of propagation of forest fires is so complex that it is desirable to point out its main features, even at the risk of repetition, in order to gain a perspective for a qualitative approach to the phenomenon as a whole.

Let us assume a steady, one-dimensional propagation of the fire in a fuel bed of constant height h. The fire front velocity u is measured along a direction referred to as the x-axis (Fig. 5a). We will assume that the front is a plane perpendicular to the x-axis, and that its height h in the combustion zone is constant. This is merely an approximation, as shown on Fig. 5b, but it helps to derive a model.

In accordance with the very schematic theory of Emmons [5], let us assume that the fuel is preheated only by the heat flowing along the x-axis and through the flame front, and moreover let us assume that this heat flow is due only to the radiation from the burning fuel and embers. The heat flow rate per unit width of the front is then

$$\Phi_o = \Phi_b \left(1 - e^{-\alpha c}\right) \tag{1}$$

where Φ_b is the heat flow generated per unit time by a burning bed whose length c is infinite, and where α, inversely proportional to length, depends specifically on the mass M of fuel burned per unit area of forest floor (here, per unit length, as we take a front of unit width).

3.2 In the steady state, the fire front velocity being u, a mass Mu is heated per unit time from the initial temperature T_o to the ignition temperature T_i. This mass Mu receives per unit time a heat quantity $MuC_p(T_i - T_o)$, equal to $\Phi_o = \Phi_b \left(1 - e^{-\alpha c}\right)$. Thus:

$$Mu\,c_p\,(T_i - T_o) = \Phi_b\,(1 - e^{-\alpha c}) \qquad (2)$$

If w is the average burning rate per unit area of forest floor (here per unit length) and per unit time,

$$Mu = cw \qquad (3)$$

where

$$c = \frac{Mu}{w}$$

Let us introduce w and α in Eq. (2). We obtain

$$\frac{w\,c_p(T_i - T_o)}{\alpha\,\Phi_b}\,\frac{\alpha\,Mu}{w} = 1 - e^{-\alpha Mu/w} \qquad (4)$$

The dimensionless fire parameter P

$$P = \frac{w\,c_p(T_i - T_o)}{\alpha\,\Phi_b} \qquad (5)$$

is a function of the dimensionless quantity

$$\cdot\,\frac{\alpha\,Mu}{w} = \alpha c \qquad (6)$$

Equation (4) may be written as follows:

$$P = \frac{1 - e^{-\alpha c}}{\alpha c} \qquad (7)$$

3.3 This function is sketched on Fig. 6. When P is small, either because the ignition temperature is low or because the heat flow rate Φ_b is high, the rate of propagation u becomes high and the combustion zone c is long. But when the parameter P increases, u and c decrease, and at the limiting P of unity the propagation rate reduces to zero. At P larger than unity, the fire would not be able to propagate under the conditions prevailing in the forest, either because the temperature T_i is too high, or because the heat flow Φ_b is too low (as, for example, when the moisture content of the forest is too high).

All the already given qualitative results concerning the propagation of fire may be defined with the aid of Eq. (6).

3.4 One of the advantages of this simple calculation is that it readily yields the width g of an efficient fire break, that is, a fire break such that fire would not reach the right side prior to dying out on the left side (Fig. 8). A very simple calculation shows that, with the assumptions made at the beginning of paragraph 3.1, the total energy supply to the left side of the fire break per unit time

$\Phi_b c/u$. Only a fraction B of this heat passes through the fire break, the other part being lost by radiation to the sky or to the ground in the fire break. Then, if we introduce again the α ratio the energy supplied per unit time to a slab of fuel of thickness dx, situated near the right side of the fuel mass, is α B Φ_b c dx/u. The heat rate required to ignite the slab dx is M cp(T_i - T_o)dx. The fire break will be efficient if:

$$Mc_p(T_i - T_o) > \frac{\alpha B \Phi_b c}{u} \qquad (8)$$

or if,

$$B < \frac{Mu/c_p(T_i - T_o)}{\alpha \Phi_b c}$$

By using Eq. (3) this can be written as:

$$B < \frac{wc_p(T_i - T_o)}{\alpha \Phi_b}$$

or

$$B < P \qquad (9)$$

Fraction B depends upon the ratio of the width g of the fire break to the height h of the bed, and the following expression for B was found to hold to within a few percent:

$$B = \frac{1}{g/h}\,(1 - e^{g/h}) \qquad (10)$$

If we compare the expression for P [Eq. (7)] to Eq. (10), we find that both equations fall on the same curve (left side curve of Fig. 9), so that Eq. (9) may also be written:

$$\frac{g}{h} > \alpha c \quad \text{or} \quad \frac{g}{h} > \frac{\alpha Mu}{w} \qquad (11)$$

Thus an evaluation of the characteristics of a given forest in terms of its ability to propagate a fire gives an approximate value of the fire parameter P and thus also of αc. It is then possible to choose the size of g in such a way that B will be smaller than P.

In practice, P and consequently αc vary from point to point in the forest, as shown by the right-side curve of Fig. 9. At some points, P may even become smaller than the value of B computed from the chosen g/h ratio. On Fig. 9, where g/h has been taken equal to 3, the shaded section indicates the fraction of the fire break that the fire will cross.

It is understood that the choice of g/h is a compromise. If g is too low, too many people and too much equipment will be needed to stop the fire. But too large a value of g will bring about much too large fire break maintenance costs.

3.5 A model of fire propagation based only on the heat transfer by radiation is too simple, as the flame and the convection phenomena are also important, even if we consider the combustion of the ground cover.

The fuel is not only preheated by the radiation from the incandescent solids but also by radiation from and convection by the hot gases.

As shown on Fig. 9, and already explained, a flame spreading in the direction on the wind will be wider and larger than a flame spreading in the opposite direction. If the slope of the flame is considered, the radiation from and convection of heat by this flame to the unburned fuel will be more important in the first than in the second case. The same applies to the case of sloping soil, depending on whether the fire is running up or down hill.

3.6 But a forest fire can spread in other ways than via the ground bed. Thus the one dimensional propagation we have considered until now is seldom the case, so vertical distribution of the fire should be also taken into account. This is more difficult as the trunk size distribution of the forest trees can change from place to place, even in well-planned forests (Fig. 10). Moreover this distribution differs in various forest tracts. The particular relationship between the trunk size of the trees and their height in an evergreen forest (Fig. 11) shows that young trees grow very quickly to reach sunlight, and that little foliage exists under a certain height. In the present case, this height is 7 meters and the foliage is densest between 11 and 18 meters.

In any event, there are several combustion levels. The first one, considered in the preceding paragraph, is the underbrush bed of pine needles, dry leaves, dead branches and so on, with a highly variable height h. This is the pyrophil level, where the fire generally starts. Over this are either brushes which easily catch fire, or young trees which, on the contrary burn very little or not at all. Finally, at the very top are the tree crowns which may catch fire individually, either because combustion is activated by the wind or because of hot gas explosion. If the trees are close together, or if a strong wind is transporting away firebrands, fire may spread directly from crown to crown. But it would be very unusual for this crown fire to propagate without a fire in the underbrush.

3.7 Finally, the shape of the fire and the forest residue left after the fire depend upon many parameters such as the moisture content of the fuel, the relative humidity of the air, the wind, the slope of the ground, the species of the trees and their health and age, and the composition of the underbrush.

It is not always true that the violent fire, which appears terrifying, is the most dangerous. For example, in a fire spreading either downhill or in the direction opposite to the wind, combustion is slower and more complete in the lower layers. Thus the temperature of the seeds and superficial roots is lower, but dangerous for a long time. It is known that the time t over which living tissue can withstand a given temperature T is a function of this temperature. Figure 12 shows a semilogarithmic plot of the variation of T with time t [6], given also by Eq. (12).

$$T = a - b \log_e t. \tag{12}$$

4.   SOME EXPERIMENTAL RESULTS

I think some numerical data are in order as a conclusion to this paper.

4.1 I selected the laboratory research work by Anderson and Rothermore [7], as they used fuels which are approximatively the same as those encountered in the Mediterranean forests. The test took place either in a wind tunnel, where the temperature, humidity, and velocity of air were measured, or at zero wind velocity on a stand called a "combustion laboratory." Both experimental setups were parts of the same test facility (Fig. 13).

The wind tunnel test section was a square, 3 meters on a side. A fuel bed 2.4 meters long, 0.45 m wide and 7.5 mm thick was placed in this test section. It consisted of two types of pine needles, ponderosa and white pine needles. The air conditions at the inlet were measured with a resistance thermometer and a wet bulb hygrometer (Fig. 14). The two thermocouple wells were placed about 6 mm above the bed surface and the rate of spread was deduced by recording the temperature as a function of time. The time interval separating the temperature increase at the two thermocouples was measured (one thermocouple was 0.6 m and the other 2.2 m from the beginning of the bed). Anderson and Rothermore found a steady state in the bed between these both thermocouples. The time τ_0 separating the increase and decrease in temperature at a given thermocouple gave the length of the burning zone c:

$$c = \tau_0 u$$

The weight loss of the bed in time was measured with strain gages mounted on the supports of the base plate.

4.2 We give at first the results at no wind. The principal result is that a steady state does exist: the burning length varies linearly with time, and the slope of the line, that is, the rate of spread, decreases with an increasing moisture content E of the fuel (Fig. 16). The rate of spread u_0 (subscript zero indicates no wind) varies linearly with E:

$$\text{Ponderosa Pine} \quad u_0 = 1.04 - 0.044\,E \quad \text{ft/mm}$$
$$\text{White Pine} \quad u_0 = 1.12 - 0.051\,E \quad \text{ft/mm}$$

The two formulae may be approximated by one, if the law of Curry and Fons [8] is used. This law says that the rate of spread u_0 is proportional to the square root of r, the ratio of the void volume to the fuel volume in the layer under-going combustion.

The two formulae then give

$$u_0 = \sqrt{r}\,(0.272 - 0.012\,E) \quad \text{ft/mm} \tag{13}$$

4.3 At no wind, the thermocouple measurements show that the length of the burning zone c is proportional to the rate of spread u_0, which is normal for steady state. But for ponderosa pine this length c is much larger (4.5 times) than for white pine, given the same rate of spread (Fig. 16). However, photographs of the flame yield significantly smaller values of c. This is because the thermocouple data take into account the length of the embers zone which does not show up on photographs.

4.4 Figures 17 and 18 show the rate of spread u versus the wind velocity v for three different moisture levels, and for ponderosa pine and white pine, respectively.

In the first case (Fig. 17) the rate of spread varies exponentially with the wind velocity,

$$u = u_0 e^{av}$$

The value of a varies slightly with the moisture content. We may take it as 0.0038, and rewrite Eq. (13) for ponderosa pine as

$$u = \sqrt{r}\,(0.272 - 0.012\,E)e^{0.0038v} \tag{14}$$

The curves on Fig. 17 plot this equation. Experimental points are pretty well distributed around them. These curves are also valid for v = 0. However, it should be noted that Eq. (14) holds only at low velocities not higher than a few meters per second. At high velocities, owing to the moisture content of the fuel, the rate of spread decreases more than indicated by the equation, and the linear function of E no longer holds.

In the second case (Fig. 18), a power law is obtained

$$u = Kv^n$$

which does not hold at v = 0. Moreover the exponent n and coefficient K depend on the moisture content via to an empirical relationship which is too involved to be interesting.

The difference between the functions u = f(v) for the two fuels may be explained in terms of the ratio α. Thus, with the thinner white pine needles the irradiated surface is larger and the spread is faster. We find here again a justification for Emmons' theory. For example, in a bed of stems 12 mm in diameter, the rate of spread will be about 15 times smaller than in a bed of white pine needles. The influence of the moisture content on the rate of spread is seen better on Fig. 19.

4.5 The percentage of burned fuel is given by the weight loss of fuel during combustion. It is a function of the air velocity and the moisture content of the fuel (Fig. 20). It always decreases with an increasing air velocity, which is understandable because a fast flame passing over does not reach the layers close to the ground.

4.6 To close this paper, I think it is good to reproduce the so-called characteristic curve of Anderson and Rothermore. This curve gives the energy E_e released per unit area and per unit time as a function of the rate of spread (Fig. 21). It is independent of the moisture content and nature of the fuel, and is therefore interesting.

Weak rates of spread occur at large E_e and high rates of spread at low E_e. Near both ends of the curve, only one quantity varies, the other becoming approximately constant.

4.7 To sum up, here are some practical conclusions. On the one hand, a high wind velocity increases the rate of spread and lets the fire extend. On the other hand, the percentage of burned fuel then decreases, as does the energy released per unit area and per unit time. Added water decreases the rate of spread, but has little influence on the percentage of burned fuel, on the energy released per unit area and time, on the height/width ratio of the flame and on the slope of the flame.

REFERENCES

1. Pierre and Roger Molinier. La foret mediterraneenne en Basse Provence (The Mediterranean Forest in the Lower Provence). Bulletin du Museum d'Histoire Naturelle de Marseille, XXXI (1971).
2. Mario Casari. Conservazione, miglioramento, polenziamento del patrimonio forestale. Atti del Convegno Nazionale di Bergamo, p. 37 (1967).
3. M. Laug. Description et mode d'utilisation du detecteur experimental d'incendies de forets (Description and operation of an experimental forest-fire detector). Centre d'Etudes et de Recherches de Toulouse, 2 av. Edouard Belin (1973).
4. Relations et Conjoncture - 21, rue Daniel Casanova Paris (ler). Rapport final d'etude sur l'accroissement d'efficacite de la lutte contre les feux de foret du Sud-Est (mai 1972)   [Final report of a study of incremental efficiency of firefighting measures employed in forest fires in south-eastern France (May, 1972)].
5. H.W. Emmons. Fire in the Forest (Air, Space and Instruments, S. Leco, ed.).
6. Fire as a physical factor, R.E. Martin, C.T. Cusluva and R.L. Miller, Forest Fires, 9, p. 271 (1969).
7. H.E. Anderson and R.C. Rothermore: U.S. Forest Service, Northern Forest Fire Laboratory, Missoula (Montana), 7th International Symposium on Combustion. The Combustion Institute, p. 1009 (1965).
8. J.R. Curry and W.L. Fons: Mech. Engin. 62, 219 (1939).

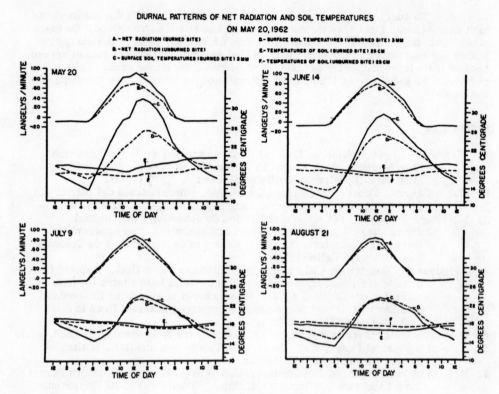

DIURNAL PATTERNS OF NET RADIATION AND SOIL TEMPERATURES
ON MAY 20, 1962

A.- NET RADIATION (BURNED SITE)   D.- SURFACE SOIL TEMPERATURES (UNBURNED SITE) 3 MM
B.- NET RADIATION (UNBURNED SITE)   E.- TEMPERATURES OF SOIL (BURNED SITE) 25 CM
C.- SURFACE SOIL TEMPERATURES (BURNED SITE) 3 MM   F.- TEMPERATURES OF SOIL (UNBURNED SITE) 25 CM

FIGURE 1

NOTE: Figure captions for this chapter are listed on page 38.

FIGURE 2

Detection in the
visible spectrum

Detection in
infrared spectrum

Adjustment of the sight

Turning disc

Motor

Fixed support

Visible signal
Infrared signal
End of travel signal
220 V

FIGURE 4

FIGURE 3

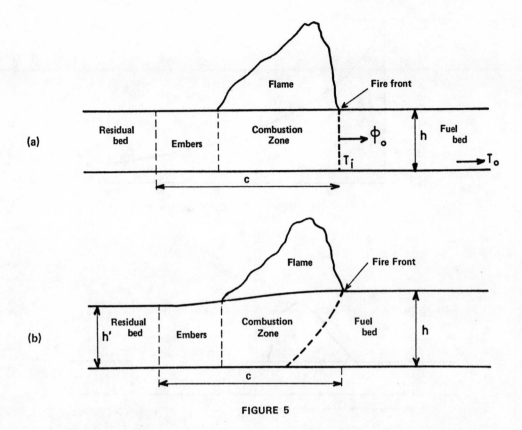

FIGURE 5

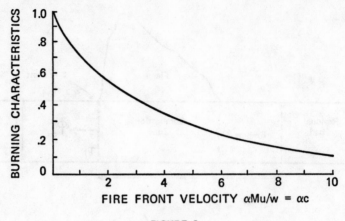

FIGURE 6

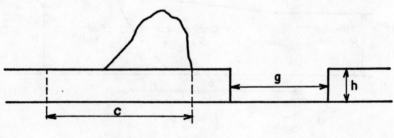

FIGURE 7

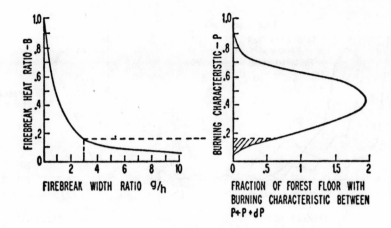

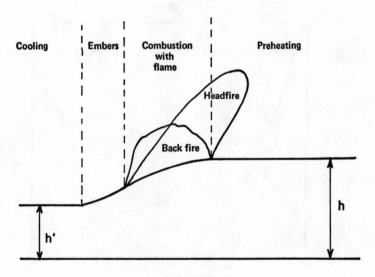

FIGURE 8

FIGURE 9

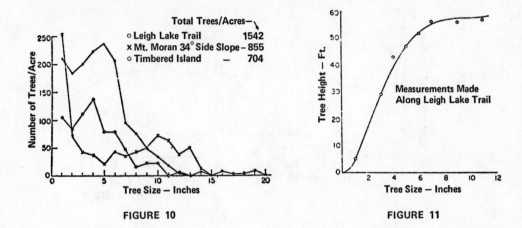

FIGURE 10

FIGURE 11

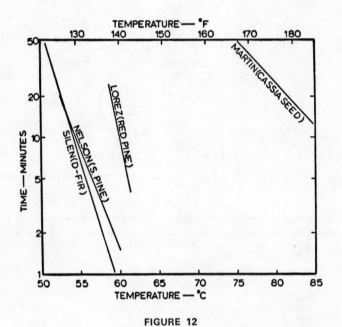

FIGURE 12

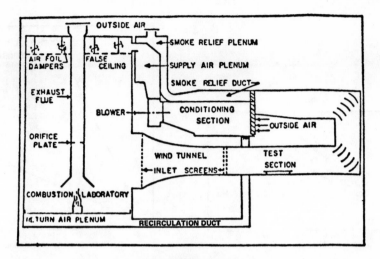

FIGURE 13

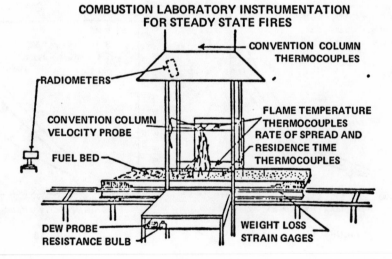

FIGURE 14

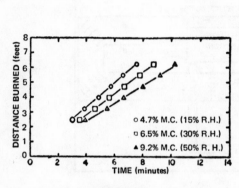

FIGURE 15

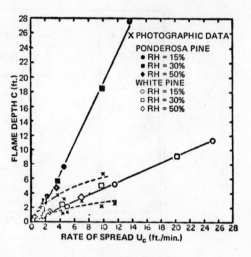

FIGURE 16

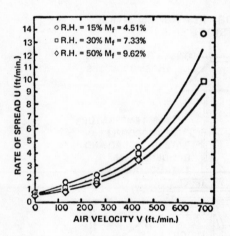

FIGURE 17

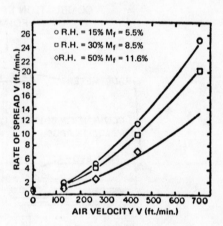

FIGURE 18

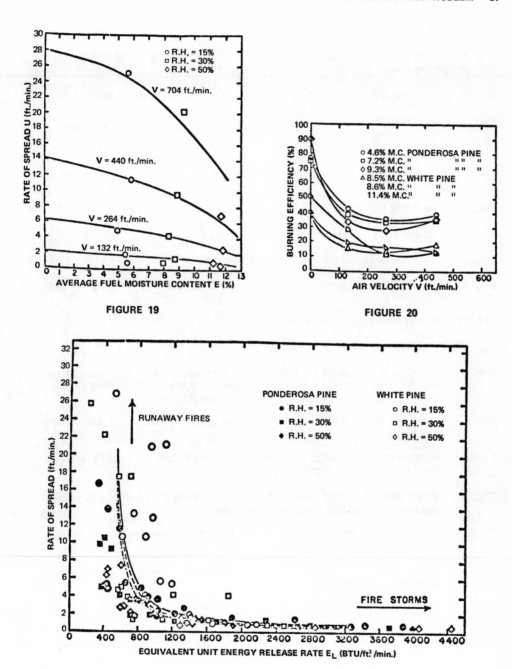

FIGURE 19

FIGURE 20

FIGURE 21

FIGURE CAPTIONS

Fig. 1. Net radiation and soil temperatures for 4 days during the growing season on the Curtis Prairie.

Fig. 2. A meadow before burnoff (left) and one year after burnoff (right).

Fig. 3. Automatic fire detector.

Fig. 4. General scheme of the detector.

Fig. 5. Scheme of the spread of fire: a) as assumed in the calculation, b) closer to the reality.

Fig. 6. Fire-front velocity.

Fig. 7. Notation used in describing the fire break.

Fig. 8. Method of estimating fire break performance.

Fig. 9. Slope of the flame as a function of the wind direction or the slope of the ground.

Fig. 10. Forest tree size distribution.

Fig. 11. Tree height – trunk size distribution.

Fig. 12. Lethal time-temperature factors for plant tissues as determined by several investigators. Time is plotted logarithmically and the relationships for any investigation approach a straight line.

Fig. 13. Combustion research facilities at the Northern Forest Fire Laboratory in Missoula, Montana.

Fig. 14. Instrumentation used in the combustion laboratory. A similar arrangement was used in the wind tunnel testing.

Fig. 15. Steady rate of spread through beds of ponderosa pine needles at different moisture contents in the absence of wind.

Fig. 16. Comparison of thermocouple and photographic measurement of flame depth.

Fig. 17. Effect of air velocity upon rate of spread at three fuel moisture levels in ponderosa pine needles.

Fig. 18. Effect of air velocity upon rate of spread at three fuel moisture levels in white pine needles.

Fig. 19. Effect of moisture content on rate of spread in white pine needles at four air velocities.

Fig. 20. Effect of moisture and air velocity upon burning efficiency measured by 100 × pounds lost/total pounds.

Fig. 21. Fire characteristic curve. Points represent individual fires.

Chapter 3

SOCIAL AND ECONOMIC IMPACT OF FIRE

P. H. THOMAS
Fire Research Station, Borehamwood
Herts, England

INTRODUCTION

It has long been recognized that uncontrolled fire is damaging to society, can waste resources, be they material or human, and requires society to take measures to prevent it and mitigate its consequences. These actions will themselves incur costs and penalties and we may then be faced with the problem of evaluating both kinds of costs so that the cost of the remedy is no greater than the cost saved. As we shall see, this task is sometimes very difficult and establishing priorities between, say, expenditure on fire safety and on road safety requires us to study fire costs in a wide context.

In evaluating the costs of protecting a single building against fire the designer has to consider the protection of the structure, the escape routes and, in some occupancies, sprinklers and detectors. What cost criteria should be adopted and how important are such costs in relation to the costs of the building? Such questions have become the talking point of government and of management in recent years and what follows is but a brief introduction to some of the issues involved.

THE SOCIAL COST OF FIRE

What is meant by the social cost of fire? Fry [1] and Silcock [2] have published estimates for the United Kingdom under various headings (see Tables 1 and 2), each of which represents some use of resources to provide protection against fire that might occur, and the loss of resources as a result of the fires that do occur. Lie [3] has given some estimates for 4 countries in W. Europe and N. America [4]. Material losses in the United States as a fraction of 'reproducible property' appear to have decreased by about 8 times in the last 150 years and little or none of this improvement occurred in the last 30 years [5]. There are

NOTE: All figures quoted in text are at the end of the chapter.

significant differences between countries which result in different research poli-
cies. An example of this is the greater relative life loss in N. America than in
Europe which makes it easier, for example, for home detectors to be cost effective. If, provisionally, and we shall consider the question later, we ascribe an
average loss as high as £ 50,000 for the cost of a fire death then we find that life
loss in fire in monetary terms is less than material losses. This pattern differs
from that for road accidents; the question of priorities between safety areas of
differing loss patterns is one of the reasons for raising the question of the cost of
life.

Materials and Property Loss

I have deliberately not attempted to quote accurate figures in Tables 1 and
2 nor corrections for inflation or currency changes. This could create a false sense
of reality to figures about which there is some doubt as to their derivation and
meaning as well as to their accuracy.

How are such figures arrived at ? Direct losses appear to be relatively
easy to evaluate. Figures are frequently quoted but in America and W. Europe
they are estimated from insurance data; it is less easy to evaluate the cost to
an economy by losing a particular plant which is a large part of a strategically
important industry. Consequential losses beyond the direct material losses present something of a problem [4]. Some organisations insure against consequential
loss to themselves and on the basis of some very limited data for large fires estimates varying up to nearly half the direct loss have been made [6], but this, it
must be emphasized, refers to insured losses only and based on few data. What
may be a loss to one company or organisation may be a gain for another. From
a national point of view there is one answer if these are in the same country and
another if they are not: It is possible sometimes to calculate what is the cost of
implementing the regulations or the law on fire control measures in buildings.
If there are limitations to the area which can be built under one roof so that the
designer is required to put in an extra wall, the cost of doing this can be calculated.
The cost of providing sprinklers, fire detectors, and means of escape can also be
estimated. It would not appear difficult to perform such calculations and those
that have been done suggest that the costs are between 1 and 10 percent of the
building cost depending on what kind of building and what kind of use it is put to.
However, do we really know how designers would build buildings if all restrictions
were removed ? Are some restrictions implicit in the design methods that are
taught to people rather than explicit in the prohibitions of regulations ? To what
extent are some fire prevention rules and practices, customs rather than the result of direct conscious activity ? A simple example will illustrate what is involved here. In the United Kingdom there are no fire requirements for baths (as
opposed to bathrooms), but as soon as somebody manufactured a bath with its
fittings made of plastic we became aware that we might need additional restrictions.
Why is this ? Because traditionally we have used incombustible materials and the
problem has not arisen and therefore there has been no requirement for a prohibi-
tion. It is part of our custom to build bathrooms in this way and when the custom
changes, new restrictions may have to be introduced.

It is tempting to talk solely about the cost of fire as if there were no benefits, though planning officials have been known to be apparently delighted to hear
that a particular building was burnt to the ground, so shortening an otherwise
lengthy and costly legal process. If certain improvements in building regulations

Table 1. Annual cost of fire protection to buildings (Silcock)

Occupancy	Construction 1965 (£million)	Average fire protection cost as a percentage of building costs	Total protection cost (£million)
Private sector			
Housing	693	1	7.0
Schools and universities	8	3	0.2
Garages	31	1	0.3
Offices	104	2	2.0
Shops	100	2	2.0
Entertainment	81	5	4.0
Industry	431	7	30.2
Miscellaneous	60	1	0.6
Public sector			
Housing	510	1	5.1
Schools and universities	169	3	5.1
Health	91	3	2.7
Offices and factories	64	5	3.2
Miscellaneous	106	1	1.0
Total	2448		63.4

Table 2. Annual cost of fire (U.K. 1967) excluding consequential loss
(after Fry and Silcock)

	£million
Loss of property (direct loss)	90
Fire fighting (public and private)	95
Fire protection in buildings	70
Hospital treatment of injury	0.5
Research (government and insurance)	0.5
Insurance (administration and technical services)	65
Total	≈ £320 million
Deaths in fire (Fire brigade reports – 1972)	1070

These exclude some in fires not attended by fire brigades and deaths after lapse of time.

have contributed to improvements to the design of a city, with benefits to the water supply, say, and to public health, then we cannot treat fire as wholly detrimental. It is no easy task to evaluate costs and benefits of this kind when many of them are long term. The practice of discounting future benefits and costs automatically defines bodies such as insurance companies as conservative with respect to changes in fire protection practice since they seek evidence on which to act from statistically rare losses and can hardly respond to 'short-term' cost benefit thinking with its possibly wide uncertainty tolerances.

Some improvements effected by building regulations are also liable to the same limitations since the rate of renewal of domestic buildings (in the United Kingdom) is of order about $1/20$ year^{-1} or less and the discounting of long term benefits tends to make any changes in building and in certain appliances of low rather than high cost effectiveness.

In one attempt to evaluate cost benefits of R and D in fire, many benefits were evaluated as minima on the basis of improving the cost effectiveness of existing plant and equipment at current levels of usage. It was concluded that additional benefit would accrue from consequent reductions in direct fire loss, but these had proved impossible to quantify with the information currently available. It follows that there is benefit to be obtained from better reporting of information, particularly on losses.

We can assume that society wishes to minimize the total costs of fire but once we embark on this task we have to discuss the interaction of fire costs with other costs and our attitude to safety in general. The builder, for example, is not concerned with minimizing only his direct fire costs if other costs are incurred, e.g. if higher heating costs result from not using materials to be preferred on grounds of lower flammability. Production can suffer if production areas are sub-divided to restrict fire spread. The evaluation of these indirect costs are not necessarily simple. It is not only the interaction of costs that is involved; there may be an interaction between performance requirements. How does one balance higher heating standards against a long term loss of fire safety?

In general, the question of how certain consequences of fire are to be regarded and costed is basically a cultural question. As such it is liable to change and to argument. This is not to say that one cannot rationally attack the problem nor that quantitative procedures are out of place, only that we recognise the problem for what it is and we are suspicious of easy answers.

The Cost of Life and Injury

Most of the lives lost in fire are lost in small domestic incidents. About three-quarters of the 800-1000 total fire deaths in the United Kingdom are lost in fires in which there was only one fatality [7].

There is much debate on whether it is proper to even discuss such a question as putting a social monetary value on life but it is desirable to have some way of relating life loss to material loss. If we do not, how can one build a large building and determine the escape facilities to be provided? Many modern designs do not allow the evacuation of the building rapidly enough and the building must be provided with a protected zone. Our choice from these designs must reflect some de facto valuation of how much one can spend to save lives.

There are, however, certain fundamental problems of meaning and method.

So long as driving a motor car was regarded as a sport, the safety of the driver and his passengers was a long way second to that of other road users. This is not the case now when some societies regard the use of a motor car as a necessity.

To what extend should society as a whole regard the danger from smoking in bed in the same way as the dangers from faulty equipment ? One has some element of a voluntary risk whilst the other has not. Whatever method we employ to obtain an economic value to life, we must surely do it with a methodology which is consistent with our purpose and with other declarations about life and death. Over-population is said to be a social problem in some places! This point is not facetious; one method which essentially examines the economic production by individuals and their consumption of resources as a function of age is already employed in certain economic assessments and it leads to average values of order £ 20, 000 or less [8, 9]. This method has been used for costing road accidents but because of the effect of age it leads to the conclusion, when applied to fire, that it is almost worthwhile allowing people to burn! It has already been remarked here that even with figures higher than this and differently dependent on age the loss of life in fire does not become the major part of the total apparent economic loss. Most people feel there is something wrong with doing sums this way but I would suggest not. It is in answer to a question which is probably of limited relevance — how to maximize material production gains. We should, however, ask a different question, not necessarily instead of, but in addition to, this, because in society we pursue more than one objective.

Many people expect to be rewarded if they take risks (for example to be paid danger money) and to pay extra for something safer; we can examine how much they are prepared to pay for how much extra safety, that is, we find what 'value' corresponds to people's behavior and preferences as consumers. An important difference between these methods is that in the first it is death and injury that attracts a valuation irrespective of cause; in the second, people may prefer to spend more (or less) on reducing the risk from fire than from some other hazard. We have already mentioned the difference between voluntary and involuntary risks which is highly relevant here.

It is customary to quote a figure such as £ 5, 000 to £ 20, 000 or, as Melinek [10] suggests, on the basis of the above method, a higher one of £ 50, 000. However, even stating the 'answer' this way begs a question of whether these figures obey the simple commutative law that the cost of 10 deaths is 10 times the cost of one. People's reactions to a disaster may well differ from their reactions to the perhaps greater losses incurred as a total of many small incidents. There is social effort expended in dealing with preventing the recurrence of a disaster. All too often people say of fire, 'It can't happen to me'. If they switch their attitudes to one of apprehension (sometimes justified, sometimes unjustified) public pressures can result in demands for safety.

Problems raised by these questions are of increasing importance now that the sizes of buildings and building complexes are such as to put thousands of people at risk at one time.

Before leaving this topic we must reaffirm that costing life in any of these ways is pertinent to investment decisions by organizations in an economic context not to personal decisions in a human one.

OPTIMIZATION

Let us retreat from these general problems and look at the limited task of finding the optimum design for fire resistance.[1]

We can in principle make a simple balance between expenditure and expected loss as follows. Following the arguments of Saito [11] and Maskell and Baldwin [12], we assume that the cost of providing extra fire resistance to a structure may be expressed as

$$c = c_0 + aR \tag{1}$$

where c_0 is the basic cost of the structure and R is the fire resistance provided (e.g., 1 h, 2 h, etc.).

In slightly different ways Saito [11] or Baldwin [13] examine probabilities of fires in offices lasting longer than an effective duration ii. We shall follow Baldwin [13] and use for offices

$$p = \alpha e^{-\gamma R} \tag{2}$$

where α is the annual probability of a fire (estimated as 6×10^{-4} year^{-1}) and $\gamma = 0.06$ h^{-1} together with a simplified model for loss, namely that if the fire lasts less than the time for which fire resistance is provided, there is no loss, but if it lasts longer than the time there is a loss D, while so long as there is no fire there is a net benefit of b each year. We discount future benefits and expenditure at a rate of interest r then the sum of immediate costs and expected losses during the lifetime, n years, of the building discounted to the commencement of the period, may be written as approximately

$$C_{net} = \left[\frac{Dp - b(1-p)}{r+p} \right] \left[1 - \left(\frac{1-p}{r+1} \right)^n \right] + C$$

Saito's result is somewhat different in form as a result of a different treatment of discounting.

Since in general $p \ll r \ll 1$

$$C_{net} = \left[(D+b)\, p - b \right] \left(\frac{1 - e^{-rn}}{r} \right) + C_0 + aR \tag{3}$$

i.e., $C_{net} - c_0' = Epn + aR$
where $E = D' + b' \simeq D'$
where D' and b' are discounted mean loss and annual benefits, and $c_0' = c_0' - b'n$.

This equation may be conveniently rewritten as

[1] We have to neglect the important qualifications that fire resistance can at present be evaluated only for single elements (whilst we seek a performance of a framework) and that the fire resistance required is not the same as the duration of a fire.

$$\frac{\gamma(C_{net} - c_o')}{a} = \gamma R + \left[\frac{\alpha\gamma En}{a}\right] e^{-\gamma R}$$

which is shown in Fig. 1.

The optimum value of R giving the minimum of C_{net} can be found as

$$(\gamma R)_{opt} = \ln\left(\frac{\alpha\gamma En}{a}\right)$$

and the minimum C_{net} given by

$$\left[\frac{\gamma(C_{net} - c_o')}{a}\right]_{min} = 1 + (\gamma R)_{opt}$$

No optimum exists (other than R = 0) if

$$\frac{\alpha\gamma En}{a} < 1$$

A problem arises from imprecise information about the parameters a, D, $\approx$, etc., and the sensitivity of the result to the formulation of the model. It is not much good having an answer such as that the optimum fire resistance is 2 h ± 1 h and the usefulness of this approach depends very much on the adequacy of the basic data.

The statistical data on which the results depend must be available (whether or not the cost and frequency probabilities follow the simple law described here does not matter in more sophisticated computations). The optimum is well defined when the damage D and the annual frequency are high in relation to the extra cost per unit of extra fire resistance provided and ill defined when they are low (see Fig. 1). For this latter condition extra safety is relatively inexpensive and the optimal economic criteria as Saito points out may well not be the decisive design criterion. The points A and B are sub-optimal designs but are not equivalent. In the event of a fire the cost A is a certain cost for an unsafe design; B is a probable cost for a safer one. Amongst other comments one can make an important one is that the constancy of the fire frequency (taken here as 6×10^{-4}) is by no means assured. Such calculations are probably to be applied to buildings of new types of design for which there is not much past experience and our energy consumption and our choice of materials will change, possibly considerably, over the expected lifetime of the building. A 'solution' to this problem (which is all too often dismissed) is to accept that whilst there are difficulties about changing passive protection after a building has been built, active protection systems (e.g., sprinkler, detector) can be installed later if justified.

"TRADE-OFF"

We turn our attention to the problem of 'trading-off' passive (e.g., structural) fire protection against active protection methods [14]. If passive and active protection were designed to achieve precisely the same objective then one can imagine that they would be interchangeable without restriction but there is in effect an important difference between them in the world as described by fire reports. Sprinklers reduce the frequency of damaging fires; fire resistance reduces the damage from damaging fires.

Figure 2 shows diagrammatically a typical distribution of fire duration. Let the fire resistance of the building correspond to the time R so that the shaded area of the frequency distribution corresponds to the risk of a fire occurring which would cause significant damage. Changing the value of the structural fire protection changes the value of R. However, if we were to introduce sprinklers, we would expect that many fires would be suppressed before they did any damage whatsoever. On the other hand some small number of fires might overwhelm the sprinklers and their subsequent behavior would then be little influenced by the provision of the sprinklers. Accordingly we see the action of sprinklers as reducing the frequency of fires that last longer than R and so cause damage. In short, structural fire protection controls the abscissa and active fire protection the ordinate of the graph. In the physical sense neither active nor passive protection affects the incidence of the embryo fire. It is the statistically reported fire for which the distinction applies but it is on the statistical data that the theory and the data rest. A second difference between active and passive protection is that the cost of introducing active measures after the building has been built is a cost that has been incurred many times in the past when owners have installed sprinklers or detectors to existing buildings. This cost, however, is quite different from the probably unacceptable cost of changing the structural fire protection of the loadbearing framework, something which is not normally entertained. A third difference is that it is not easy to turn off structural fire protection. Sprinklers and detectors are much more vulnerable to malicious behaviour which in some countries is increasing markedly. In 1970 about 30 percent of the direct loss in fires costing over £ 10, 000 and having a known cause was attributed to malicious or intentional ignition (see also Fig. 3). Given such distinctions between active and passive fire protection, it is only to be expected that in some fire situations depending on the relative cost of the two methods of protection, one or other or neither is to be preferred whilst there are only some situations where there is an optimum mix. However, because of the relatively low cost of introducing active measures after the building has been built, it will be economical in some situations to introduce them some time after the building has been built, i.e., the optimum design may be to defer the introduction of active measures until sometime in the future. What one does at the time of building design depends on what one envisages to be the future trend in fire frequency. These considerations are quite general because we can in principle consider the cost as lives or compound lives and material losses through some economic evaluation of life.

THE ROLE OF RESEARCH IN REDUCING FIRE LOSSES

All losses, be they direct loss or consequential, can be reduced either by reducing the number of fires initiated or by reducing the extent of the fires that do take place. This latter means of reducing loss may be effected either by the use of passive protection, i.e., the use of safe designs, compartmentation, and less flammable material, etc., or by active fire protection which will include detection, sprinklers, and fire brigades.

Initiation

There are, in principle, three ways in which the initiation of fires can be reduced in frequency:

1) energy sources can be made safer and more reliable
2) threshold levels at which combustible materials are ignitable can be raised by treatment or protection, and
3) the conjunction of an energy source and a combustible material can be made less likely.

Since fire and heat are an essential part of every day life, it is doubtful if research can influence fire initiation to anything but a limited degree, particularly if one bears in mind the number of fires started deliberately or from avoidable causes, such as children playing with matches or smoking materials; research cannot completely overcome negligence and accidental damage. Research on energy sources such as electric appliances will take many years to diffuse through the existing stock of equipment, though clearly in special instances dramatic results can be achieved. Figures 4 and 5 refer to well-known examples in the United Kingdom of the effects of research and legislation. Figure 4 shows the numbers of fires attributed to space oil heating most of which are associated with portable oil heaters. The stopping of the upward trend is associated with the stronger control on design and sale introduced about 10 years ago.

This is part of a larger story of reducing burns from the ignition of clothing (see Fig. 5). Analysing the effects of the various measures introduced during the 1950's and 1960's is complicated. There is a changing pattern in the production of fabrics of different properties. The open coal fire is disappearing in England and there has been a trend in recent years to shorter skirts both of which can be presumed to have beneficially affected the accident incidence. On the other hand there is an increase in accidents to young males, thought to be associated with the greater use of cotton in trousers and jeans.

These comments serve to emphasize the importance of social and cultural aspects of fire problems.

Raising the threshold at which materials can be ignited may be very effective for those products which are consumed rapidly, such as cooking oil. Household furniture, on the other hand, remains in existence for a relatively long time so that improvements diffuse slowly throughout the existing stock. Because only a small fraction of any one of these materials is involved in fire, the costs of protecting the whole of an output must be a very small proportion of the material cost if it is to be cost-effective. Reducing the accidental conjunction of an energy source and a combustible material (item 3 above) is possibly more within the scope of publicity and education than in that of research, and introduced problems of a rather different kind. The reduction in initiation is not always a benefit! Control burning is a recognised forestry practice. Small fires in forests are deliberately started over large areas in Western Australian forests to prevent accumulation of fuel becoming too large so causing disastrously large accidental fires [16].

Detection of Fire

In 1970, 57 percent of the fires in the United Kingdom were detected between 0600 and 1800 h but the proportion of large fires[2] detected between the same times was only 39 per cent, and a similar situation has existed for many years [17]. It follows from this that if the detection standards at night could be improved to match those during working hours, considerable savings — up to 20 percent of the total fire bill — might be effected. This, of course, is not simply a matter of a detector and a detection system; one difference between daytime and night time is the smaller number of people available to detect a growing fire at night [17].

There is much activity in many laboratories to develop cheaper and simpler modes of fire detection. One of the great obstacles to their widespread use in industry and in the home is their cost which at present would appear to be too high for the universal use to be cost effective. Also, whilst the current usage produces a tolerable level of false alarms, greater usage would make this problem embarrassing to fire brigades. In reference 5 it is suggested that a highly sensitive detection system universally in homes in the United States might save about 2,800 lives a year, and this would be worthwhile if costs were less than £ 100 a household. The Holroyd Committee in the United Kingdom suggested [18] 1/2 percent of home capital costs but no basis for this figure is given which may well be too high for the United Kingdom since the lower relative rate of casualties in the United Kingdom imposes greater constraints on the development of a cost-effective fire detector than in the United States. As detectors become cheaper so their use will extend in industry from the specially hazardous or high value risks on which they are used at present to areas of less hazard and lower value.

A recent study [19] in the United Kingdom showed that the average chance of being a casualty in a hotel fire was little different from the chance of being a casualty in a domestic fire, given the existence of the fire. However, the chance of being involved in a fire in a given time was perhaps 10 times greater in hotels than in domestic buildings. Although the similarity of the averages as such (they are not averages over the same population) is not an argument against improvements in hotel escape systems it may be more cost-effective to improve the detection of fires in their early stages in the hazardous areas, namely kitchens and public lounges.

Spread of Fire

We have just discussed detection and we should remember that installing a detector does not normally interfere with any of the building's normal functions nor necessarily with any other building system. Reducing the flammability of materials, or restricting the use of materials to those which are less flammable rather than more flammable, does introduce conflicts with other requirements. Very frequently materials have to serve several functions and some of these may be in competition. A wall lining is primarily chosen for its decorative and acoustic properties, washability and thermal insulation, before a choice on fire safety grounds is made. These interactions of the requirements of fire safety with other building functions is possibly greatest in the area of spread of fire. Wider separations between stored goods, reduced heights of stacks of goods, compart-

[2]Defined as costing over £ 10,000 (direct loss).

mentation in production areas, usually impose considerable economic penalties. Greater demands on land in urban areas leads to smaller separations and more hazardous conditions. The size of buildings and automatic warehouses has become considerably larger in recent years and in very large buildings it is economically impracticable to design escape routes to evacuate the building in, say, 10 or 20 min, and it becomes necessary to design the building so that people can safely remain within it while there is a fire in some part of it.

This mixing of the requirements for property and life safety demands that rational safety criteria be evolved. It is unlikely that existing regulations and design procedures are optimal and there is a possibility that some of them are over-safe rather than under-safe, so that savings can be effected by a rationalization of design. A recent example of an attempt at such profitable rationalization in the United Kingdom may be quoted. Both physical and statistical studies show that the risk of a serious fire in a well-ventilated, above the ground, multi-storey car park and assessments of the likely fuel content, does not warrant the conventionally high requirements for fire resistance [20]. Savings could be achieved by allowing unprotected steelwork in the construction. Such developments have taken place in a number of countries though progress in the United Kingdom is retarded by the classification of such car parks as 'Petroleum Storages' which are subject to controls outside building regulations.

Extinction and Fire Control

Sometimes, rapid developments take place in producing new extinguishing agents as new hazards appear, but their cost precludes them for general use for which water is by far the most economical — delivered either by sprinklers in fixed installations or by the fire brigade as a mobile force. In protecting shopping precincts under cover, there is a tacit recognition of the value of sprinklers as a means of reducing casualties and this is perhaps the forerunner of other developments in this area.

Fire brigades have made use of computers in storing information about hazards and buildings at risk, and in forest fires they have been used to store strategies [21], but the process of tackling a fire with water carried in a hose line has not changed very much. Experiments have been conducted with inert gas generated in large quantities and with high expansion foam, but their role is limited to special risks. Nevertheless, fire brigade operations have become the area in which much operational research is being undertaken, both to ascertain the optimum levels of cover and protection in a community, the siteing of fire stations [22], and other organizational matters, and no doubt soon we may see some operational research on the fire field to appraise and hopefully improve fire-fighting tactics, at least for certain 'difficult' fires. There is evidence that most large fires are large not because fire-fighting is poor but because the fires are already large when discovered [17].

However much research is done and however effective it may appear to be, its effect can only be measured if suitable statistical data is available. The United Kingdom is fortunate in having had national statistics for several years; even so, one cannot expect that sufficient data for all research purposes can be obtained from fire brigade reports. Special surveys are needed and, both in the United Kingdom and United States, survey teams are visiting fires and reporting details of fire and building behaviour of well-defined samples of fires. Only by such activities can adequate 'feedback' be obtained on technical matters.

ACKNOWLEDGEMENT

The paper is Crown Copyright reproduced by permission of the Controller, H.M. Stationery Office. It is contributed by permission of the Director, Building Research Establishment. (The Fire Research Station is the Joint Fire Research Organization of the Department of the Environment and Fire Offices' Committee.)

REFERENCES

1. FRY, J.F. The cost of fire. Fire, 1964, 56 (706) 591-4; Fire Internat., 1964, 1 (5) 36-45.
2. SILCOCK, A. Protecting buildings against fire: background to costs. Archit. J., 1967, 146 (24) 1515-8; Fire protection of buildings — what does it cost ii Fire, 1968, 60 (752) 461-3.
3. LIE, T.T. Fire and buildings. London, 1972. Applied Science.
4. A study of fire problems, held at Woods Hole, Mass. U.S. National Academy of Sciences, National Research Council Publication 949. Washington.
5. A program for the Fire Research and Safety Act. Report of the Sub-committee of the National Bureau of Standards of the Committee on Science and Astronautics. U.S. House of Representatives. 91st Congress, 2nd Session, July 1, 1970. Washington, 1970. U.S. Government Printing Office.
6. CHAMBERS, E.D. Private communication.
7. United Kingdom fire and loss statistics 1970. Department of the Environment and Fire Offices' Committee Joint Fire Research Organization Borehamwood, 1973. Building Research Establishment.
8. DAWSON, R.F.F. Cost of road accidents in Great Britain. Road Research Laboratory Report LR79. Crowthorne, 1967.
9. ABRAHAM, C. and THEDIE, J. Le Prix d'une vie humaine dans les decisions economiques. Revue fr. Rech. oper., 1966 (16) 157-67.
10. MELINEK, S.J. A method of evaluating human life for economic purposes. Joint Fire Research Organization Fire Research Note 950/1972.
11. SAITO, H. Fire duration for design of fire resisting structure. Conseil International du Batiment colloquium on 'Safety of building structures in fire'. Paris, 1971.
12. MASKELL, D.V. and BALDWIN, R. A relationship between the fire resistance of columns and the cost of construction. Joint Fire Research Organization Fire Research Note 934/1972.
13. BALDWIN, R. Some notes on the mathematical analysis of safety. Joint Fire Research Organization Fire Research Note 909/1972.
14. BALDWIN, R. and THOMAS, P.H. Passive and active fire protection — the optimum combination. Joint Fire Research Organization Fire Research Note 963/1973.
15. BULL, J.P., JACKSON, D.M. and WALTON, C. Causes and prevention of domestic burning accidents. Br. med. J., 1964, 2 1421-7.
16. BAXTER, J.R., PACKHAM, D.R. and PEET, G.B. Control burning from aircraft. Commonwealth Scientific and Industrial Research Organisation Chemical Research Laboratories. Melbourne, 1966.
17. DUNN, JENNIFER E. and FRY, J.F. Fires fought with five or more jets. Fire Research Technical Paper No. 16. London, 1966. H.M. Stationery Office.
18. HOLROYD, Sir RONALD (Chairman). Report of the Departmental Committee on the Fire Service. Home Office and Scottish Home and Health Department Cmnd 4371. London, 1970. H.M. Stationery Office.

19. FRY, J.F. An estimate of the risk of death by fire when staying in an hotel. Instn Fire Engrs Q., 1970, 30 (77) 69-70.

20. BUTCHER, E.G., LANGDON-THOMAS, G.J. and BEDFORD, G.K. Fire and car park buildings. Fire Research Station Fire Note No. 10. London, 1968. H.M. Stationery Office.

21. STOREY, T.G. Fire control, planning sumulation and the Forest Service. U.S. Department of Agriculture Forest Service Pacific Southwest and Range Experiment Station. Berkeley, Calif., 1971.

22. HOGG, JANE M. The siting of fire stations. Opl. Res. Q., 1968, 19, 275-87.

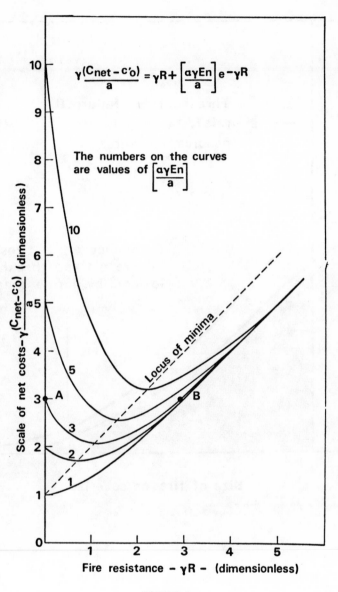

$$\gamma\frac{(C_{net}-c'_0)}{a} = \gamma R + \left[\frac{a\gamma En}{a}\right]e^{-\gamma R}$$

The numbers on the curves are values of $\left[\frac{a\gamma En}{a}\right]$

Locus of minima

Scale of net costs — $\gamma\frac{(C_{net}-c'_0)}{a}$ (dimensionless)

Fire resistance — γR — (dimensionless)

FIGURE 1

NOTE: Figure captions for this chapter are listed on page 58.

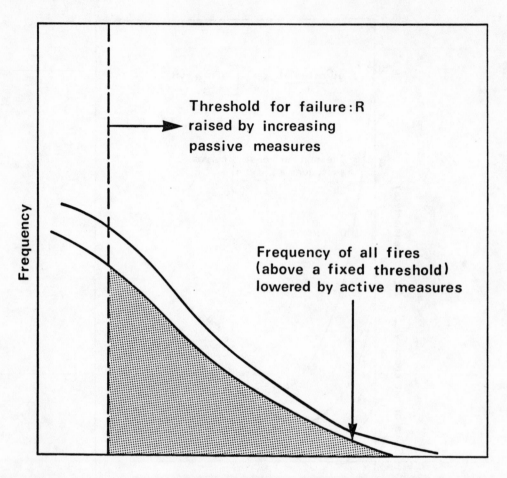

Size of fire or severity

FIGURE 2

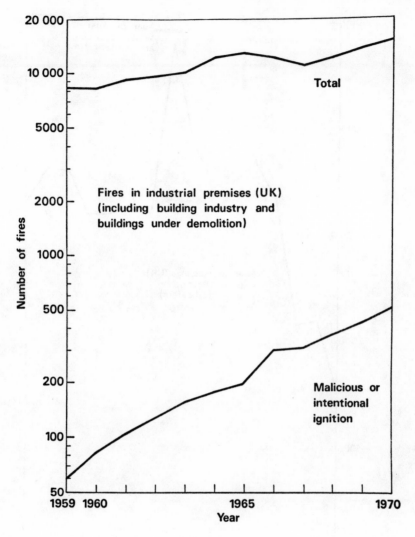

FIGURE 3

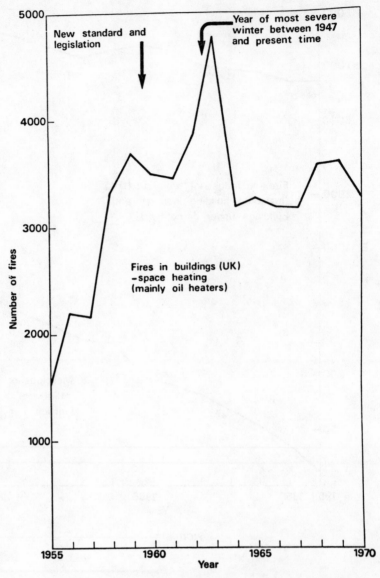

FIGURE 4

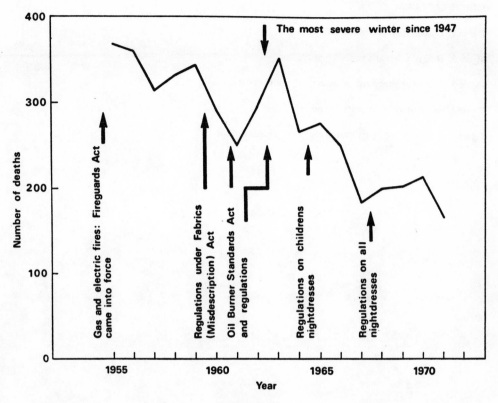

FIGURE 5

FIGURE CAPTIONS

Figure 1. Optimization of fire costs

Figure 2. Role of active and passive measures

Figure 3. Increasing importance of malicious or intentional ignition

Figure 4. Reduction in fires caused by oil heaters

Figure 5. Deaths caused by ignition of clothing (Registrar—General)

PART II

GEOMETRIC PARAMETERS FOR CLASSIFYING FULL-SCALE FIRES

Chapter 1

EFFECTS OF FUEL GEOMETRY ON FIRES IN SOLID FUEL ARRAYS

P. H. THOMAS
Fire Research Station, Borehamwood
Herts, England

INTRODUCTION

IN EVEN THE SIMPLEST ATTEMPTS to classify fire risks, in forests or in various types of building occupancy, some implicit recognition is given to the amount of fuel present. The more fuel there is the more likely is an ignition source to come into hazardous proximity to it, and although, for example, surface area and volume are separate geometric properties of the fuel they are likely to be correlated with the total amount, at least for a given type of building occupancy. The amount of fuel is also more easy to estimate than are other properties and even today, surveys of fire loads in buildings perforce concentrate on total quantity; specific surface and uniformity of distribution over a floor are less adequately reported.

We shall discuss some of the roles in fire behaviour of the geometric properties of fuel. It will emerge that some of our knowledge is very unevenly distributed. A great deal of work, for example, has been done on the ignition of vertical surfaces by radiation, very little on any other orientation, and little if any on the effects of departure from a flat surface.

The differences in ignition behaviour may not be very large, the difference between the tendencies of hollows, say, and plane surfaces to <u>continue</u> burning after the ignition source is removed can be much more important.

We shall examine the ignition, some selected aspects of spread and the steady burning of solid fuel beds; we shall omit problems peculiar to liquid fuels, viz the role of ullage in tank fires and the role of thickness in the burning of liquid films spreading along a surface.

NOTE: All figures quoted in text are at the end of the chapter.

1.1 IGNITION

Ignitability may have to be assessed by two properties — the threshold condition for ignition and the speed of response to an igniting source above the threshold. Neither are entirely separate from geometric effects.

There is, in general, no ignition temperature which is a property of the material as such. For example, in ignition by self-heating the heat generation occurs, albeit non-uniformly, within the volume of the material and is lost by cooling from the surface. Critical temperatures are therefore size dependent. Spontaneous ignition of the flammable gases emitted by pyrolysis in solids may also involve self-heating within the gas stream, so that the thickness of the external boundary layer will be relevant.

The thermal response of a material heated by an external source can clearly be influenced by its thickness, thin materials heating more rapidly than thick ones. The effect is relevant not only to the onset of ignition but also to whether the ignition leads to sustained burning on withdrawal of the original igniting source. Thick wood can be ignited, but is less likely to continue burning than thin wood once thermal exposure stops. The effect results from the cold interior acting as too strong a heat sink.

Extensive reviews of the ignition, particularly of wood and related cellulosic material, have appeared in the literature [1,2].

1.2 FIRE SPREAD

Thin Fuel

Figure 1.1 shows some typical rates of spread of fire systems. Spread along paper was studied in Japan [13] in the 1920's and it has long been recognized that thin materials burn fater than thick ones —for the same reason that they ignite sooner—they are weaker heat sinks with respect to heating from outside the surface.

Papers and fabrics of a given substance burning in air in various thicknesses follow an inverse law [3-5] between the rate of spread V and the thickness usually observed as

$$mV = \text{constant} \qquad (1.1)$$

where m is the mass per unit area.

Because the mass rate of involvement mV of thin material by an advancing fire is independent of m the pre-heating of the solid fuel to ignition is itself independent of m — as one would expect if the source of heating is external to the fuel. This is not so for fires in porous beds which are not too shallow. For these the rate of spread is independent of the thickness of the fuel bed and the inverse relation (for thin fuel elements) is between V and bulk fuel density ρ_b (see Fig. 1.2) based on data given by Byram et al [6].

$$\rho_b V = \text{constant} \qquad (1.2)$$

This is a consequence of the pre-heating of unburnt fuel to ignition by a flux within the fuel bed [7,8].

Thick Fuel

For thick fuels, the effective thickness is the depth heated, viz kt where the characteristic time t is proportional to $1/V$ where 1 is the distance scale of the heating process, e.g. the mean free path through the fuel bed if the spread is by radiation through a bed. For vertical spread on the surface of a semi-infinite thick solid driven primarily by the flame, 1 is flame length L.

Hence a first approximation for a heat balance is

$$\Delta H_i V p \sqrt{\frac{kL}{V}} \approx \left(q'' - \alpha \theta_i\right) L \tag{1.3}$$

We have neglected longitudinal conduction: this is permissible for fast spread.

More elaborate models which include the coupling between the heat producing reaction (or a simple form of it) and the aerodynamics of the thermal feedback by convection have recently been studied by de Ris [9] and Tarifa et al [10] and Lastrina et al [11], who have discussed the role of diffusion at the base of the flame.

The Distribution of Specific Surface

The variation of specific surface within a fuel bed has an importance as well as the total surface. Thick fuel is likely to be burning when all the thinner fuel has burnt out and if the thick fuel is too isolated it may be unable to sustain continued burning. This effect will depend on the previous moisture changes because moisture changes have time constants dependent on fuel size. Thus in forests thin fuel can dry out more quickly after rain than thick fuel.

Effect of Fuel Depth

The constancy of $\rho_b V$, Eq. (1.2) implies that heating from outside the fuel bed is of secondary importance but if the fuel bed is shallow the contribution of the flames above the fuel bed may no longer be negligible. This may be seen by examining the relationship between flame size and fuel burning rate R which is nonlinear [12]

$$\frac{L}{D} \propto \left(\frac{R^2}{D^s}\right)^n \tag{1.4}$$

for a two-dimensional source of width D, n is less than $\frac{1}{2}$. If V is independent of the depth h_c of the fuel bed, R/h_c is constant, then

$$L \propto h_c^{2n}$$

For a spreading fire D is largely controlled by the thickness of the fuel elements and the rate V, and is largely independent of h_c (if ρ_b V is constant). Thus $\dfrac{L}{h_c}$ is small for deep fuel beds: burning buildings have flames of a height of the same order as the building height and the flames do not contribute much to the spread (unless deflected by a wind). L is absolutely smaller for a grass fire, it is relatively larger and its orientation more affected by the wind and the role of the flame is more important. Also, in a wind, 'flying brands' may determine the spread of a grass fire particularly in dangerous windy conditions and a conflagration produces its own winds.

Despite these qualifications one can see that there may be at least two regimes of behaviour associated with differing scales of fuel depth.

1.3 THE BURNING OF STICKS

The classic problem of interpreting how fast a stick of wood burns raises many problems as yet incompletely solved though this is probably less the result of deficiencies in knowledge of combustion than in our knowledge of the properties of wood both virgin and pyrolised.

The duration of burning varies roughly as the three-halves power of the diameter or thickness of sticks. Various writers have described the relation as

$$t_B \propto b^n \tag{1.5}$$

with $n \approx 1.6 \overset{+}{-} 0.2$ approx.

It has been pointed out by Gross [13] and Kanury [14] and others, that this result is close to that expected from considerations of thermal diffusion into the sticks. If the heating condition imposed on an inert material, say, a given temperature on the surface, the propagation of an isotherm to describe pyrolysis leads to the well known diffusion square law ($n = 2$) but a more realistic surface condition with surface cooling leads to a dependence of t_B on b reasonably well described by the 1.6 power law as a first approximation. Although not a description of the pyrolysis process this relation illustrates the important role of thermal diffusion, which is responsible in part for the scale effect. Similar relations have been reported for the burning times of assemblies of widely spaced sticks (open cribs) [13, 15]. Diffusion considerations alone will not predict the limiting conditions in which burning is just possible. It is commonplace that a single stick may be too thick to sustain steady burning alone but assemblies of such sticks may do so, presumably owing to the retention of the radiated heat.

Various computed solutions exist of mathematical models of pyrolysis though there are many uncertainties in the physical properties of wood and char at high temperatures. A recent model includes some considerations of the effect of these [16].

1.4 RATE OF BURNING CRIBS

Gross [13] distinguished between cribs with large spaces between the sticks (open cribs) which burnt at a rate determined by the stick size and cribs which had small values of A_v/A_s (densely packed) which burnt at rates determined by A_v/A_s.

Block [15] developed a theory for densely packed cribs which leads to a relation different from, but apparently as satisfactory as Gross's empirical one. Block argued that the flow conditions within the crib could be approximated by vertical flow up the shafts without much horizontal flow so that he could solve the flow equation as for a duct under buoyancy. He assumed a constant friction factor, and applied the Reynolds Analogy. He treated the wood as endothermic with the

vapour having total heat of 1.750 MJ/Kg* and the solid a surface temperature rise of 370°C and obtained a good agreement with his data.

On the basis of an analysis of an extensive set of data Smith and
Thomas [17] had expressed the burning of cribs by

$$m'' = \frac{R}{A_s} \propto \sqrt{h_c \frac{A_v}{A_s}} \qquad (1.6)$$

which also accounts for all but the smallest of Gross's cribs. The
data they correlated are not so well fitted by Block's theory.

However if the Reynolds Analogy does not hold for cribs and the
heat and mass transfer coefficients are less than assumed by Block
better agreement is obtained [18].

1.5 FUEL GEOMETRY IN ENCLOSURES

Much work has been done with cribs in compartments but unlike
most others Nilsson [19] raised his cribs above the base. This makes the
assumption of vertical flow more realistic. Some of his data refer to
a regime controlled by the crib porosity, not by the fuel surface (for
which m" would be more nearly constant) nor by the window which has
only a secondary effect for that regime. These data are now being
analyzed [20] and one can now define crib designs to be avoided in experiments where the crib is meant to represent an amount of fuel of given
specific surface.

1.6 STOCHASTIC MODELS OF SPREAD

We have just referred to non-uniformity in the fuel bed and such
variation, natural to forests but also present in buildings, has led
to the consideration of stochastic models of spread [21].

These of course are appropriate for considering large collections
of data, not for any particular fire. We consider here only a highly
idealized linear spread and we consider the early stages of a fire
propagating across a series of cells or strips.

Let $P_s(t)$ be the probability that the fire is burning in the
(s)th cell, but not burning beyond it. We consider a much over-simpli-
fied model that there is a probability of $\lambda \delta t$ of spreading to the
(s + 1)th cell in the time interval t to $t + \delta t$ and we neglect burn-
out.

The basic equation is then

$$P_{s+1}(t + \delta t) = P_s(t) \lambda \delta t + (1 - \lambda \delta t) P_{s+1}(t) \qquad (1.7)$$

in particular

$$P_1(t + \delta t) = (1 - \lambda \delta t) P_1(t)$$

We define the initial condition as

*Values reported for this property range from small negative values to
over twice that assumed by Block.

$$P_1 (0) = 1$$

$$P_s (0) = 0 \quad (s \neq 1)$$

it readily follows that

$$P_s (t) = \frac{(\lambda t)^{s-1}}{(s-1)!} e^{-\lambda t}$$

The mean distance of spread is

$$\sum_{s=1}^{\infty} (s-1) \, Ps(t)$$

which is λt so λ can be related to the rate of spread. Phung and Willoughby [21] include burn-out where the probability of burn-out is related to the duration of burning of a fuel element and the discuss two-dimensional spread. The solution of all but such trivial problems as above, in general, defy analytic procedures.

Two-dimensional spread models are relevant to conflagrations and, insofar as the mechanism of spread is radiation between buildings and excludes for example spotting by 'flying brands', the probabilities of spread are related to the fraction of all buildings closer to their neighbour than the critical separation distance for that part of the building. In such ways, physical processes and statistical modelling are linked together to allow for the variation in real fuel beds.

1.7 CONCLUSION

A brief reference has been made to some of the more obvious effects of fuel bed depth, porosity, and fuel thickness and its variation on ignition, rate of spread and steady rate of burning.

Practical grading systems for fire risk in buildings (if not in forests) recognize effects of the fuel other than its total density per unit area of floor (fire loading) more implicitly than explicitly and this emphasises that some aspects of fire behaviour may have to be treated by statistical modelling. For this the fire technologist is called upon to supply input data, both physically meaningful and practically available.

1.8 NOMENCLATURE

A_s — fuel surface area

A_v — total area of cross-sections of vertical shafts in crib

b — stick thickness

D — a characteristic linear dimension

h_c — fuel bed height

ΔH — fuel enthalpy

k — thermal diffusivity

l — characteristic distance of heat transfer

L — flame length

m — mass of fuel per unit area

m'' — mass rate per unit area $(= {}^{R}\!/_{A_s}$)

n — a coefficient

P — probability (Eq. 1.7)

q'' — heat flux per unit area

R — rate of burning

s — an integer

t — time

t_B — duration of burning

V — linear rate of spread

α — heat transfer coefficient

θ — temperature rise

λ — probability of spread per sec

ρ — density

ρ_b — bulk density

Subscript

i — ignition

1.9 ACKNOWLEDGEMENT

The paper is Crown Copyright reproduced by permission of the Controller, her Britannic Majesty's Stationery Office. It is contributed by permission of the Director, Building Research Establishment. (The Fire Research Station is the Joint Fire Research Organization of the Department of the Environment and Fire Offices' Committee)

1.10 REFERENCES

(1) SIMMS, D. L.   Ignition of cellulosic materials by radiation. Combust. Flame, 1960, 4 (4) 293–300.

(2) MARTIN, S. B.   Diffusion-controlled ignition of cellulosic materials by intense radiant energy. 10th Symposium (International) on Combustion. pp.877–896. Pittsburgh, 1965. Combustion Institute.

(3) KINBARA, T.   A survey of fire research in Japan. Fire Res. Abstr. Revs, 1961, 3 (1) 1–12.

(4) STOTT, J. B.   A study of flame spread over cellulosic materials. PhD Thesis, Leeds University, 1949. (Unpublished)

(5) LAWSON, D. I., WEBSTER, C. T. and GREGSTEN, M. J.   The flammability of fabrics. J. Text. Inst., 1955, 46 (7) T453–463

(6) BYRAM, G. M. et al.   An experimental study of model fires. U.S. Department of Agriculture, Forest Service, South Eastern Forest Experimental Station, Technical Report No 3. Macon, Georgia, 1964.

(7) A study of fire problems (1961). U.S. National Academy of Sciences, National Research Council, Publication 949. Washington, DC, 1961.

(8) THOMAS, P. H. and SIMMS, D. L.   A study of fire spread in forest fires. Report on Forest Research, 1963, pp. 108–112. London, HM Stationery Office.

(9) de RIS, J. N.   Spread of a laminar diffusion flame. 12th Symposium (International) on Combustion. pp. 241–252. Pittsburgh, 1969. Combustion Institute. (see also discussion to ref. (11)).

(10) TARIFA, C. S., PEREZ del NOTARIO, P. and MUNOZ TORRALBO, A.   On the process of flame spreading over the surface of plastic fuels in an oxidizing atmosphere. idem. pp. 229–240.

(11) LASTRINA, F. A., MAGEE, R. S. and McALEVY III, R. F.   Flame spread over fuel beds: solid phase energy considerations. 13th Symposium (International) on Combustion. pp. 935–948. Pittsburgh, 1971. Combustion Institute.

(12) THOMAS, P. H.   The size of flames from natural fires. 9th Symposium (International) on Combustion. pp. 844–859. New York, 1965. Academic Press.

(13) GROSS, D.   J. Res. Natn. Bur. Stand., 1962, <u>66C</u> 99.

(14) MURTY KANURY, A.   Burning of well-ventilated wood cribs. Factory
     Mutual Research Corporation. Serial No. 19721-1, 1970.

(15) BLOCK, J. A.   A theoretical and experimental study of nonpropa-
     gating free-burning fires. 13th Symposium (International) on
     Combustion.pp. 971-978. Pittsburgh, 1971. Combustion Institute.

(16) KUNG, H. and KALELKAR, A. S.   On the heat of reaction in wood
     pyrolysis. Combust. Flame, 1973, <u>20</u> (1) 91-101.

(17) SMITH, P. G. and THOMAS, P. H.   The rate of burning of wood
     cribs. Fire Technol., 1970, <u>6</u> (1) 29-38.

(18) THOMAS, P. H.   (To be published)

(19) NILSSON, L.   The effect of the porosity and air flow factor on
     the rate of burning for fire in enclosed space. Statens Institut
     för Byggnadsforskning. Rapport R22, Stockholm, 1971.

(20) THOMAS, P. H. and NILSSON, L.   (To be published)

(21) PHUNG, P. V. and WILLOUGHBY, A. B.   Prediction models for fire
     spread. Pyrodynamics, 1965, <u>2</u> (1) 39-52.

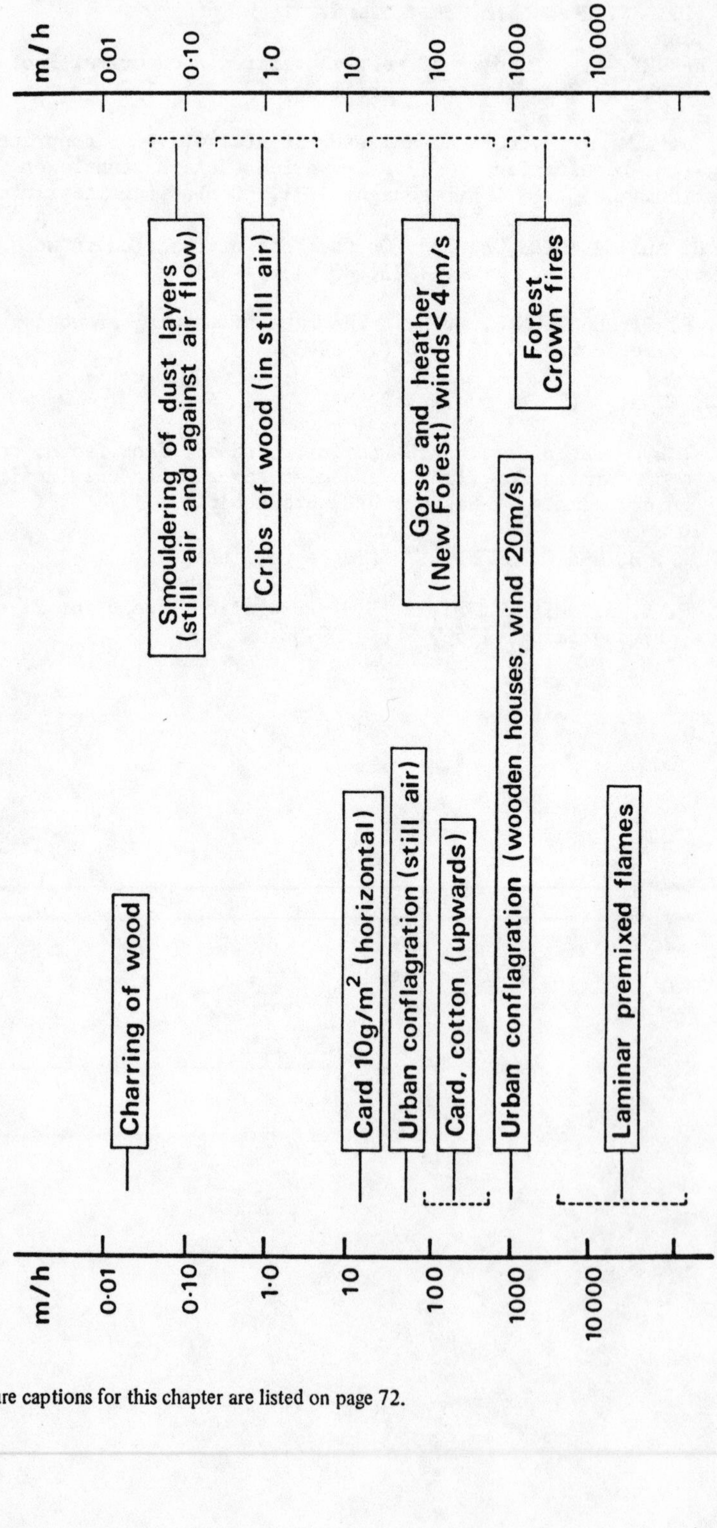

m/h
0·01
0·10
1·0
10
100
1000
10 000

Smouldering of dust layers (still air and against air flow)

Cribs of wood (in still air)

Gorse and heather (New Forest) winds < 4 m/s

Forest Crown fires

Charring of wood

Card 10g/m² (horizontal)

Urban conflagration (still air)

Card. cotton (upwards)

Urban conflagration (wooden houses, wind 20m/s)

Laminar premixed flames

m/h
0·01
0·10
1·0
10
100
1000
10 000

FIGURE 1

NOTE: Figure captions for this chapter are listed on page 72.

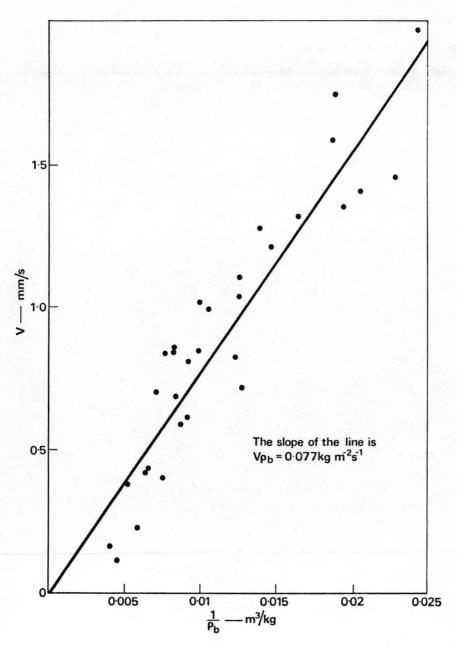

FIGURE 2

FIGURE CAPTIONS

Figure 1. Some typical rates of spread

Figure 2. The effect of bulk density on fire spread
(Based on data from Byram et. al. [c])

Chapter 2

FIRES IN ENCLOSURES

P. H. THOMAS
Fire Research Station, Borehamwood
Herts, England

INTRODUCTION

MOST OF THE INJURIES AND DEATHS and most of the financial loss from
fire are the result of fires in buildings and the principal traditional
design feature of buildings employed to reduce these losses is sub-
division by horizontal and vertical partitions. These contain the fire
and limit the size of the maximum likely loss. Such partitions are subject to perhaps the oldest of fire regulations, and have their parallel
in cities and forests in the sub-division of the fuel by the use of fire
breaks.

The advent in building of new materials and structures having high
strength to mass ratios leads to load-bearing partitions having lower
thermal resistance and capacity than in traditional materials and structures. The thermal impedence required to protect the load-bearing element of a structure may then have to be specially provided at additional
cost.

Some modern buildings are so large and have so many occupants that
total evacuation within say 10 min of a fire alarm is impossible and
therefore the containment of a fire is necessary, not only to limit
economic loss, and to ease the difficulties in fire-fighting, but also
to allow occupants to remain safely in some part of the building which
itself does not suffer structural damage during the course of the fire.
It is no surprise therefore that an important area of the fire protection of structures is the testing of structural elements, floors, walls
and columns in furnaces which simulate the action of a fully-developed
but confined fire at temperatures typically in the range 800–1300°C.
Such testing is done using an internationally agreed standard heating

NOTE: All figures quoted in text are at the end of the chapter.

program [1] and the property measured is the time for which the structural element is able to carry the load with limited distortion and in the case of partitions without allowing the passage of flame or too high a temperature rise on the surface away from the fire. The main problems in such structural testing are:

1. The variability in fire behaviour and the necessity for one or very few heating programs and the consequent problems of relating furnace and fire behavior.

2. The absence of a realistic representation of interaction between, say, beams and columns.

3. The variation between national furnaces and the consequent difficulties of standardization between them.

Much research has been devoted to examining the relationship between the transient fire and the standard used in furnaces [2]. If, for example, the failure of a particular structural element can be defined by the attainment of a particular temperature on some element, such as the prestressing wires in reinforced concrete, it is a relatively straightforward matter to compute the equivalence between the two temperature/time characteristics. Various methods have been reviewed by Lie [3]. The accuracy or inaccuracy of such proceedings must be assessed against the strict impossibility of defining failure in terms of a particular temperature at a particular depth in a structural element. In principle it is not possible to decouple from each other, the fire, the structural element and the way it is supported, and the loading condition, because the fire temperatures depend on the heat transfer into the structure, and failure depends on the stress which in turn depends on the loading. Also the accuracy needed depends on whether one is examining a particular structure or seeking some generalization appropriate to a class of structures. Whatever method is employed adequate material property values at high temperatures are required. Before examining in more detail the behavior of fully-developed fires we shall follow the development in general terms of a fire in a vented enclosed space.

IGNITION

One usually thinks of fire in a building as starting with ignition and then spreading, but the process of ignition is important even after the fire has started because its spread often involves further ignitions. Sometimes fire in part of a room will ignite a combustible item some distance away by radiation or hot gas, and even the continuous spread of flame along a surface may be regarded theoretically as a rapid succession of ignitions. Usually when we speak of ignition we do not really mean the ignition of a solid but the ignition of the gases evolved by the solid when it is heated. Combustible materials evolve volatiles more rapidly as they are heated and if oxygen is present the flammable vapors may ignite. If insufficient oxygen is available, as inside a room which is burning fiercely, and in which the only air available for combustion comes through the window, some of the fuel burns outside in flame emerging from the window.

Much of the description of the ignition of solid materials is a discussion on how they respond to heating [4, 5]. If the material is a bad conductor the surface warms up more quickly than if it is a good conductor. Also, materials having low density or low specific heat warm up quickly. Conventional heat conduction theory can be used to predict the rate of rise of the surface temperature under various heating conditions. When this temperature approaches the region of 300°C for untreated cellulosic materials such as wood, sufficient flammable vapors are emitted to be ignited by a small pilot flame. If there is no flame present, then the material must be heated up somewhat more, to about 500–600°C, before the gases will spontaneously ignite.

These temperatures which are not necessarily basic properties of a material, can be dependent on the duration of heating, higher temperatures being necessary for higher heating rates and shorter times. Nevertheless if we need to describe ignition among a large set of other processes the simplicity of assuming an 'ignition temperature' is very convenient and not very inaccurate. Although not a common cause of fires in buildings, spontaneous ignition resulting from slow heating over a long period is known to occur in certain solid materials which generate heat as they decompose. This heat can usually be disregarded if the material is exposed to a strong heat flux externally, though of course self-heating in the emitted volatile material may well be relevant to the ignition process itself.

FIRE SPREAD

In practice most combustibles in a building are not single materials but composites, with joints, and corners. The arrangement and amount of furnishings varies in time and in space (e.g., between one hotel room and another). There is therefore a stochastic element in the risk of ignition in an occupancy class. When some heating has taken place the solid surfaces and walls and ceilings themselves begin to emit radiation. Because radiation increases more rapidly with a rise in temperature than does convection or conduction, it is often the most important process at a later stage in spreading fire throughout a building, or from one building to another.

If we think of a large space with a relatively small part of that space containing burning materials, the floor on which those burning materials are resting becomes heated by conduction. The ceiling immediately above the burning materials and some distance away will become heated by a convection plume and since these convection currents travel at a few meters a second, materials a long way from the fire will be heated by the hot gases, an effect which of course can be exploited to operate fire alarms. However, near to the fire other combustible materials just surrounding the burning area will be heated by radiation from the flames and the heated ceiling. These three processes of heat transfer have different relative importance according to the situation and the stage of development of a fire.

What are the major points of interest about fire spread in buildings that concern us? Briefly there are two. Firstly, will it spread at all? Secondly, if it will, what is the rate of spread? We are interested in the latter because this determines how big the fire will become by the time the fire brigade arrives, and the fire-fighting and rescue effort required. The hazard from smoke of course deserves separate attention.

Flame Spread Along Continuous Surfaces

If the flammable surface is vertical, on a wall, say, convection may play an important part in heating the unburnt fuel above the burning zone. This heating leads eventually to further ignition which occurs when enough volatiles are being released fast enough to sustain a flame - at about 300°C for wood. In this way the burning zone extends.

Horizontal spread, along a floor, say, is much slower because more of the heat is convected away from the unburnt fuel ahead of the fire front. Conduction of heat through the material does not play an important role in spread along 'thin' materials.

Spread along 'thick' materials is little affected by the total thickness. Heat is lost into the interior of the material by conduction - unlike 'thin' materials - so that the thermal conduction properties of 'thick' fuels are important. If the fuel is a poor insulator, heat lost into the interior by conduction and spread will be slow; for a good insulator heat will be conducted away from the burning zone more slowly and spread will be quicker.

The thermal properties of a base material will influence the fire behavior of a coating which may itself be 'thin'. Rates of fire spread have been correlated with the thermal properties of a base and different thicknesses of an applied coating [6]. Comparisons between different coatings are, however, at present a matter for testing rather than theory, some aspects of which will be pursued in greater detail below.

Spread in a Room

If there are no extensive readily flammable surfaces in a room, fire must spread from one item of furniture to another for fire to propagate and in the absence of continuous flammable floor surfaces this requires the fire to 'jump' across the intervening gaps: there are certain similarities with urban conflagrations. A sparsely furnished room is less likely to produce a fully-developed fire than a densely furnished one. However if the fire can spread enough for flame to extend under the ceiling, the radiation emitted downwards extends over a wider area (see Fig. 2.1) and fire then spreads more readily through the fuel standing on the floor.

Plate II.I demonstrates this important aspect of fire spread in compartments. Here the fuel bed is tall in relation to the compartment and approaches the ceiling (as do goods in warehouses). One such pile burning freely would produce flames much taller than the compartment but here such flames are deflected horizontally beneath the ceiling and lengthen because of the lower rate of entrainment between the heavy air and hot light fuel steam. The ceiling is heated and the area heated by radiation from above is considerably increased. In an extreme situation as shown here the fire spreads over the tops of nearby piles of goods. High piles of fuel and tall furniture beneath low ceilings are hazardous because of this effect which is due to the presence of a ceiling as well as its nature.

An analysis of this combination of spread along a surface in one part of a room and spread by jumping in other parts is clearly not simple and perhaps the best approach is a statistical one which gives, say, the probability that fire will spread and not die out, or the probability that the whole room will be involved in some given time [7]. We have not referred much to the effect of the room as such, largely because during the initial spread the fire is small enough to behave as if it were in the open. Work on open fires, forest fires, is relevant here. However when the fire grows large enough its enclosure in a confined space becomes important. Growth often accelerates until it is stopped, either because there is no unburnt fuel for the fire to spread to, or because the rate of air supply becomes a limiting factor.

Flash over

If fire growth continues unimpeded it is customary to speak of development leading to 'flashover' - but it is far from clear whether this word is always used to mean the same phenomenon. Firstly there are various types of fire development in a compartment and secondly there are various types of fully-developed (or post-flashover) fires.

The following table is not comprehensive but it includes the main types of fire and the most general use of the word 'flashover' is to denote a transition from A to B.

Table 2.1

Type of fire and factors controlling behavior

A Developing fire	B Fully developed fire
1 Adequate ventilation Fire spreads by heating of unburnt fuel which when ignited burns freely	Fire burns almost as if in the open and is strongly influenced by the amount of surface area of fuel
2 Restricted ventilation Spread is slower. The fire may occasionally become choked by lack of oxygen	Conventional windows can restrict air supply to fire. Rate of burning depends primarily on area of window opening

All fires starting from a small source begin with sufficient ventilation: ie (A1). They may then change to (A2) if the windows are not broken. A fire can go from (A2) to (B1) if windows break or a door is opened at a crucial time and one may use the term 'flashover' for what is observable as the ignition of unburnt gases. Fully-developed

fires are often of the type (B2) (until the roof or floors collapse).
Thus even within this limited classification of fires we can see that
there is more than one way for a fire to grow and reach the fully-
developed stage. Some (A2) fires are self-extinguishing and never
become fully-developed - this is sometimes the course to be followed
by a fire in a small room if the windows do not break and all doors are
closed.

'Flashover' is usually spoken of as a 'time' and because the factors controlling the behavior of the fire are often different before and
after 'flashover' eg the arrangement and action of fuel before, and the
air supply afterwards, the 'flashover' time divides the fire into two
nearly independent phases. Even breaking windows may not provide enough
air to enable the fire to burn completely freely but the rapid change in
the rate of burning even from (A1) to (B1) - the rapid engulfing of large
areas of fuel by flames and the emergence of flames out of windows - has
led to this transition also being called 'flashover'. It is tempting
to define 'flashover' as the moment when all exposed flammable surfaces
contribute to the production of gaseous fuel, but this may not be
possible in very deep compartments where all the fuel cannot be involved
at the same time. It is difficult to make a definition that is both
useful and general.

Prior to 'flashover' life is at risk in a building from flame,
smoke and toxic products. The rapid increase in the rate of burning and
the temperatures accompanying 'flashover' (see Fig. 2.2) generally leads to
much greater quantities of toxic products and puts a larger space into a
hazardous condition although the character of the combustion may change
from (A2) to (B1) or (B2) to produce subsequently lower concentrations
of toxic products.

If the discovery of a fire depends upon its size - not how quickly
it is changing its size - a short 'flashover' time will mean that there
is a greater likelihood that the fire brigade will be confronted on
arrival with a big, rather than a small, fire. Although life is
endangered before 'flashover', conditions in those parts of a building
remote from a fire rapidly worsen at 'flashover'. Design and construction features, such as flammable linings, which assist fire growth - ie
hasten 'flashover' - may be subject to building controls.

Fire Spread from Room to Room

The spread of fire within a building has rarely been studied. Smoke
movement has been computed, largely in terms of plug flow driven by
buoyancy. Labes [8] has proposed a formula of exponential growth on the
basis of some of the very limited work done on real buildings subjected
to complete burn out.

Some Other Relevant Studies of Fires Developing in Enclosures

Most of the detailed studies of heat and mass transfer on growing
fires have been made with the view to designing detection and sprinkler
systems. Thus Alpert [9] has recently studied the details of the interaction of a rising plume and a ceiling. Orloff and de Ris [10] have studied
steady-state ceiling fires and have derived theoretical relations between
dimensionless mass transfer rates and the Spalding mass transfer number B

for the fuel (radiation omitted). Torrance et al [11] and Larsen [12] have made calculations of the flow from steady thermal sources in enclosures and the present development of computer studies of flow combined with equations for chemical reaction are the forerunners of major developments in this area [13]. Major problems are to validate adequate descriptions of turbulence in the fire system, to predict theoretically the formation and size of solid particles in flames and hence the radiation from luminous flames, and to accommodate the variation of the conditions in practice which can only be determined in a statistical manner. Certain particular features of fire spread in enclosures have been studied in some detail; thus Thomas and Hinkley [14] devised design rules for automatic venting in large single-story industrial buildings based on simple buoyancy scaling. Much research has been devoted to fires in one-dimensional systems because of the current concern with escape routes in large buildings and with fire spread through service ducts which often perforate fire resisting partitions and are themselves liable to be made of flammable material. Fires in mine roadways, driven by forced ventilation, have been studied by Roberts and Clough [15] and de Ris [16] who developed a theory in terms of quasi steady propagation. De Ris wrote the overall energy relation as:

$$M_1 m_{ox} \frac{\Delta H}{r} = PVQ_1 + q_w \qquad (2.1)$$

and calculated q_w to be about 36 per cent of the heat released by combustion and based on Roberts' and Clough's data proposed a value for Q_1 of 3.4 MJ/kg. In mines there are many possible ignition sources in or near the main roadway but in corridors and escape routes in buildings there are few and these are mostly in rooms off the escape route itself. The emphasis in research is likely to be on preventing spread into the escape route, ie on self-closing doors, pressurization of corridors to prevent smoke leakage and on suppression systems, rather than on the behavior of the propagating fire.

We now return to the confined fire and consider the fully-developed fire in a single compartment in order to estimate how hot it becomes and for how long it burns.

FULLY DEVELOPED FIRES

By fully-developed we mean that stage of the fire when it has become as large as it can be within the enclosing partitions. The fire will be limited either by the availability of air or by the total fuel surface. The problem appears first to have been examined by the US National Bureau of Standards in the 1920's [17]. Early studies concentrated mainly on those fires in which the burning was controlled by the flow of air in through the window and it is with this that we shall deal first.

Air Flow

The air flow into an enclosure through a single window is conventionally calculated using a formula which was originally derived by Fujita [18] and which became the basis of work by Kawagoe [19] and others:

$$M_1 = \frac{2}{3} \frac{C_D P (2g\theta/T_2}{[1 + (T_2/T_1)^{1/3} (M_2/M_1)^{2/3}]^{3/2}} \qquad (2.2)$$

The equation is derived assuming that the temperatures in the enclosure
are uniform and that there is no internal vertical acceleration or
friction. For most wood fires it is sufficient to put M_2/M_1 as 1.2 on
the right hand side and for temperatures above about 500^oC one can take

$$M_1 \approx 0.5 \, A_w H_w^{\frac{1}{2}} \qquad kg \; m \; s \qquad (2.3)$$

This approximate relationship breaks down for very small window openings
because the temperature falls and the approximation of high temperature
is no longer valid. It also fails as the windows [20] become large,
firstly, for experimental fires using cribs, because the resistance of
the crib fuel bed to the air flow may become more important than the
resistance of the window, and secondly because the temperature ceases
to be uniform within the enclosure and the flow of air into the
combustion zone is controlled by entrainment processes rather than by
the chimney effect produced by the buoyant stack. Despite these and
other limitations this relationship plays a central part in fire
engineering. Clearly for fires in which the heat release is controlled
by the air supply and hence by the size of the window one can calculate
a first approximation to the temperature attained allowing for heat loss
to the walls, ceiling etc., and one can include the transient element
associated with the thermal capacity of the structure [21-26]. However
this relation alone cannot tell us how long the fire will last and for
this we turn to discussion of the rate of burning.

Rate of Burning

The values of the rate of burning are usually given as mean rates
over a period omitting the first and last stages of a fire. Typically
the rate is given over the time that the fuel mass falls from 80 to 30
per cent of its initial value. Temperatures are frequently quoted as a
temperature/time relation and recently data have been reported as mean
values similarly defined over the 80/30 period. Various writers
including those referred to in references 21-26, have used the following
experimentally derived relations or some minor variant of them. The
work of Fujita [18] and Kawagoe [19] led them to correlate the burning rate
'R', of wood fuel in compartments as proportional to the air flow and so
by equation (2.3) to a proportionality between R and $A_w H_w^{\frac{1}{2}}$ viz

$$R = 5.5 \, A_w H_w^{\frac{1}{2}} \qquad kg/min \qquad (2.4)$$

where A_w and H_w are in m^2 and m respectively.

For fires in compartments with large enough windows the values of
R tend to be controlled by the fuel bed and independent of $A_w H_w^{\frac{1}{2}}$. Use
is then frequently made of the nominally constant charring rate of wood
observed in fires. We summarize the appropriate mean values of R based
on the majority of data (see Harmathy [26])

(a) For the window-controlled regime (equation 2.4)

$$R \approx 0.095 \, A_w H_w^{\frac{1}{2}} \quad kg \; m \; s \qquad (2.5)$$

Recent experiments have shown even for window-controlled fires that $R/A_w H_w^{\frac{1}{2}}$ is only effectively constant within a limited range of design conditions, albeit typical ones for real buildings [27]. A few values of nearly 0.2 for $R/A_w H_w^{\frac{1}{2}}$ (about twice that in equation (2.5)) have been reported [28,29] and recent work has established its systematic nature [27] (see Fig. 2.3).

(b) For the fuel-surface-controlled regime

$$R \approx 0.006 \, A_s \quad kg \; m \; s \qquad (2.6)$$

(This omits the well known variation of R/A_s with stick size; the coefficient 0.006 is typical for, say 2-4 cm). Tsuchiya and Sum [24] have performed computations of the transient temperature employing refinements of these relations using whichever one gives the lower rate of burning in a particular case. Although these relations or refinements of them have appeared extensively in the literature there is one serious drawback: these relations are empirical, and are based almost exclusively on wood fires. Before giving an outline of a suitable theory we should note that a first approximation for free-burning cribs has been given so that in experimental fires we approximate in addition to regimes (a) and (b) a regime dependent on the porosity of cribs.

(c) For cribs controlled by their porosity [30]

$$R \approx 0.07 \, \sqrt{h_c A_v A_s} \quad kg \; m \; s \qquad (2.7)$$

from which first approximations to the boundaries between the three regimes can be obtained.

Energy Equation

An energy equation for the window-controlled regime may be written as:

$$M_1 \left(m_{ox} \frac{\Delta H}{r} \right) = M_2 \, \Delta H_2 + q_w + I_w A_w + R \left[Q_1 + \left(\frac{1}{w} - 1 \right) Q_2 \right] \qquad (2.8)$$

where

$$M_2 = M_1 + R$$

We have here neglected any oxygen in the flow out of the window and the small transient changes in the enthalpy of the gases within the enclosure. We have included the energy Q_1 required to produce unit mass of volatiles from the solid wood at its initial temperature and also the sensible heat Q_2 of unit mass of residue at its mean temperature. If R/M_1, is taken as constant the last term can be accommodated in a redefinition of the rate of heat release as implied by equation (1.3) and (1.4) and equation (2.8) written with

$$m_{ox} \frac{\Delta H'}{r} = m_{ox} \frac{\Delta H}{r} - \left(\frac{R}{M_1} \right) Q_3 \qquad (2.9)$$

where

$$Q_3 = Q_1 + \left(\frac{1}{w} - 1\right) Q_2 \qquad (2.10)$$

It is essentially this form of heat balance using $\Delta H'$ which has been used by Kawagoe [23], Sekine [24], Odeen [22], Magnusson and Thelandersson [24] and most recently by Harmathy [26] in using 'best fit' values for $R/A_w H_w^{\frac{1}{2}}$ and for an effective calorific value of solid wood of 10.8 MJ/kg. With equation (2.3) these give

$$m_{ox} \frac{\Delta H'}{r} = \frac{0.095 \times 10.8}{0.5} \approx 2 \text{ MJ/kg}$$

By coupling the above relation to the transient heat transfer through the walls, ceiling, etc, transient temperature relations have been computed by these authors.

Values of $m_{ox} \dfrac{\Delta H}{r}$ based on the calorific value of wood volatiles given by Roberts [31] on the basis of his own and other data are equivalent to over 3 MJ/kg for wood with less than 15 percent initial moisture and 3.5 MJ/kg for the volatiles from dry wood.

The term $(\frac{1}{w} - 1)Q_2$ is small (approximately 0.1 MJ/kg) as is the heat lost by radiation through the window, but the term Q_1 may be important and is the subject of much interest. Estimates vary from small negative values (an exothermicity of the order of 0.4 MJ/kg) to values of the order of 4 MJ/kg (endothermic) (c.f. de Ris's estimate above). The above equation is almost useless for determining Q_1, since small variations in the relation between R/M_1 and $R/A_w H_w^{\frac{1}{2}}$ are magnified in the estimate of Q_1.

Several authors have sought to describe the pyrolysis of wood including recently Murty Kanury and Blackshear [32], Roberts [33] and Kung and Kalelkar [34]. There is still uncertainty concerning the value of Q_1, partly because of uncertainties in the property values of wood volatiles and char residue.

The Fuel Surface Equation

As well as a total energy balance there is also a local energy balance for each part of the fuel surface and in an approximate integrated form this can be written as:

$$I_f A_s + h_c A_s (T_2 - T_s) = R(Q_3 + Q_4) \qquad (2.11)$$

We assume that the exit temperature and the mean gas temperature inside the compartment are the same T_2.

The last term on the right hand side represents the sensible heat flow in the flux of volatiles leaving the fuel and entering the gas phase. If the heat transfer within the pyrolyzing solid is high the volatiles will emerge at the surface temperature of the residue i.e., T_s but if the transfer between the convective flow and the solid residue is poor the volatiles will emerge at a temperature little above the median temperature of their formation, T_p. The flux, I_f, depends on the gas temperature (moreover on its 4th power) and the fuel surface temperature and, for flames which are not fully emissive, on the wall temperatures which are present in subsidiary relations determining q_w. The dependence on flame emissivity itself raises considerable difficulties for anything but a simplified treatment because of the difficulties in predicting the soot particle size and the emissivity of the flame in terms of other variables. This may well be less of a problem for real fires which are large than for small experimental fires where the emissivities would be expected to be lower. These two relations are themselves not sufficient to determine the properties of a fire; one needs either to assume that the fuel surface temperature is some particular value or to describe in greater detail the process of pyrolysis within the solid. Of course for a liquid fuel T_s can be taken as in effect the boiling point and perhaps for some plastics which melt readily a similar approximation is possible. With some assumptions and approximations one can obtain from equations (2.9) and (2.11) values of R and T_2 as functions of other terms either as mean values or if q_w is coupled to the thermal transfer through the boundary surfaces, as a transient. It emerges that the value of $R/A_w H_w^{\frac{1}{2}}$ is not constant but increases somewhat as $A_w H_w^{\frac{1}{2}}/A_T$ falls, which is in accordance with recent observations [29], but experimental data are not sufficiently precise and extensive to test the theory in detail.

Fires with Very Low Levels of Ventilation

If the size of the openings is reduced so that the opening factor $A_w H_w^{\frac{1}{2}}/A_T$ falls below say 0.01 $m^{\frac{1}{2}}$ one must envisage low temperatures and the possibility of kinetic limitations to the combustion, oscillation, flame extinction and smouldering. Little work has been done in these regimes which are important primarily because the total production of toxic carbon monoxide can rise to a maximum in these conditions [35]. Although the mass flow of combustion products is lower, the amount of air which can be contaminated to toxic levels is highest at low levels of ventilation. A feature of the very low ventilation regimes is that periodic behavior or instability can arise and a real fire in this state could present a danger to firemen who enter it.

Tewarson [36] regards

$$\frac{A_w H_w^{\frac{1}{2}}}{A_T} \approx 0.0036 \text{ m}^{\frac{1}{2}}$$

as the critical region for periodic explosion in a compartment roughly twice as long as its width and height.

FULLY DEVELOPED FIRES — EXTERNAL HAZARDS

In the above energy equation there are two terms of importance to the hazard of neighboring buildings viz I_w and $(M_1 + R) \Delta H_2$.

Radiation

The radiation from a window is a relatively small part of the total rate of energy release but in still air the main hazards to nearby buildings are the radiation from the walls and other surfaces behind the open windows and to a lesser extent the radiation from the flames. These of course can and do sometimes spread fire up the outside of the facade of the building. As a first approximation one can define the radiating area by the window area and evaluate whether a fire can spread across a street. This is mostly a geometrical problem and for the purpose of regulating building separation nominal radiation fluxes are defined. The spread of conflagrations in an urban area is determined largely by the mean probability of any one building being closer to its neighbor than the critical separation. If one fire is likely to start more than one other fire before burning out a large fire can develop.

Convection

Very roughly about $\frac{1}{4} - \frac{1}{3}$ of the energy released by a fire is emitted as radiation from the visible flame and $\frac{2}{3} - \frac{3}{4}$ convected upwards thereby presenting little hazard to people or buildings nearby. Putting any cover over the spaces between buildings, turning streets into arcades may prevent this rapid convection and so raise by a factor of 3 to 4 the energy flows at lower levels near where people are. The effect of such covers is similar to that of a ceiling in a room; a large part of the energy is convected to other areas at risk instead of to where it would be comparatively harmless.

Flames from Openings

Yokoi's [37] extensive study, mostly with models, was the major work on the trajectory and temperature distribution of 'hot upward currents' for plumes from openings in the vertical plane, e.g. windows. He gave correlations of a dimensionless temperature rise

$$\theta \left(\frac{g C_p^2 P^2 r_o^5}{q^2 T_1} \right)^{1/3}$$

as functions of position, the window shape and r_o the characteristic size of the window $\left(= \sqrt{A_w / 2\pi} \right)$.

Measurements at low temperatures could thereby be used to discuss isotherms of higher temperatures and, with approximations, the size of flames. Thomas and Law [38] have discussed the relation of this work to that of Webster and Raftery [39] and Siege [40] who included measurements at higher temperatures and of flame length, but who studied very few geometries. Allowance has to be made for radiation in any problem involving, say, the design of buildings with external steel frames. So far the flow pattern of the flame and the temperature distribution can only be described in terms of the flow out of the window. Relating this to the fire in order to determine if a flame will impinge on the structure requires determining the term $(M_1 + R) \Delta H_2$ with more precision than is required for determining the thermal input to the structure as discussed above.

Consequently certain approximate treatments which are satisfactory in practice for determining fire resistance of the enclosing structure are inadequate for determining certain important characteristics of flames from openings. This is but one of several problems where the practical designer of buildings requires assistance in areas of concern to the heat and mass transfer engineer.

NOMENCLATURE

A_s	–	fuel surface area
A_T	–	total area of walls, ceiling and floor of compartment
A_v	–	total area of cross sections of vertical shafts in crib
A_w	–	window area
B	–	Spalding mass transfer No.
C_D	–	discharge coefficient
C_p	–	specific heat at constant pressure
D	–	characteristic dimension
g	–	acceleration due to gravity
ΔH	–	heat release per unit mass of fuel
$\Delta H'$	–	net heat release (see equation (9))
ΔH_2	–	enthalpy of gases leaving compartments
ΔH_{vol}	–	enthalpy of volatiles leaving pyrolyzing wood
h_c	–	height of crib
H_w	–	height of window
I_f	–	mean net radiation flux onto fuel
I_w	–	mean radiation flux from window
m_{ox}	–	oxygen concentration
M_1	–	inlet flow (of air)
M_2	–	exit flow
P	–	perimeter
Q_1	–	heat required to produce unit mass of volatiles from solid wood at T_p
Q_2	–	enthalpy of residue at mean temperature
Q_3	–	$(= Q_1 + (1/w - 1)Q_2)$
q_w	–	rate of loss of heat to walls, etc
q	–	rate of heat release
r	–	stoichiometric ratio of oxygen/fuel
r_o	–	effective linear size of window $\left(= \sqrt{\dfrac{A_w}{2\pi}}\right)$
R	–	mean rate of burning (fuel mass)
T_1	–	temperature of inlet (ambient) flow
T_2	–	temperature of exit flow (assumed equal to mean temperature of gases within compartment)

T_s –   temperature of fuel surface

T_p –   mean temperature of pyrolysis

V –   linear rate of spread

w –   mass fraction of solid wood pyrolyzed

θ –   temperature rise

ρ –   density

ACKNOWLEDGEMENT

The paper is Crown Copyright reproduced by permission of the Controller, H.M. Stationery Office. It is contributed by permission of the Director, Building Research Establishment. (The Fire Research Station is the Joint Fire Research Organization of the Department of the Environment and Fire Offices' Committee).

REFERENCES

(1)   Fire resistance tests of structures. International Organization for Standardization, Recommendation R834, 1968.

(2)   LAW, M. and ARNAULT, P. Fire loads, natural fires and standard fires. ASCE and IABSE Joint International Conference on Planning and Design of Tall Buildings. Lehigh University, Bethlehem, Pa, U.S.A. 21–26 August 1972.

(3)   LIE, T. T. Fire and buildings. London, 1972. Applied Science Publishers Ltd.

(4)   SIMMS, D. L. Ignition of cellulosic materials by radiation. Combust. Flame, 1960, $\underline{4}$ (4) 293–300.

(5)   MARTIN, S. B. Diffusion–controlled ignition of cellulosic materials by intense radiant energy. 10th Symposium (International) on Combustion. pp 877–896. Pittsburgh, 1965. Combustion Institute.

(6)   GROSS, D. and LOFTUS, J. J. Surface flame propagation on cellulosic materials exposed to thermal radiation. J. Res. Natn. Bur. Stand., 1963, $\underline{67C}$ (3) 251–258.

(7)   VODVARKA, F. J. and WATERMAN, T. E. Fire behaviour, ignition to flashover. Illinois Institute of Technology Research Institute (IITRI) Project No. N6059 Summary Research Report, June 1965.

(8)   LABES, W. G. The Ellis Parkway and Gary dwelling burns. Fire Technol., 1966, $\underline{2}$ (4) 287–297.

(9)   ALPERT, R. L. Fire induced turbulent ceiling jet. Factory Mutual Research Corporation, Serial No. 19722–2 (1971).

(10) ORLOFF, L. and de RIS, J. Cellular and turbulent ceiling fires. Combust. Flame, 1972, $\underline{18}$ (3) 389–401.

(11) TORRANCE, K. E., ORLOFF, L. and ROCKETT, J. A. Experiments on natural convection in enclosures with localized heating from below. J. Fluid Mech., 1969, 36 (1) 21-31.

(12) LARSEN, D. W. Analytical study of heat transfer in an enclosure with flames. School of Mech. Eng. Heat Transfer Lab. Purdue University, 1972.

(13) PATANKAR, S. V. and SPALDING, D. B. A computer model for 3-dimensional flow in furnaces. 14th Symposium (International) on Combustion. State College, Pa. U.S.A., 1972. (To be published).

(14) THOMAS, P. H. and HINKLEY, P. L. Design of roof-venting systems for single-storey buildings. Department of Scientific and Industrial Research Fire Offices' Committee Joint Fire Research Organization. Fire Research Technical Paper No.10. London, 1964. H.M. Stationery Office.

(15) ROBERTS, A. F. and CLOUGH, G. The propagation of fires in passages lined with flammable material. Combust. Flame, 1967, 11 (5) 365-376.

(16) De RIS, J. Duct fires. Combust. Sci. Technol., 1970, 2 (4) 239-258 (See also ROBERTS, A.F. idem. 1971, 3 (6) 263-262).

(17) ROBERTSON, A. F. and GROSS, D. Fire load, fire severity and fire endurance. pp. 3-29 In: Fire test performance. ASTM Special Technical Publication 464. Philadelphia, 1970. American Society for Testing and Materials.

(18) FUJITA, K. Characteristics of fire inside a non-combustible room and prevention of fire damage. Japanese Ministry of Construction, Building Research Institute, Report No.2(h).

(19) KAWAGOE, K. Fire behaviour in rooms. Japanese Ministry of Construction, Building Research Institute, Report No.27, 1958.

(20) THOMAS, P. H. Studies of fires in buildings using models. Part 1. Experiments in ignition and fires in rooms. Research 1960, 13 (2) 69-77.

(21) SEKINE, T. Room temperature in fire of a fire-resistive room. Japanese Ministry of Construction, Building Research Institute, Report No.29. Tokyo, 1959.

(22) ÖDEEN, T. Theoretical study of fire characteristics in enclosed spaces. Swedish Royal Institute of Technology, Bulletin 10, 1963.

(23) KAWAGOE, K. Estimation of fire temperature-time curve in rooms, third report. Japanese Ministry of Construction, Building Research Institute, Research Paper No.29, 1967.

(24) MAGNUSSON, S. E. and THELANDERSSON, S. Temperature-time curves of complete process of fire development. Theoretical study of wood fuel fires in enclosed spaces. Acta Polytechnica Scandinavica Civil Engineering and Building Construction Series No.65. Stockholm, 1970.

(25) TSUCHIYA, Y. and SUMI, K. Computation of the behaviour of fire in an enclosure. Combust. Flame, 1971, 16 (2) 131-140.

(26) HARMATHY, T. Z. A new look at compartment fires Part 1. Fire Technol., 1972, 8 (3) 196-217.

(27) THOMAS, P. H. and HESELDEN, A. J. M. Fully-developed fires in single compartments. A co-operative research programme of the Conseil International du Batiment (CIB Report No.20). Joint Fire Research Organization Fire Research Note 923/1972.

(28) SALZBURG, F. and WATERMAN, T. E. Studies of building fires with models. Fire Technol., 1966, 2 (3) 196-203.

(29) GROSS, D. and ROBERTSON, A. F. Experimental fires in enclosures. 10th Symposium (International) on Combustion. pp.931-924. Pittsburgh, 1965. Combustion Institute.

(30) SMITH, P. G. and THOMAS, P. H. The rate of burning of wood cribs. Fire Technol., 1970, 6 (1) 29-38.

(31) ROBERTS, A. F. Comments on a letter (Calorific values of the volatile pyrolysis products of wood) by J. J. Brenden. Combust. Flame, 1967, 11 (5) 439-441.

(32) MURTY KANURY, A. and BLACKSHEAR Jnr., P. L. Some considerations pertaining to the problem of wood burning. Combust. Sci. Technol., 1970, 1 (5) 339-356.

(33) ROBERTS, A. F. Problems associated with the theoretical analysis of the burning of wood. pp.893-903. 13th Symposium (International) on Combustion. Pittsburgh, 1971. Combustion Institute.

(34) KUNG, H. and KALELKAR, A.S. On the heat of reaction in wood pyrolysis. Combust. Flame, 1973, 20 (1) 91-101.

(35) RASBASH, D. J. and STARK, G. W. V. The generation of carbon monoxide by fires in compartments. Joint Fire Research Organization Fire Research Note 614/1966.

(36) TEWARSON, A. Some observations on experimental fires in enclosures Part 1: Cellulosic materials. Combust. Flame, 1972, 19 (1) 101-111.

(37) YOKOI, S. Study on the prevention of fire spread caused by hot upward currents. Japanese Ministry of Construction, Building Research Institute, Report No.34, 1960.

(38)   THOMAS, P. H. and LAW, M. The projection of flames from burning buildings. Joint Fire Research Organization Fire Research Note 921/1972.

(39)   WEBSTER, C. T. and RAFTERY, Monica M. The burning of fires in rooms. Pt.II. Tests with cribs and high ventilation on various scales. Joint Fire Research Organization Fire Research Note 401/1959.

(40)   SIEGEL, L. G. The projection of flames from burning buildings. Fire Technol., 1969, 5 (1) 43-51.

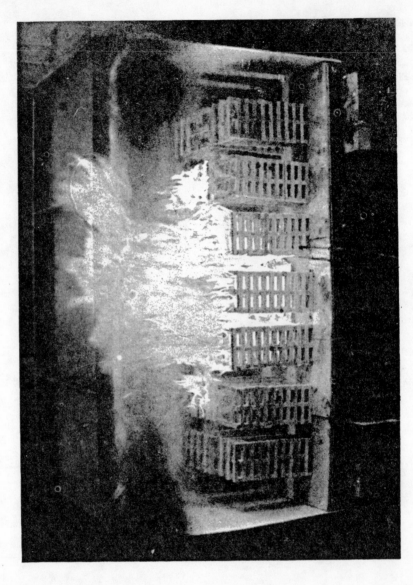

PLATE 1

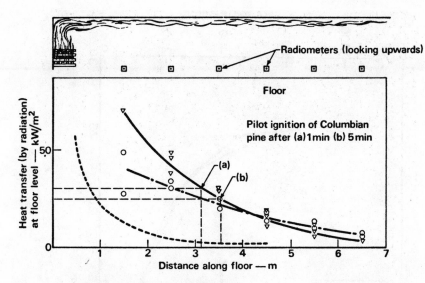

Radiometers (looking upwards)

Floor

Pilot ignition of Columbian
pine after (a)1min (b) 5min

(a)

(b)

Heat transfer (by radiation) at floor level — kW/m²

.50

0

1   2   3   4   5   6   7

Distance along floor — m

▽ Crib base 970mm beneath ceiling
o Crib base 660mm beneath ceiling
-----Calculated line for flames rising vertically from the same
    fuel bed (no ceiling)

FIGURE 1

NOTE: Figure captions for this chapter are listed on page 94.

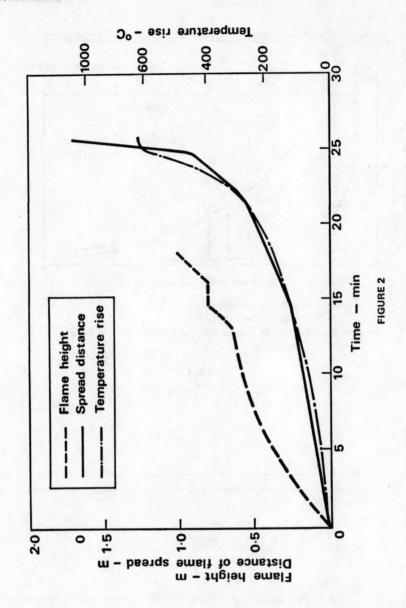

FIGURE 2

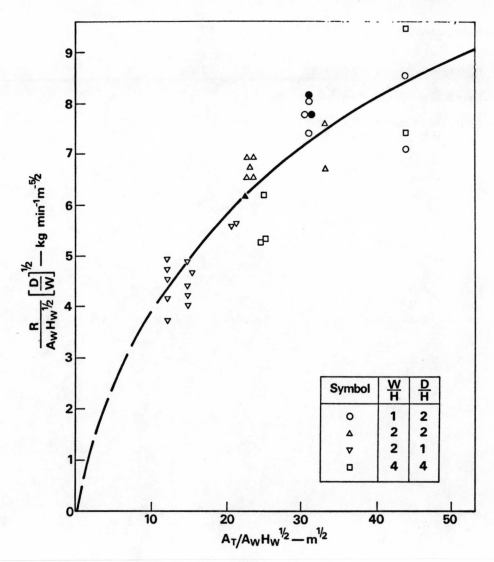

FIGURE 3

FIGURE CAPTIONS

Figure 1. Rates of heat transfer downwards

Figure 2. Some typical data for spread in a compartment

Figure 3. Variations of mean $R/A_W^{1/2}$ with shape and $A_T/A_W H_W^{1/2}$ (Solid points are means of 8–12 data)

Chapter 3

ON THE COMBUSTION AND HEAT TRANSFER IN FIRES OF LIQUID FUELS IN TANKS

P. G. SEEGER
Forschungsstelle für Brandschutz Technik, Germany

*[Research Conducted at the Research Institute for Fire Prevention Techniques
at the University of Karlsruhe]*

INTRODUCTION

In solving problems in the fields of fire prevention and fire defense, it is important to understand the initiation, the course of events and the heat transfer in the combustion process.

The Research Institute for Fire Prevention Techniques, therefore, has during the last 20 years conducted several research projects with the goal of clarifying the course of events in fires as influenced by several factors that affect fire spread, first, in storage tanks of combustible liquids, and second, in rooms or buildings with a more or less pronounced fire hazard.

Since experimentation in fires of natural size and under natural conditions would be far too costly and the difficulties of conducting the experiments would be hard to control, and since, in addition and as a rule, influences based on climate and exterior environment could not be excluded, scaled models were used under carefully controlled conditions. Experiments on the combustion process in model rooms containing fires of solid materials have been conducted within the framework of two international research programs coordinated by the Fire Research Station in Borehamwood, England. The evaluation of test results of 10 participating research institutes was also conducted by the above institute. Since Dr. Thomas, who represents Borehamwood, will report in detail during the course of this workshop on the results of these experiments, my report will be limited to the experimental

NOTE: All figures quoted in text are at the end of the chapter.

work done at the Research Institute for Fire Prevention Techniques
on tank fires.

THE BURNING PROCESS IN TANKS

During fires in storage tanks containing liquid fuels, combustion
occurs, as shown on Fig. 3.1 in such a way that the heat necessary for the preheating and evaporation of the top layer of the liquid
is transferred from the flame through radiation of the flame itself,
through conduction through the tank wall, and through convection in
the combustible liquid stored in the tank. Initially, the very slowly
rising fuel vapor reacts with the surrounding air above the tank
opening. The hot combustion gases cause an updraft that results in
locally reduced pressure in the tank; as a result surrounding air is
drawn into the tank. This air mixes below the tank opening with the
rising fuel vapor in such a way that the initial stage of the flame
reaction will move deeper and deeper into the tank, in consonance with
the lowering of the level of the liquid.

The heat transferred into the liquid supports the evaporation of
the top layer of the liquid, and raises the temperature of the liquid
remaining in the tank. Depending on the time and positionwise process
of heating the layer below the surface level of the liquid, we distinguish
liquid fuels that form a preheated zone, and others that do not. As is
shown in Fig. 3.2, the heated layer is very thin in the case of
fuels without a heated zone. After having formed in the beginning
phase of the fire, its thickness hardly changes provided a constant
burning rate is given. Then heat transferred into the liquid goes altogether to evaporate fuel at a given rate and at the same time serves
in heating fuel at the same rate from the initial to the temperature of
the uppermost layer of the liquid.

As can be seen furthermore from Fig. 3.2 liquid fuels that
form a heated zone display a layer of constant temperature under the
surface of the liquid. The transition from this heated zone to the fuel
unaffected by the fire takes place in a relatively thin transitional zone.
The thickness of the heated zone steadily increases during the fire and
may encompass, after a considerable duration of the fire, all of the
stored liquid.

The speed with which the lower border of the heated zone penetrates into the unaffected fuel, is after a certain starting transient
relatively constant. It depends on the type of fuel and has to be determined by experimentation. The simple heat balance that occurs in
liquid fuels without heated zone does not occur here.

In order to avoid these difficulties, the Institute limited its research to liquid fuels without the heated zone.

Burning Rates and Temperature Levels Below and Above the Surface of the Liquid

The first experiments of the Institute for Fire Prevention Techniques were conducted with fires of liquids stored in tanks, the goal to gain closer insights into the relationship of burning rate and temperature levels below and above the surface of the liquid and the tank dimensions and type of liquid fuel involved   [3, 4, 5, 6, 7]

In fires of liquid fuels without heated zones, the burning rate is understood to be the rate of lowering of the surface level through evaporation of the topmost layer of liquid in which process heat flux from the flame supplies the heat of prewarming and of vaporization.

$$Va \quad \frac{\dot{Q}_Z}{b \, [r + c \, (t_s - t_u)] \, F_T} \tag{3.1}$$

Which means:

Q_Z the total of the heatflow added

b the fuel density

r the evaporation temperature of the fuel

c the specific heat of the fuel

t_s the boiling point of the fuel

t_u the temperature of the fuel unaffected by the fire

F_T the cross surface of the tank

For the investigation model tanks were used whose diameters varied between 12 cm and 120 cm. Each individual tank height was selected in such a way that the relationship of height to diameter for all model tanks was a constant 4:3. Later on the experiments were also extended to tanks with a height-diameter relationship of 3:4. (4, 5) The fuels were ethyl alcohol and gasoline for motorized vehicles. During the course of the experiments the burning rates were determined partially through weighing, and partially--in the cases of model tanks with the large diameters—through measurements of the changes of the levels of the tank liquids. In addition, the local temperature changes above and below the surface of the liquid were determined at regular intervals during the burning by recording readings of thermoelements that had been arranged in the vertical axis of the tanks. Fig. 3.3 shows the temporal burning process of the two fuels in three of the model tanks examined. It can be clearly recognized that on the one hand the burning rate increases with the tank diameter, with gasoline showing a greater increase than ethyl alcohol; on the other hand, the rate during the burning period does not remain constant, but rather, after an initial increase at the start of the experiment, decreases constantly with the lowering of the level of the tank liquid, with gasoline again registering a stronger decrease. The condition at the start of the experiment can be related to the fact that first the total upper layer of the liquid has to be brought to the boiling point, before a stationary condition between evaporation and burning of the fuel can be achieved. The differing duration of this, as related to the two fuels, could be explained by the higher specific heat as well as the three times greater evaporation temperature of ethyl alcohol. In addition, in contrast to alcohol with its homogeneous composition, the inhomogeneous mixture of gasoline has a tendency towards fractional distillation, which sets in at temperatures far below the boiling point of ethyl alcohol.

These differing physical qualities of the two fuels are also responsible for the greater burning-off speed of gasoline in the upper half of the tank. On top of that, gasoline flames radiate much greater heat quantities into the liquid fuel contained in the tank. than alcohol flames, since gasoline burns with a strongly luminous flame, due to its greater carbon content and the incomplete combustion based on greater demand for air. This, in turn, causes a more or less intensive formation of soot. In contrast to gasoline, ethyl alcohol burns only with a weakly luminous flame, e.g. heat can be transferred only through the infrared radiation of carbon dioxide and steam, and the effect of heat transport through convection as well as the radiation and conduction of the tank wall can be ignored.

The reduction in the burning rate, as the level of the liquid falls, can be explained by the increasing distance between flame and liquid, and as Drotschmann (8) points out, by the fact that the necessary air supply for the combustion process becomes progressively poorer; in the case of gasoline due to the higher air demand this effect is stronger than would be observed with ethyl alcohol.

The temperature range on the tank axis, read at regular intervals during the experiments, has been graphed for the two fuels used as seen in Figures 3.4 & 3.5 for a model tank of diameter 30 cm and height 40cm and in Figures 3.6 & 3.7 for a model tank of diameter 120 cm, and height 160 cm. As can be seen, the temperature distributions all follow a similar course. Below the surface of the liquid the temperature quickly falls within a relatively thin layer from the boiling to the initial temperature of the liquid. This temperature range in the upper layer of the liquid remains almost constant during the total duration of the burn. Due to the lesser specific heat and the higher boiling point of gasoline, the thickness of this layer is greater during a burn in gasoline than in ethyl alcohol. Additionally, during a burn of gasoline a certain fractional distillation can be observed in the beginning phase as figure 3.5 shows, with an increase in the boiling temperature at the surface of the liquid over the duration of the burn-off. In this particular figure, each surface of the liquid is marked by corresponding triangles.

Due to the flame the temperature rises steeply above the level of the liquid and drops again relatively fast after reaching a maximum. Since this temperature maximum retains, during the total duration of the burn, an almost constant distance from the surface of the liquid, it shifts deeper and deeper into the tank, as the surface of the liquid becomes lower. To be sure, the size of the temperature maximum is subject to change depending on the height of the tank liquid, reaching its greatest reading, when the liquid has decreased by one half. The general decrease of the maximal temperature accompanied by the sinking of the level of the liquid in the lower part of the tank can be explained as was the reduction in burning rate, by the fact that the necessary air supply for combustion becomes progressively worse because of the existence of the tank wall. On comparing the temperature curves of gasoline and ethyl alcohol it is seen that the maximum temperature in the gasoline flame is only slightly above that of the alcohol flame. The difference should yield relatively higher flame temperature for gasoline since the heating value of gasoline is almost twice as large as that of ethyl alcohol. On the other hand, the theoretical air demand for combustion of gasoline is almost twice as large as that of ethyl alcohol; it can be assumed that the combustion process in the gasoline flame is considerably more incomplete than that of the ethyl alcohol flame, which fact is confirmed by the formation of soot in the gasoline flame. This incomplete combustion is the cause of the marked reduction in the flame temperature.*

Research performed with three model tanks with respective diameters of 1.28m, 1.6 m, and 2.0 m, and an equal height for all three tanks of 2 m, using a gasoline as fuel, led to similar results [1].

* Editor's Note: An alternate explanation for the small differences in temperature may be found in the fact that

    1) The adiabatic flame temperatures are very nearly the same, and

    2) The turbulent structure of the flame may be nearly the same so that the sensors average the turbulent temperature fluctuations to yield nearly identical results; all large gasoline fires smoke.
(see Spalding, Some Fundamentals of Combustion)

**Heat Transfer From the Tank Wall to the Liquid Fuel in the Area
Immediately Below the Surface of the Liquid**

With the aid of a heat balance, Hottel [9] was able to prove that in the case of small tank diameters considerable quantities of heat were transferred by conduction through the tank wall to the liquid, which, in turn, caused high burning rates. Conversely, the portion of heat conducted by the tank wall in the case of large tank diameters becomes negligibly small, so that a constant heat quantity, ignoring a certain convective heat transport, is exclusively radiated from the flame into the liquid, as a consequence of which the burning rate assumes a constant value as tank size increases further. In the transition region, the fraction of heat transferred by conduction through the tank wall into the liquid decreases with increasing diameter on the one hand, whereas, on the other hand, the heat transferral by radiation from the flame due to a low flame density attains its constant maximal value only after the tank diameter increases. This yields relatively low burning rates [in intermediate size tanks*]. In order to demonstrate the significance of the heat conducted through the tank wall into the liquid fuel in the region immediately below the surface of the liquid, the Research Institute for Fire Prevention Techniques conducted experiments with model tanks with an interior diameter of 225 mm and height of 300 mm with wall thicknesses between 1 mm and 22 mm (1, 10) The fuel was ethyl alcohol. The burning rate was established by measuring the temporal decrease of the level of the liquid in the tank, using pressure taps. The temperature distribution in the tank wall was measured by means of thermocouples firmly built into the wall. The heat conducted through the tank wall was measured at a place on the tank wall whose distance between the level of the liquid and the tank rim amounted to 140 mm. This was done in order to guarantee a sufficient wall height above the level of the liquid as well as an adequate amount of liquid in the tank.

The heat conducted through the wall can then be determined from the measured temperature gradients in the wall in the direction of the tank axis, the known heat conductivity and the diameter of the tank wall; the contribution of heat conduction in the wall to the total burning rate can then be established. This contribution could also be determined directly and experimentally by the following method. If the flame is extinguished in a tank, the liquid will evaporate because of the heat supply from the tank wall until the tank has cooled down. If the temporal process of the evaporation rate after extinction is extrapolated back to the point of extinction, the contribution to the total burning rate due to heat supplied from the tank wall results. Figure 3.8

*Editor's addition

shows the values of evaporation rates measured, the values both of the total as well as conduction heat determined from Eq. (3.1) and the corresponding contributions to burning rates as they depend upon the wall thickness to tank diameter ratio. It can be seen from Figure 8 that the share of heat flow (by conduction through the tank wall into the liquid fuel) can be a considerable portion of the total heat flow depending on the thickness of the tank wall, and that it may not always be neglected.

Position of the Lower Flame Boundary in the Tank and the Length of Flames

The flames occurring in tank fires are buoyant diffusion flames which fall in the transition region between laminar and purely turbulent flames and in which in most cases, the momentum of the evaporating fuel vapor may be neglected in comparison to the momentum caused by gravitational effects. As Thomas (11) has stated, the median flow speed increases so rapidly with height owing to the strong buoyant forces in the flame that turbulence occurs at a distance constituting only a fraction of the flame height, even if the Reynolds number based on the flow of the fuel vapor at the surface of the liquid, corresponds to a laminar flow. It can therefore be assumed with Thomas (12) that flame length relating to tank diameter depends only on the Froude number.

There is a series of theoretical and experimental papers extant (13, 14, 15, 16, 17, 18) which conclude that the relationship between flame length and Froude number can be described by an exponential function. Thus Sunavala (14) derives the following equation for the length of the buoyant diffusion flame; here he assumes a cosine distribution of the flow velocity in the flame, and neglects the initial momentum.

$$ L/d = 5 \left(\frac{1}{c_s}\right)^{2/15} \cdot \left(\frac{\rho_0}{\rho_f}\right)^{2/15} \cdot \left(\frac{f}{\rho_0 - f}\right)^{1/15} \cdot Fr^{1/15} \qquad (3.2) $$

Where in:

L   flame length

d   interior tank diameter

c_s mass concentration of the fuel on the flame axis for a stoichiometric mixture

ρ_0 the density of the fuel vapor

ρ_f the density of the flame gases

$Fr = \dfrac{u^2}{g \cdot d}$ the Froude number

u   median flow speed of the fuel vapors

g   acceleration due to gravity

The valid range for flames of liquid fuels is a tank diameter of 30 mm and above, and for gaseous flames a burner diameter of 20 mm and above.

A similar equation was found by Steward (15) in which the Froude number also appears with the exponent of 0.2. Hess (16) experimented with buoyant diffusion flames of various gaseous fuels on burners with diameters of up to 22 mm. He established that a flame obeys an exponential law, in which the Froude number appears with the exponent 0.232, for all fuels. Kremer (17), too, found in his experiment with a buoyant city gas diffusion flame, the exponent of the Froude number is 0.2. Putnam and Speich (18) investigated the process of free burning liquid fuel and solid fuel flames by means of a burner out of which buoyant diffusion flames were emitted through a number of jets. A theoretical analysis of the results, a gain, yields a Froude number with the exponent 0.2.

In order to gain insight into the position of the lower flame border in the tank and into the flame length during tank fires, the Institute for Fire Prevention Techniques conducted experiments with model tanks with interior diameters at 25.8 cm, 17.9 cm, and 8.6 cm, all with a height of 40 cm, (19). The walls of the model tanks were water cooled in order to exclude the possibility of wall radiation and conduction and their potential influence on mixing and combustion processes. The rising vapor develops through heat from the flame which heats and evaporates the liquid. In order to be able to fix the flow rate as required, the experiment was conducted with a gaseous fuel. Propane was the choice since it is also heavier than air as is the case with most liquid fuels.

The Froude numbers of real tank fires [in the case of tanks with diameters from 10 to 100 m and velocities of the vapors of ca. 1 cm/sec., (corresponding to a burning rate of the liquid fuel of 0.4 cm/min)] are order of magnitude 10^{-6} to 10^{-7}. Since the velocity of the gas stream in experiments with model tanks with Propane as the fuel cannot be substantially decreased below 1 cm/sec. and guarantee stable combustion, the Froude numbers with experiments on model tanks of the sizes quoted here range between 10^{-3} and 10^{-5}. The assumption was here also, that in this range of the Froude number the flames are almost exclusively influenced by buoyancy effects which were verified

by the experiments. Since below a gas velocity of 0.75 cm/sec. a
stable combustion no longer occurred in the tank with the smallest
diameter with a measuring probe in it, the experiments were, therefore, conducted with gas velocities between 0.75 cm/sec. and 1.5 cm/
sec.

The process of the entering of surrounding air and the exact spot
of the start of the first combustion phenomena in the tank were recorded
through measurements of the local concentration of the two gas components in the fuel-air mixture, propane and carbon dioxide, by utilizing instruments for the analysis of infrared absorption, and by
temperature measurements with an iron-constantan thermocouple.
The length of the flames was determined via photography using time
exposures of 20 sec.

Figure 3.9 represents the process of the time averages of the local
propane and carbon dioxide concentration on the tank axis, as it applies
to the four tank diameters used. As this figure shows, the measured
values can be, as far as the individual gas velocities are concerned,
quite well approximated over a large part of the measured range by
straight lines. Since the exact point of the start of the decrease of
the 100% propane concentration, e.g., the point of the first appearance
of carbon dioxide is extremely difficult to record, under the given
circumstances, these points were determined through extrapolating
the extension of each individual concentration profile. In the execution
of this mapping, it was discovered that the points of the start of the
decrease of the propane concentration on the tank axis coincided very
well with those of the first appearance of the carbon dioxide within the
limits of measuring accuracy.

It can be further deduced in the figure that corresponding measured
values will shift downstream towards the tank opening and beyond with
decreasing tank diameter and increasing flow speed of the combustible
gas. Whereas in the model tank with the greatest diameter, the surrounding air penetrates relatively deeply with the fuel flow moving
in the opposite direction, the model tank with the smallest diameter
will show the mixing and combustion with the surrounding air above
the tank opening exclusively, for all speeds used in the experiments.
This dependency of the position of the lower flame boundary in the
tank upon the diameter of the tank and the gas velocity can be seen also
in Fig. 3.10, in which the onset of the drop in propane concentration is
plotted over the tank diameter for several gas velocities. Due to the
paucity of measured values, the extrapolations to greater diameters
than those used, should be done with caution. Only additional experiments will show whether this linear dependance fits tanks of greater
dimensions than those studied here.

The axial and radial temperature profiles, and fluctuations in
these measured values serve to confirm that the fuel stream flowing

counter to the penetrating air stream mixes with it below the tank
opening all the way to the lower flame border (as was established by
concentration measurements;) and that combustion occurs simultaneously. An attempt was made to determine the position of the lower
flame boundary in the tank by means of temperature measurements.
The assumption was made that the lower reaction zone is defined by
those points in which during 70% of the measuring time no temperature
changes occurred. The temperature changes during the rest of the
measuring time were not attributed to combustion reactions, but to
the convection of pockets of hot gas since the temperatures never
exceeded a maximum reading of 300°C. The results of this experiment are shown in Fig. 3.11, which represents the position of the
lower flame border in the model tank with the greatest diameter, and
for the various gas velocities. The values on the tank axis obtained
from the extrapolation of the measured propane concentrations have
been recorded also and correspond well with those values obtained
through temperature measurements. The fact that the lower flame
boundary obtained from concentration measurements lies somewhat
farther downstream than the boundary established through temperature
measurements may be due to error introduced by the extrapolation
procedure. In this figure, too, it is seen that the flame shifts towards
the tank opening as the fuel flow rate increases.

The Research Institute for Fire Prevention Techniques is currently running yet another long range research project which is concerned with the penetration of the surrounding air into the tank interior
and its consequent mixing with the rising fuel vapor. The experiments
are conducted with water cooled model tanks having interior diameters
of 36.0 cm, 50.8 cm, 72.0 cm and with heights of 55 cm, 70 cm, and
90 cm respectively. The varying height gradation takes account of
the increasing penetration depth of the flame which occurs with an
increase in tank diameters. The experimental gas is propane as in
the earlier experiments. In order to examine the effect of density on
the process, additional experiments with methane as the burn fuel
have been planned. Whereas in the experiments described above only
the time averaged values of the local propane and carbon dioxide concentration along the tank axis have been determined, in the experiments
now in progress the whole field will be covered, meaning that not only
time averaged values of the propane, methane, and carbon dioxide
concentration will be measured, but also those of the remaining significant combustion gas components, e. g., of carbon monoxide and
oxygen. Furthermore the time averaged values and the fluctuation
amplitudes of temperature and pressure will be established, in order
to gain closer insight into the flow and mixing processes.

The tentative results of the measurements obtained to date of the
concentrations of the individual combustion components and of the
temperatures in various cross-levels both inside and outside the

model tank, (with diameter 36. 0 cm and height 55 cm) suggest that
the main share of the penetrating air enters the tank in the range of
the tank axis. (For given height) the oxygen concentration diminishes
symmetrically from the tank axis towards the tank wall. Correspondingly, the components propane and carbon monoxide show their
minimal concentration in the tank center. At the tank wall their concentrations are at a maximum. The highest volume fraction of carbon
dioxide is found, where the stoichiometirc mixture of propane and
oxygen occurs, e. g., (for constant and) in a small circular zone between
tank axis and tank wall, where the highest temperatures are also found.

The measured gas concentrations and temperatures are consistent
with a picture of the lower flame boundary which resembles on the
average a paraboloid of rotation open towards the top with axis of rotation congruent with the tank axis. It appears that the combustible gas
flowing upwards in the vicinity of the tank axis burns up more or less
completely while still inside the tank depending on its flow speed,
whereas the overwhelming share of the combustible gas flowing upwards in the border zone reacts only outside the tank.

The log of flame length (l) divided by tank diameter (d) has been
plotted in Fig. 3.12 against the log of the Froude number, flame
lengths having been established photographically. Also recorded on
the chart are the flame lengths established photographically by Hess
in propane burners of small diameter. As can be seen in Fig. 3.12,
all the values obtained within the limitation of this subjective measuring method are grouped around a line which obeys the following
equation:

$$l/d = 40 \, Fr^{0.2}$$

(3.3)

In order to show the validity of the connection between the relative
flame length and the Froude number as it applies to other fuels, too,
the findings of Hess for methane and those of Putnam and Speich (18)
for natural gas, again for experiments with burners of small diameter, were also included in Fig. 3.12. As can be seen these values
lie around a line subject to the following equation:

$$l/d = 29 \, Fr^{0.2}$$

(3.4)

A similar equation has been already mentioned by Putnam and Speich
in their paper (18). The flame lengths established by Blinov and
Khudiakov (20) for tanks with diameters up to 22. 9 m and with gasoline
as fuel, also fit the equation above.

It might be added here that, according to Sunavala's equation (3.2),
the following relationships exist for the flame lengths of propane and
methane at a flame temperature of $1000^{\circ}C$:

Propane flame: $l/d = 23 \, Fr^{0.2}$ (3.5)

Methane flame: $l/d = 21 \, Fr^{0.2}$ (3.6)

Based on these experiments it can be established that diffusion flames in the range of the Froude number from 10^{-5} to 10^4 are largely controlled by buoyancy. But Fig. 3.12 also shows that--and this was also observed by Hess (16) and Kremer (17)—that the related flame length approaches a constant value above a certain Froude number. This is, however, characteristic of the purely turbulent diffusion flame, in which case the buoyancy in relationship to the initial momentum no longer exerts any influence on the flame length.

HEAT TRANSFER IN TANK FIRES THROUGH RADIATION TO THE SURROUNDINGS

The heat released during a fire in storage tank of inflammable liquids is transferred to the surroundings through radiation as well as convection. The fraction of heat transferred by convection is larger only in the close vicinity of the burning tank. With distance the chance of contact between flames and combustion gases decreases steadily so that heat will be transferred by radiation only.

Due to the strong soot formation which generally occurs as a consequence of incomplete combustion, the flames in this case are of the luminous variety. In the case of flame thickness of more than 1 m one could by way of approximation consider such flames as black or grey bodies with an emissivity between 0.9 and 1, and then consequently fall back on the Stefan-Boltzmann law for the determination of the radiation level. Utilizing this assumption, the radiation intensity occurring on any randomly located plain element in the vicinity of the tank can be determined according to the following relationship:

$$\dot{q} = \varphi \, \varepsilon \, \delta \, T^4$$ (3.7)

Which is:

$\dot{q}$ radiation intensity

φ relationship of the angles

ε emissivity

δ the Stefan-Boltzmann constant; $\delta = 5.77 \cdot 10^{-12} \, W/cm^2 \, K$

T the absolute temperature of the flame

If the shape of the flame is changed to a cylinder whose diameter equals the tank diameter and whose height equals the flame length, the relationship of the angles can be established for various cases according to the relationships in the literature [21,22,23].

In order to examine how well the equation reflects the radiation intensities that occur in the surroundings during real tank fires, the Research Institute for Fire Prevention Techniques conducted fire experiments in model tanks with diameters of 1.28 m, 1.6 m, and 2.0 m, respectively, and an equal height for all tanks of 2.0 m. (1, 24). The walls of the model tanks could be cooled by water. Fuel oil served as the fuel with a boiling range of 167 - 367° C and with a residue of 1.5/vol %. At three average fuel levels in the tanks of 1.8 m, 1.0 m, and 0.2 m radiation intensities at various points in space were recorded by total radiation pyrometers. The flame temperature was determined by a radiation pyrometer and the flame height by photography.

Figure 3.13 shows the experimental recordings in the case of a tank with a diameter of 1.6 m. The radiation intensities were measured at tank height at a right angle to the tank axis and are plotted against ratio of distance to tank diameter, on a semi-logarithmic plot. Figure 3.14, from the same experiment, shows the radiation intensities measured at a distance of 3 m from the tank axis, along a line parallel to the axis, verses height (from tank top) over diameter. A comparison with the radiation intensity according to Eq. (3.7) (the solid line in figure 3.13) shows good agreement in the case of the full tank experimental results. In this comparison the values inserted in Eq. (3.7) were emissivity = 0.9, and flame temperature = 900°C. (This latter value agrees with the measurements in the experiments described above, as well as with the reports from references (1, 25).) For the establishment of the relationship of the angles the author fell back on the photographically determined flame length. Since the determination of flame length is extremely difficult, due to the cluser-like structure of the flame and the recorded by short duration exposures, the maximum value was used in order to be on the safe side. This value is found to be related to tank diameter by l/d = 1.5.

Further details concerning these experiments will be presented during the next International Seminar of 1973, entitled "Heat Transfer from Flames".

SUMMARY

This paper is concerned with research conducted at the Research Institute for Fire Prevention Techniques with the goal to gain closer insights into the combustion process and heat transfer during tank fires of liquid fuels. The research covers experiments on the dependency of burning rate and temperature distribution on the tank dimensions and on the type of liquid fuel, the determination of the heat portion of flux which is transferred via conduction through the tank wall to the liquid fuel, the determination of the position of the lower flame boundary in the tank and length of flames during tank fires, and finally, the determination of radiation intensities as they occur due to the radiation of the flame in the vicinity of a burning tank. At this time another research project is in progress in order to get to know more closely the mechanism of penetration of the surrounding air into the interior of a tank as well as the process of the mixing of air and rising fuel vapor.

REFERENCES

1. Werthenbach, H. G.: Fires of Natural Oil Products in Tanks-- Experiments and Scale Model
   Research Report 21, Workshop Fire Prevention (AGF), 1971.

2. Werthenbach, H. G.: Fires of Natural Oil Products in Tanks-- Experiments and Scale Model
   VDI Journal, Vol. 115 (1973), 5, pp. 383-388.

3. Kruger, A. and Radusch, R.: Investigations for the Determination of Fire Processes in Storage Tanks with Combustible Liquids
   Natural Oil and Coal, Vol. 8 (1955) 7, pp 482-486.

4. Magnus, G.: Tests on Combustion Velocity of Liquid Fuels and Temperature Distribution in Flames and Beneath Surface of the Burning Liquid.
   International Symposium on the Use of Models in Fire Research, National Academy of Sciences -- National Research Council, Washington, D. C., Publication 786, 1961, pp. 76-90.

5. Magnus, G.: Burn-Off of Liquids in Model Tanks
   First International Seminar for Fire Prevention of the VFDB.
   VFDB Journal, Vol. 9 (1960), Special Issue 3, pp. 101-104.

6. Hinrichs, B. R.: Problems of the Model Fire Technique
   VFDB Journal, Vol. 14 (1965) 2, pp. 42-50.

7. Seeger, P. G.: Experiments on Models in the Technique of Fire Prevention
   VFDB Journal, Vol. 19 (1970), 1, pp. 37-44; 2, pp. 91-94.

8. Drotschmann, H.: The Questions of Burn-Off of Liquids in Tanks
   First International Seminar of Fire Prevention of the VFDB.
   VFDB Journal, Vol. 9 (1960) Special Issue 3, pp. 106-108.

9. Hottel, H. C.: Certain Laws Governing Diffusive Burning of Liquids

Fire Research Abstracts and Reviews, Vol. 1 (1958/59), 2,
pp. 41-44.

10. Werthenbach, H. G.:
    Model Experiments on Tanks, About the Influence of the Thickness
    of Tank Walls
    VFDB Journal, Vol. 16 (1967), 2. pp. 43-49.

11. Thomas P. H., Webster, C. T., and Raftery, M. M.:
    Some Experiments on Buoyant Diffusion Flames
    Combustion and Flame, Vol. 5 (1961), 4, pp. 359-367.

12. Thomas, P. H.: Some Studies of Models in Fire Research
    First International Fire Protection Seminar of the VFDB.
    VFDB Journal, Vol. 9 (1960) Special 3, pp. 96-101.

13. Thomas, P. H.: The Size of Flames from Natural Fires
    9th Symposium (International) on Combustion, Academic Press,
    New York - London, 1963, pp. 844-859.

14. Sunavala, P. D.: Dynamics of the Buoyant Diffusion Flame
    Journal of the Institute of Fuel, Vol. 40 (1967) pp. 492-497.

15. Steward, F. R.: Linear Flame Heights for Various Fuels.
    Combustion and Flame, Vol. 8, (1964), pp. 171-178.

16. Hess, K.: Flame Length and Flame Stability
    Doctoral Dissertation, Technical Academy, Karlsruhe, 1964.

17. Kremer, H.: The Expansion of Inhomogeneous, Turbulent Free
    Rays, and of Turbulent Diffusion Flames
    Doctoral Dissertation, Technical Academy, Karlsruhe, 1964.

18. Putnam, A. A. and Speich, C. F.: A Model Study of the Inter-
    action of Multiple Turbulent Diffusion Flame
    Ninth Symposium (International) on Combustion, Academic Press,
    New York - London, 1963, pp. 867-877.

19. Seeger, P. G., and Werthenbach, H. G.: Diffusion Flames
    with an Extremely Low Flow Speed
    Chemie-Ingenieur-Technik, Vol. 42 (1970) 5, pp. 282-286.

20. Blinov, V. and Khudiakov, G. N.: Certain Laws Governing
    Diffusive Burning of Liquids
    Doklady Akademii Nauk SSSR, Vol. 113, 5, (1957) pp. 1094-1098.

21. VDI Heat Atlas, Statistical Charts for Heat Transfer
    VDI Press, Dusseldorf, 1963

22. Perrperhoff, W.: Temperature Radiation
     Dr. D. Steinkopf, Darmstadt, 1956

23. McGuire, G. H.: Heat Transfer By Radiation
     Fire Research Special Report, 2. H. M. Stationery Office,
     London, 1953.

24. Seeger, P. G: Investigation of Heat Transfer Through Radiation
     from a Burning Object to the Environment
     Research Report 22, Workshop on Fire Prevention (AGF),
     (yet to be published)

25. Atallah, S. and Allan, D. S.: Safe Separation Distances from
     Liquid Fuel Fires
     Fire Technology, Vol. 7 (1971), 1, pp. 47-56.

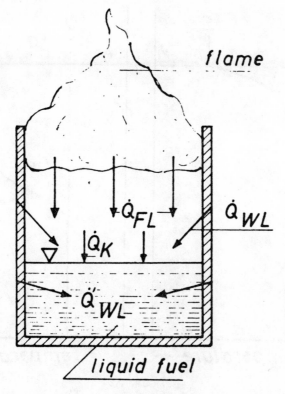

FIGURE 3.1

NOTE: Figure captions for this chapter are listed on page 124.

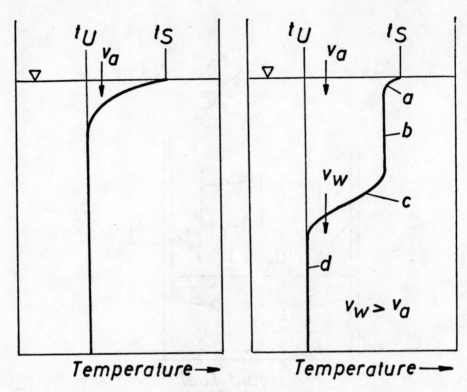

FIGURE 3.2

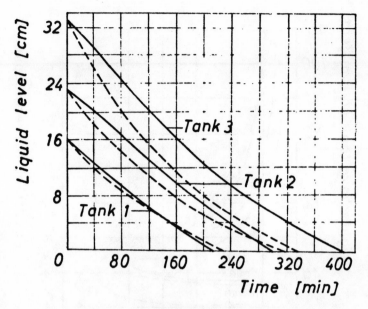

FIGURE 3.3

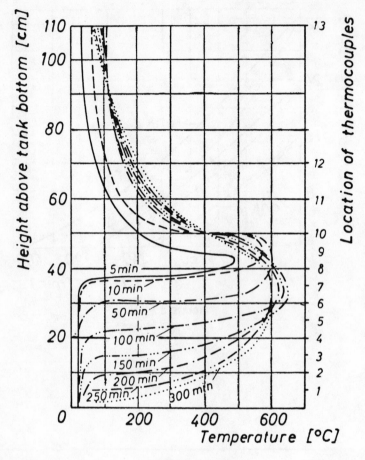

FIGURE 3.4

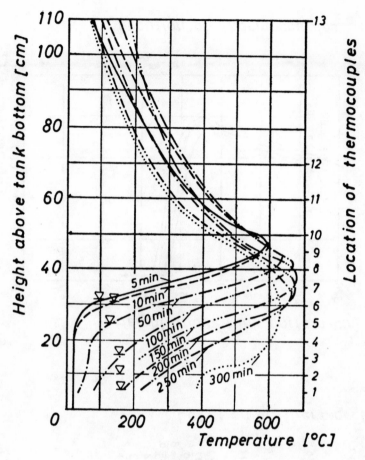

FIGURE 3.5

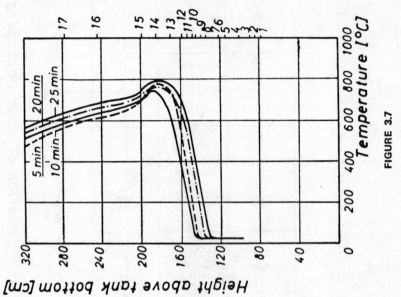

Location of thermocouples

FIGURE 3.7

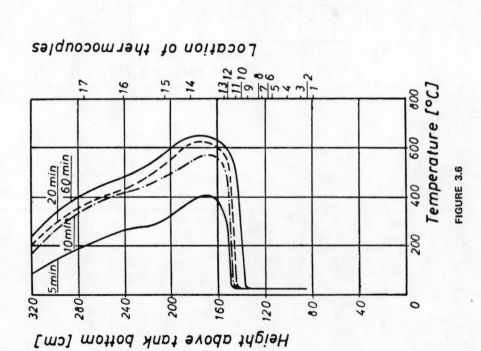

Location of thermocouples

FIGURE 3.6

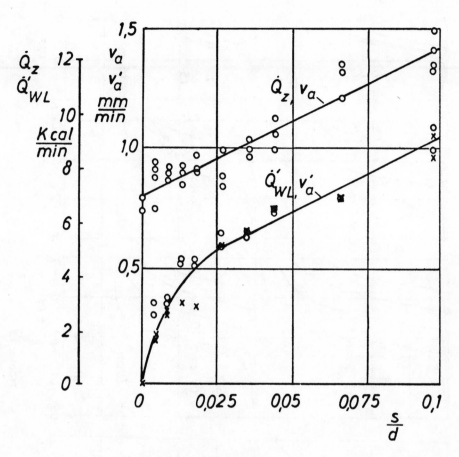

FIGURE 3.8

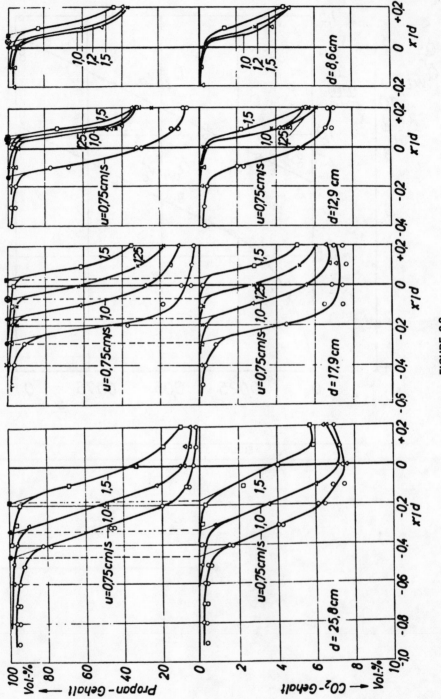

FIGURE 3.9

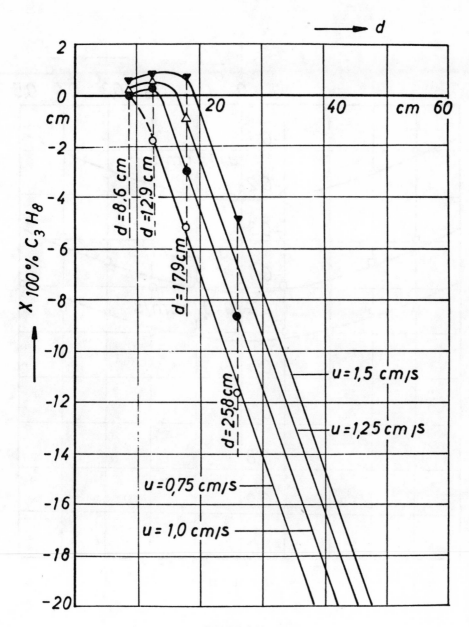

FIGURE 3.10

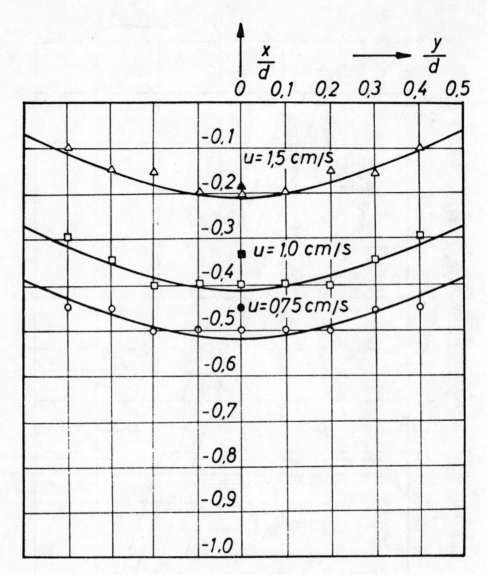

FIGURE 3.11

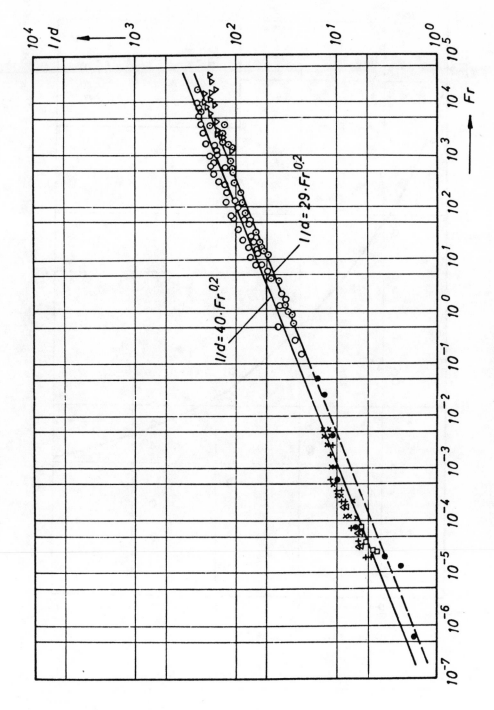

$l/d = 29 \cdot Fr^{0.2}$

$l/d = 40 \cdot Fr^{0.2}$

FIGURE 3.12

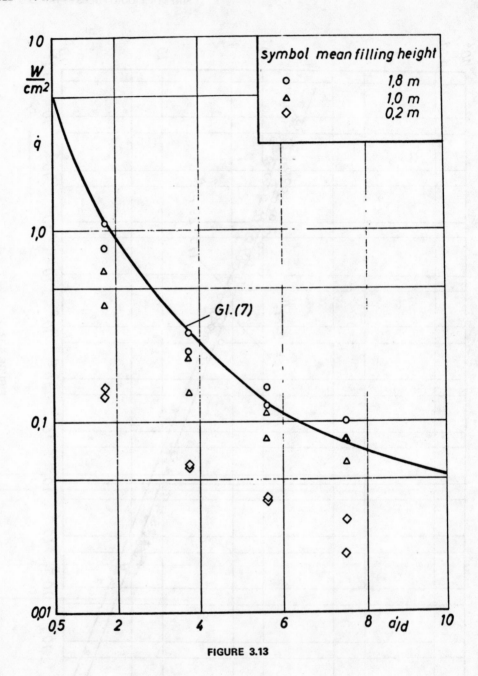

FIGURE 3.13

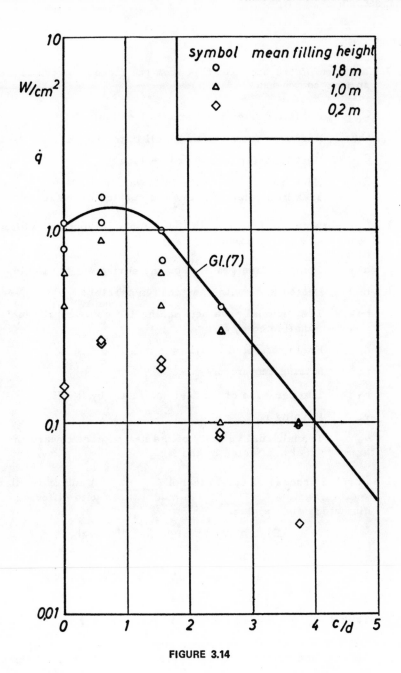

FIGURE 3.14

FIGURE CAPTIONS

Figure 3.1 Division of the heat backflow from the flame into the liquid
in four components.

Q_{FL} Radiation from the flame to the liquid

Q_{WL} Radiation from the tank wall to the liquid

Q_K Heat transfer through convection

Q_{WL} Heat transfer from the tank wall to the liquid in the
area immediately below the surface of the liquid

Figure 3.2 Temperature distribution in liquid fuels without and with a
heated zone information.

a   Temperature gradient on the surface of the liquid

b   Heated zone with constant temperature

c   Transitional area between heated zone and the fuel
unaffected by fire

d   Fuel unaffected by fire

t_S Boiling temperature

t_U Temperature of the fuel unaffected by the fire

v_a Burning rate

v_w Transitional velocity of the heated zone toward the
fuel unaffected by the fire

Figure 3.3 Burning process of ethyl alcohol (_____) and of regular
motor gasoline (_____) in model tanks with differing
diameters and heights.

	Diameter (cm)	Height (cm)	Fill Level (cm)
Tank 1	12	16	16
Tank 2	18	24	24
Tank 3	25.5	34	34

Figure 3.4 Axial temperature distribution during burn-off of ethyl
alcohol in a model tank with a diameter of 30 cm, a height
of 40 cm, and a fill level of 37.5 cm, in dependence of the
burn-off time.

Figure 3.5 Axial temperature distribution during burn-off of regular
motor gasoline in a model tank with a diameter of 30 cm,

a height of 40 cm and a fill level of 31. 2 cm, in dependence
of the burn-off time. ($\triangle$: distribution of the fill level
during the duration of the burn-off)

Figure 3.6 Axial temperature distribution during burn-off of ethyl
alcohol in a model tank with a diameter of 120 cm, a height
of 160 cm, and a fill level of 149 cm, in dependence of the
burn-off time.

Figure 3.7 Axial temperature distribution during burn-off of regular
motor gasoline in a model tank with a diameter of 120 cm,
a height of 160 cm, and a fill level of 149 in dependence of
the burn-off time.

Figure 3.8 Distribution of the total (Q_Z) heat flux and that of the heat
flux by conduction through the tank wall, e. g. distribution
of the corresponding burning rates (v_a, v_a) as dependent
upon the wall thicknesses over tank diameter assuming a
distance of 140 cm between the surface of the liquid to the
mouth of the tank.
X   Establishment of Q_Z e. g. v_a via the temperature
gradient in the tank wall.
O   Establishment of Q_{WL} e. g. v_a by extrapolation after
putting out the flame
s/d = 0: through water cooling of the tank wall, with s = 1 mm

Figure 3.9 Distribution of the temporal median values of the local propane and carbon dioxide concentration along the tank axis,
with various tank diameters d and gas velocities u.

extrapolated as 100 vol/% propane = 0 vol/% CO_2

Figure 3.10 Distribution of the points on the tank axis on which the
decrease of propane concentration starts, above the tank
diameter d with the velocity u as parameter.

Figure 3.11 Position of the lower flame border in the model tank with
the diameter of 25. 8 cm according to temperature measurements with various gas velocities u.

according to measurements of the propane concentration

Figure 3.12 Flame lengths l relating to each individual tank diameter d
as a function of the Froude figure Fr.
Comparison with measurements undertaken by other researchers. Measurements by the Research Institute of Fire

Prevention Techniques:
(1.19) with propane flames;   25.8 cm,   17.9 cm, +12.9 cm,
8.6 cm burner diameter;

o   according to Blinov and Khudiakov (20), gasoline 0.6
      to 2290 cm burner diameter;

o   according to Hess (16), propane, 0.4 to 2.21 cm
      burner diameter;

⊙   according to Hess (16), methane, 0.15 to 1.8 cm
      burner diameter;

∇   according to Putnam and Speich (18), natural gas,
      0.48 to 0.95 cm burner diameter.

Figure 3.13 Radiation intensity q at tank height and at a right angle to
the tank axis in dependency of the equivalent distance a'
divided by tank diameter d, with the tank wall cooled. Tank
diameter 1.6 m.

Figure 3.14 Irradiation intensities q at a distance of 3 m from the tank
axis in dependency of the height a divided by the tank diameter d, the height being above the tank opening with the
tank wall cooled.
Tank diameter 1.6 m.

PART III

HEAT AND MASS TRANSFER IN GASEOUS AND CONDENSED PHASES

Chapter 1

INTERACTIONS BETWEEN FLAMES AND CONDENSED PHASE MATTER

R. C. CORLETT
University of Washington, USA

SOLIDS OR LIQUIDS ARE heated upon contact with flames. When subjected to thermal attack by flame gases, many substances undergo chemical changes resulting in loss of material. They are "consumed" by the flames. Such heating and material loss are universally recognized manifestations of heat and mass transfer phenomena at the interface between flame gases and a condensed-phase medium. Analysis and interpretation of this class of phenomena requires understanding of processes both within the condensed phase and within the gas-phase region of the fire, and of the manner in which these processes are coupled at the phase interface.

The first section of this chapter considers this interface coupling. It begins with a discussion of condensed-phase surface models. Corresponding to a surface model is an appropriate mathematical boundary condition formulation. Some of the more tractable boundary formulations are described.

Boundary conditions can be classified as "detailed" or "overall". Detailed boundary conditions are expressed in terms of field quantities, such as temperature or species concentrations, and their gradients, and require corresponding field solutions for application. Overall boundary conditions require a less complete fire description. They are expressed in terms of quantities relatively likely to be known. The connection between detailed and overall fire interface boundary conditions is briefly explored.

The second section contains some comments on convective transport in the vicinity of a condensed-phase surface. In this book, the word "convection" is used to describe all gas-phase transport by molecular mechanisms. Thus both the orderly macroscopic, or bulk, flow of flame

NOTE: All figures quoted in text are at the end of the chapter.

gases, and transport due to random motion of molecules on a micro-
scopic scale, are considered as convection. This usage is consistent
with the standard literature on convective heat and mass transfer.
Indeed, convective heat and mass transfer data developed for non-reac-
tive systems often can be applied directly to fire phenomena. Approximate boundary layer relations are set down. But this approach encounters some serious difficulties and there is a need for a realistic
data base for convective heat and mass transfer in fires. Data for
small pool fires, for which a reasonably clear picture has emerged, is
discussed.

In concluding this chapter, two examples of the significance of
convective heat transfer in fires are outlined. One is "energy feedback". By this is meant heat transfer to a condensed-phase primary
fuel. The coupling between energy feedback and the evolution of gasified fuel is a critical determinant of the state of the fire. In
general, a fire may exist in a steady (or quasi-steady) state, or it
may exhibit various types of unstable behavior, depending on the way
the fuel responds to energy feedback and on the energy feedback as a
function of fuel gasification rate. These ideas are illustrated by
consideration of coupling in pool and pool-like gas fires.

The second example is the split between convective and radiative
heat transfer as a determinant of thermal insulation performance. In
general, both the convective and radiative environments to which insulation is exposed must be characterized.

1.1 CONDENSED–PHASE SURFACE MODELS AND BOUNDARY CONDITIONS

Nearly all condensed-phase material undergoes some irreversible
chemical or physical change when exposed to flame. Precise description
of such changes in most cases would lead to unmanageable complications.
Simplified models which retain essential features of the problem at
hand are thus adopted. Modeling in this sense, namely the conceptual
abandonment of some aspects of reality for purposes of simplicity or
generality, is of course a necessary step in any application of physical science to the real world. But such modeling is an especially
vital element of fire studies. Often, rather sweeping simplifications
are required to render problems tractable. For practical fire problems, the selection of suitable surface models is a source of chronic
uncertainty and controversy.

In much of the combustion literature, there is reference to homogeneous and heterogeneous systems. A condensed-phase fuel burning in
an oxidizing atmosphere is in such a sense always an heterogeneous
system. In this chapter, the words homogeneous and heterogeneous
aroused in a different sense, to characterize the condensed phase alone.

In the simplest case, the condensed phase may be modeled as homogeneous and inert. As meant here, an inert substance undergoes neither
chemical reaction nor phase change. Its physical properties depend
only on the local thermodynamic state which, in the absence of significant pressure variations, is characterized by temperature alone. A
steel plate exposed to flame, for example, would normally be modeled

as homogeneous and inert. The modeling assumptions ignore the development of a surface oxide layer and solid-phase transformations. Under special circumstances, the insulating effect of the oxide layer or property variations due to solid phase transformations could be significant. If the plate were exposed for a long time to an intense fire, an appreciable amount of material might be lost by oxidation and erosion.

Surface heterogeniety is illustrated in Fig. 1.1. Heterogeneity may be a direct characteristic of the construction of the exposed material, or may be a consequence of staged decomposition. For example, many fabrics have a construction which is manifestly heterogeneous even in the absence of chemical reaction. Many building materials, although smooth-surfaced, are of heterogeneous composition; limiting cases of interest are aggregate and layered construction. Heterogeneity due to staged decomposition is exemplified by char-forming plastics. The plastic pyrolyzes, yielding char plus gaseous products, and subsequently the char either oxidizes by surface reaction or vaporizes at relatively high temperature. In reality pyrolysis consists of several stages, some of which may compete in parallel. However, a one-step pyrolysis process is a frequently adopted model.

A non-inert condensed material may react chemically, or undergo phase change. If the surface is initially homogeneous, and if the material vaporizes or reacts chemically into a uniquely constituted gas phase, it is said here to transform uniformly. Uniform vaporization is an accurate characterization of single component liquids with boiling points too low for significant chemical reaction. It is often a useful model even for mixtures such as gasoline. Uniform chemical transformation is rarely an accurate characterization but is a common model.

An initially homogeneous material which reacts in stages may sometimes be modeled one-dimensionally. All of the heterogeneity would then be accounted for in terms of material variation with depth and time, but not with position along the surface. Such a model is based on the assumptions of stable response to thermal attack. An example of unstable response is mechanical cracking, a common behavior of char-layers. Chemical instabilities can also be envisaged. For example, a material which endothermically gives up absorbed or bound water, and then oxidizes exothermically at or near its exposed surface when sufficiently high temperature is reached, might decompose at an accelerating rate in regions where heating perturbations occur.

The mathematical formulations discussed here are far from the most general conceivable, even within a one-dimensional framework. In particular, it is assumed that pressure perturbations in fires are much smaller than the absolute pressure level, so that in thermodynamic and kinetic expressions the pressure may be assumed known and constant. This means that conservation of energy and species at the surface can be uncoupled from surface stress considerations. Some exceptions do occur for free liquid surfaces and are mentioned later.

In general, diffusive transport of energy and species in the gas phase is important where a condensed phase contacts flames. Normally,

the gas phase consists of a mixture of species. Rigorous specification of diffusive fluxes would unacceptably complicate the mathematics.
Moreover, it is doubtful that enough basic data exists to apply accurate gas-phase transport theory to fire phenomena. Here, simplified
transport expressions, commonly referred to as the Fick's and Fourier
laws, are adopted. The reader is referred to works on transport
theory (e.g. Ref. 1) for more accurate formulations. These approximations are thought to be reasonable if the molecular weights and
structures of the important species are not too different. To this
level of approximation, the one-dimensional flux of an individual
species is given by:

$$J_i = -\rho \ D_i \ \partial Y_i / \partial y \tag{1.1}$$

Equation (1.1) expresses the y-direction component of the diffusive mass
flux (mass/area/time) of species i, with mass fraction Y_i. The total
mass density is ρ and D_i is the diffusion coefficient. If the
participating species are not too dissimilar, all of the D_i are of
the same rough magnitude. Introduction of a universal, but not necessarily constant, diffusion coefficient D is a useful modeling
approximation. In general D depends on the local thermodynamic state.
It increases sharply with increasing absolute temperature. It tends
to decline with increasing molecular weight. However, in view of the
inaccuracy normally inherent in the universal D modeling assumption
itself, the dependence of D on molecular weight is usually neglected
or, at most, represented crudely.

To the same level of approximation, the y-component of diffusive
energy flux q is given by

$$q = - \lambda \ \partial T / \partial y \ + \ \sum_i J_i \ h_i \tag{1.2}$$

where λ is the thermal conductivity and h_i the specific enthalpy
of the i-th species. To this level of approximation λ depends on the
local thermodynamic state only. It is related to the diffusion coefficient D through the Lewis number Le

$$Le = [\lambda \ / \ (\rho \ c_p)] \ / \ D \tag{1.3}$$

where c_p is the constant pressure specific heat, hereafter
referred to simply as specific heat. It is also possible to define
species Lewis numbers using the D_i ; but this book makes no use of
these. The group $\lambda / (\rho \ c_p)$ is called the thermal diffusivity, and
for gases is roughly equal to D . Le $\equiv$ constant, and indeed, Le $\equiv$ 1,
are widespread modeling assumptions.

Equation (1.1) could be framed with equal validity as a molar flux
in terms of a mole fraction gradient. Molar and mass fraction formula
tions each have advantages. Molar specific heats are relatively independent of molecular weight. On the other hand mass but not number
of moles is conserved in chemically reactive systems. Moreover, external flow fields require momentum analyses which are handled relatively easily in terms of mass fractions. As a practical matter, the

general level of approximation in treatment of most fire phenomena precludes exploitation of the above cited factors. A mass fraction formulation is used consistently in this book.

If the condensed surface is inert and non-catalytic, all of the J_i's are zero. Surface catalysis is thought not to be an essential feature in fire problems of current interest and is not considered further in this book. For a non-inert surface, Eq. (1.2) takes a more familiar form if a universal D is assumed and if species specific heats are taken as identical, i.e. if

$$h_i = h_i^0 + \int_{T^0}^{T} c_p \, dT \qquad (1.4)$$

where h_i^0 is the heat of formation of species i at reference temperature T^0 and the universal specific heat c_p is in general a function of temperature T. The result is

$$q = -\lambda \, \partial T / \partial y + q^0 \qquad (1.5)$$

where the term

$$q^0 = \sum_i J_i h_i^0 \qquad (1.6)$$

can be interpreted as the net diffusive flux of formation energy. Its significance is that the total flux of formation energy, $q^0 + \sum_i \rho v h_i^0 Y_i$, v being the bulk or macroscopic velocity, does not change across a surface except due to chemical reactions or phase changes which occur there. To the extent that the specific heats $c_{p,i}$ do vary among species, there would be an added term in Eq. (1.5). The assumption of a universal c_p is valid as regards energy transport if this correction term is small compared to the heat conduction term $-\lambda \, \partial T / \partial y$.

In general there is a net macroscopic mass flow component of magnitude ρv away from the condensed phase. At one atmosphere pressure the value of v just outside the surface is typically on the order of magnitude of 1 cm/sec and the value of D ranges roughly from 0.2 cm^2/sec at ambient temperature to 1 cm^2/sec at red heat. Thus the typical diffusion length D/v is on the order of millimeters.

The simplest, and by far the most widely used, boundary condition statement applies to a surface modeled as an homogeneous material which is uniformly gasifying as schematically depicted in Fig. 1.2. In this simple model there is no flux of any species other than fuel across the interface. Conservation statements are necessary for energy and species. An additional statement is required relating temperature and fuel concentration just outside the interface. These statements provide boundary conditions for the temperature and species fields and determine the mass flux ρv . Conservation of fuel can be written

$$\rho v = \rho v Y_F - D \, \partial Y_F / \partial y \qquad (1.7)$$

For any other species, say the i-th, the boundary condition reads

$$0 = \rho v Y_i - \rho D \, \partial Y_i/\partial y$$

or
$$\partial \ln Y_i/\partial y = v/D \qquad\qquad (1.8)$$

Conservation of energy can be written in terms of the increase in formation energy flux across the interface,

$$\rho v L = q_{RAD} - \lambda_- \, (\partial T/\partial y)_- + \lambda \, \partial T/\partial y \qquad (1.9)$$

In Eq. (1.9) the quantity L denotes the increase in formation energy per unit mass of gasifying fuel, and is here called the effective latent heat. If the gasification process is endothermic, L is positive. The term q_{RAD} consists of absorbed incident radiation minus radiation emitted by the surface. This contribution is dealt with in Chapters 9 and 10. The term $\lambda_- \, (\partial T/\partial y)_-$ denotes heat conduction from the interior of the condensed phase and the term $\lambda \, \partial T/\partial y$ conduction to the gas phase.

A homogeneous inert surface is a limiting case of the foregoing. Then ρv is zero and Y_F is meaningless. On the other hand, with finite ρv, this simple model need not be restricted to fuel surfaces, applying in principle to any material gasified by fire thermal attack. For brevity, the gasifying material will be referred to as "fuel" in the remainder of this section.

A great variety of statements connecting the surface fuel concentration Y_F and temperature T can be appropriate, depending on circumstances. At surface temperatures too low for significant chemical reaction, a phase equilibrium model is common. It should be emphasized that true thermodynamic equilibrium never exactly prevails at a surface where there is finite heat or mass transfer. Nevertheless at normal pressure levels, temperature equilibrium is approached virtually exactly. In the absence of chemical reaction and at typical fire mass transfer rates, phase equilibrium is also approached very closely. Then Y_F and T may be connected through a vapor pressure-temperature relationship. A good approximate equation derivable from the Clapeyron equation is

$$1/T^0 - 1/T = (R^0/L) \ln (Y_F/Y_F^0) \qquad\qquad (1.10)$$

where R^0 is the gas constant expressed in the same units as L, and Y_F^0 is the equilibrium fuel vapor fraction at reference temperature T^0. In Eq. (1.10) T is expressed in absolute units (e.g. degrees Kelvin).

When $Y_F=1$, a liquid medium is at its boiling point. Equation (1.10) is often replaced by a statement that the surface is at the boiling temperature T_B. The rationale is as follows. If the ratio $L/R^0 \gg T_B$, a wide range of Y_F corresponds to a very small variation of T. If the T variation corresponding to orders of magnitude variation of Y_F is small compared to the characteristic temperature differences of the problem (such as that between flames and surface), then it is appropriate for modeling purposes to take the surface T = constant, in lieu of Eq. (1.10). In this case the surface fuel fraction, Y_F and the gasification flux ρv are determined by simultaneous solution of Eqs. (1.7) and (1.9). For pools of low-boiling liquids, the constant surface T is frequently taken as T_B , or even as the ambient value T_∞ .

It is worth noting here that for some important problems surface models must incorporate relatively detailed variations of temperature T . Consider the problem of flame spread rate over a free liquid surface whose initial T corresponds through Eq. (1.10) to Y_F below the lean flammability limits. For purposes of this discussion, the essential physics of the problem may be stated thus: the flame will spread as fast as the liquid surface ahead can be heated to a temperature where the equilibrium fuel concentration is flammable. This temperature is fairly well-defined empirically. The argument stated above that determination of flame-to-liquid heat transfer does not require accurate determination of surface T is still true. But in this case flame-to-surface heat transfer does not by itself control the flame-spread rate. The T distribution itself is decisive. A liquid pool with a free surface can flow horizontally, and such horizontal flow can influence the T distribution. Driving mechanisms for horizontal flow are aerodynamic shear due to gas motion above the surface, and surface tension gradients. The latter have been demonstrated to control an important class of liquid fuel flame spread problems and the essential theoretical problem has been formulated.[2] Surface tension varies with T . A gradient in surface tension has the effect of a shear force applied to a free liquid surface. Thus the momentum boundary condition for liquid motion both determines and depends on the surface T distribution ahead of the flame.

Equations (1.7) and (1.9) hold also if a uniform chemical transformation is added to or replaces the phase change at the surface, provided that the effective latent heat L correctly characterizes the formation energy change for the transformation. For example, a polymer might pyrolize completely to volatile species. In this case, however, the fraction Y_F cannot be related to T through an equilibrium equation. If the controlling process were a chemical decomposition, the correct functional relationship would be based on finite rate chemical kinetics. The fuel gasification flux ρv is then a function primarily of T . Frequently, ρv is such a rapidly varying function of T that the change in T , taking ρv from negligible to overwhelming proportions, is small compared to the characteristic differential temperature of the fire. Then it is again acceptable to model T as a constant at the interface. This constant T is a so-called pyrolysis temperature.

In both of the above stated limits, that of decomposition kinetics varying rapidly with temperature and that of equilibrium vapor pressure varying rapidly with temperature, the same mathematical formulation results. The surface temperature is a constant, perhaps experimentally determined, and the flux of gasified fuel is controlled by conservation of energy.

It is important to realize that, even if the surface temperature is fixed in accordance with the limiting arguments just stated, the

model still couples the heat transfer with the flux of gasified fuel.
Of course, the general fire configuration and to some extent the thermal environment presented by the fire depend on this flux. However, there is also a very important local effect. At sufficiently high heat flux rates, there is a blockage effect which manifests itself as a reduction of the surface temperature gradient $\lambda \, \partial T/\partial y$. This type of blockage is discussed later in conjunction with overall boundary conditions.

In the limit that the mass heat flux is negligibly small, it is possible to eliminate ρv from Eqs. (1.7) and (1.9), formally taking $v = 0$. The analytical advantage accruing from such a simplification lies in the treatment of the momentum equation which determines the fire flow field. As is discussed shortly, however, a great practical advantage when the finiteness of v can legitimately be neglected is that heat transfer theory and correlations for inert surfaces can often be directly applied.

The homogeneous and uniformly gasifying surface model, whose boundary conditions are stated above, is the simplest which retains the essential coupling between energy feedback and gasified fuel evolution. Because analytical difficulty increases sharply with further complication of the surface model, there is a strong incentive to apply this relatively simple model to a maximum extent. If the medium is an agglomerate of different substances, or is otherwise not properly describable in one dimension, efforts are made to select average parameters with which to apply a one-dimensional model. The rigorous approach to doing this is to perform an analysis with appropriate multi-dimensional features and then to carry out averaging processes over the surface. In practice, such an approach is usually not feasible. A more reasonable approach is a semi-empirical one, in which parametric analyses based on a one-dimensional model are used in conjunction with experiments to deduce values of the parameters. In this book, no attempt is made to develop or analyze any multi-dimensional surface formulations.

Within a one-dimensional framework, attention is now briefly directed to non-uniform gasification of an initially homogeneous material. Here a number of gaseous species leave the surface toward the flame. Other species approach the interface from the interior. More generally, some species could migrate into the condensed phase from the flames. A completely general mathematical formulation would require an energy conservation statement, definitions of all the relevent fluxes in terms of flow velocities and concentration gradients, and a number of equilibrium or kinetic equations relating the various concentrations and the temperature (and perhaps also formation or destruction rates) at the interface. In this book, such general boundary conditions are not pursued. Instead, extension to two special cases, themselves of considerable generality, will be indicated.

The first special case is that of a homogeneous multi-species condensed phase gasifying to a vapor phase of the same set of species but of generally different fractional composition. If there are N species,   then Eq. (1.7) must be replaced by N definitions of component fluxes m_i ,

$$m_i = \rho \ v \ Y_i - D_i \ \partial Y_i / \partial y \qquad (1.11)$$

where subscript i denotes the ith species. The left hand side of the energy equation must be replaced by a term $\sum_i m_i \ L_i$ where L_i is the effective latent heat of the i-th species.   Eq. (1.10) must be replaced by a set of equations describing multi-species equilibrium, decomposition kinetics or both. In general, if multi-species features are to be treated at all realistically, then a variable temperature T must be retained.

The above model could be further complicated by allowing for chemical reaction at the interface. For example, oxygen could diffuse to the interface, react at the surface with some or all of the evolving gas species or directly with the condensed phase, creating new gas-phase combustion products. Surface oxidation generally is associated with high surface temperatures. A more likely model for inclusion of surface oxidation is the simple char-layer model.

In the simple char-layer model, the condensed-phase medium decomposes at relatively low temperature to form pyrolysis gases and porous char, each considered as a single species. The char may oxidize at some higher temperature or, in principle, may sublimate at a still higher temperature. Conceivably, the destruction of the char may be enhanced by mechanical processes such as erosion by flame gases. Also, char contains a fraction of ash whose concentration at the exposed surface may impede oxidation or sublimation there. Mechanical and ash accumulation effects are not contained in the simple char-layer model. It is doubtful that such effects can be incorporated in any quantitative surface model presently in use for fire conditions.

The char formation process can be modeled as continuous or discontinuous. Discontinuous char formation occurs at a distinct internal surface where virgin fuel transforms completely to char plus pyrolysis gases. This is indicated in the char layer sketch in Fig. 1.1. In a continuous char formation model, the transformation at any point occurs gradually, the porosity and char concentration increasing and the virgin fuel concentration decreasing with time. Further discussion of char formation models is deferred to later chapters. Suffice it to say here that at the exposed char interface, there is a mass flux of gaseous pyrolysis products m_p and a mass flux m_c into the interface corresponding to the regression rate of the surface. The total mass flux to the gas phase $\rho \ v$ is

$$\rho \ v = m_p + m_c \qquad (1.12)$$

The pyrolysis rate m_p is determined by consideration of internal processes beyond the scope of this chapter. The char regression rate m_c must be consistent with an energy balance at the interface and with appropriate statements of chemical or phase equilibrium, or of finite rate chemistry. For ordinary fuels, such as wood or plastic, burning in air, the principal reactive char constituent is carbon. Fire flame temperatures are normally too low for appreciable sublimation. It is often assumed that at high interface temperatures the gases just outside the interface are approximately in equilibrium with the solid carbon

and that the oxidation of the carbon proceeds just fast enough to maintain this equilibrium, thus controlling m_C. At sufficiently low interface temperature, m_C must be kinetically controlled but would then
be negligibly small. Hence, for practical purposes a statement of
chemical equilibrium between solid carbon (C), and gas-phase oxygen
(O_2), carbon monoxide (CO) and carbon dioxide would replace Eq. (1.10).
The flux of each participating gas species would be given by Eq. (1.11).
The left hand side of the energy equation, Eq. (1.9), would require
modification, in general

$$m_C h_C - \sum_i h_i m_i = m_C h_C^0 - \sum_i h_i^0 m_i + m_C \int_{T0}^{T} (C_{P,C} - C_P) \, dT$$

$$= q_{RAD} - \lambda_- \, (\partial T/\partial y)_- + \lambda \, \partial T/\partial y \qquad (1.13)$$

Here subscript i denotes gas-phase species only, and the integral term
accounts for the difference between the specific heat of solid carbon
$C_{p,c}$ and the mean gas specific heat C_p. Also, λ_- is the thermal
conductance of the char, depending in general upon the char structure.
If the carbon oxidation were dominated by a single reaction (e.g.
$C + CO_2 \rightarrow 2CO$) of significant rate only within a temperature range
sufficiently narrow to preclude important variation of the specific
heat term in Eq. (1.13), then an effective latent heat L can be defined, with the left hand side of Eq. (1.13) replaced by $m_C L$.

Indeed, for modeling purposes it is frequently desirable to neglect the interface T range altogether, taking T as approximately
constant, in analogy to the constant surface T limits previously
described for the homogeneous, uniformly transforming surface model.
In this case the argument would be that the relevant equilibrium constant varies so rapidly with T that a char reaction temperature T_C
can be assigned such that for $T < T_C$ the char is effectively inert
($m_C = 0$) and that otherwise $T = T_C$ with m_C determined by the energy
equation. Another possibility, which usually does not occur in fires,
is the complete absence of gas species capable of reacting with carbon,
such as CO_2 or O_2; this would render the char inert until sublimation temperatures were reached. Char inerting at relatively low T
may occur temporarily and locally if jets of gaseous pyrolysis products appear. Such jets frequently accompany wood burning.

Up to now, boundary conditions have been discussed in the context
of field solutions. The various surface models and associated mathematical formulations of boundary conditions are only useful if the
fields of the temperature T and the pertinent species fractions Y_1
are to be calculated throughout the condensed-phase and gas-phase
regimes of the fire system. Even in terms of the variables incorporated in the simplified surface models presented here, such calculations
are effectively impossible for the great majority of fire situations.
Fully numerical solution of two-dimensional laminar combustion flow
fields is only now becoming a practical research possibility. The next
best procedure exploits simplifications inherent in boundary layer
approximations. The fundamentals of boundary layer theory as applied
to combustion problems are presented in Ref. 3. In some cases, a

boundary layer formulation allows similarity transformations which reduce partial differential equations to ordinary differential equations.
In others, relatively efficient two-dimensional numerical procedures
are facilitated. For laminar boundary layer flow models multi-compo-
nent transport problems can be handled with approximately realistic
property variations.   Two typical configurations amenable to similarity transformations are indicated in Fig. 1.3. The fire flow field
for a burning flat plate can be solved in this manner if a single,
very fast reaction is assumed. The model incorporates a reaction zone
of negligible thickness, often called a flame sheet. The flat plate
boundary layer model can describe a complete, albeit idealized, fire
including free convection in the flame boundary layer. For the stagnation point external flow model, multi-step reactions with finite
rate chemistry can be treated. But the similarity transformation them
requires special external flow conditions. Specifically, the external
velocity component u_∞ parallel to the condensed phase surface must
be directly proportional to the distance x from the stagnation point.
Similar analytical principles apply to the opposed jet diffusion
flame[4], a widely exploited laboratory configuration. It must be pointed out also that successful boundary layer combustion field similarity
analysis also places restrictions on the amount of condensed-phase detail treatable. Almost all of the field solutions so far available are
for isothermal, homogeneous and uniformly vaporizing surface models.

     Clearly, there is a need for means of determining the response of
fuel from relatively crude descriptions of fire flow and energy fields.
The same problem has been faced many times in other engineering heat
and mass transfer problem areas. Overall, instead of detailed, boundary condition statements are sought. These statements inevitably
take the form of driving forces and conductions or resistances.

     For example, the heat conduction term $\lambda\, \partial T/\partial y$ has as a driving
force the temperature gradient $\partial T/\partial y$ and as a conductance λ . Here
λ is a material property which in principle is known in advance. But
$\partial T/\partial y$ cannot be rationally determined without an understanding of the
temperature field which in turn requires knowledge of the flow field.
A more convenient driving force is the overall driving temperature
differential $T_g - T_w$ where T_g denotes a characteristic flame temperature[1] and T_w the wall, or interface, temperature. The idea is
that for a given flow configuration the interface temperature gradient
should be proportional to $T_g - T_w$. The proportionality constant is
expressed in terms of a unit heat conductance,[2] α

---

[1] As is demonstrated in Chapter 2, a characteristic temperature for
fire gases does exist. This is approximately the flame temperature
that would result from complete, adiabatic, constant pressure combustion of the fuel, as provided to the gas phase, in stoichiometric
proportion with ambient air. For normal fuels in standard air,
this can approach 2000°C.

[2] The unit heat conductance is more commonly given the symbol h ,
which in this book is used consistently to denote a specific enthalpy.

$$\alpha \ (T_g - T_w) = [\lambda \ \partial T / \partial y]_w \qquad (1.14)$$

In other words, α/λ is approximately a property of the flow field only

For Eq. (1.14) to be useful both α and the differential temperature $T_g - T_w$ must be known. Usually T_w is far lower than T_g . Thus determination of T_w to within a few ten's of °C determines $T_g - T_w$ to considerably better than order-of-magnitude accuracy. This is one of the main justifications for the various constant interface temperature approximations described earlier.

Frequently, it is necessary also to characterize thermal phenomena within the condensed phase in an overall sense. The most common and useful way of doing this is to generalize the concept of effective latent heat L . As originally introduced, L is the change information energy per unit mass gasifying fuel across the interface, modeled as a mathematical surface. In writing the energy balance on a mathematical surface, heat flow into the condensed-phase must be considered. If the condensed-phase is a liquid, complicated convective mixing phenomena may have to be unravelled to determine the internal heat transfer. However, frequently the net result of internal heat transfer processes is merely to bring the liquid up to surface temperature from its state as provided to the pool. The exact manner in which this happens may influence some transient and space distribution aspects of a fire. To the extent that such details can be ignored, it is permissible to redefine L to mean the difference in formation energy of the fuel from its "as provided" state to its gasified state leaving the interface. In other words, the energy required to bring the fuel to interface temperature is merely added to the true latent heat. In many cases this is a rather minor correction. Similar reasoning applies to solid fuels, frequently even to charring ones. With increasing pyrolysis complexity, the selection of L becomes more difficult. Also, application of an overall effective latent heat L to charring fuels masks important transient effects.

In terms of α and L as just defined, the mass flux of gasified fuel $\rho \ v$ is given by reinterpretation of Eq. (1.9) as

$$\rho \ v = q_{RAD}/L + \alpha \ (T_g - T_w)/L \qquad (1.15)$$

It has been pointed out that α is roughly constant for a given surface point in a fixed flow field. But $\rho \ v$, or more generally its distribution over the condensed-phase interface, is itself part of the flow field specification. If the magnitude of $\rho \ v$ is large enough, α must have a different value than for the $\rho \ v = 0$ flow field. Indeed, it is intuitively obvious that independently increasing $\rho \ v$ eventually has a blocking effect which decreases α .

A dimensionless parameter which accounts for such a blocking effect is a special case of Spalding's[5] mass transfer parameter B ,

$$B = \bar{c}_p \ (T_g - T_w)/L \qquad (1.16)$$

where $\bar{c}_p$ is a mean specific heat. As thus specialized, B approximately determines the blocking effect through the relation

$$\alpha/(\alpha)_{(\rho\,v\,=\,0)} \cong [\ln\,(B\,+\,1)]/B \qquad (1.17)$$

In flame configurations amenable to boundary layer analysis (e.g. Ref. 6), B arises naturally out of the detailed dimensionless formulation. In other cases,[7] the group expressed as the left hand side of Eq. (1.17) serves as a convenient correlation parameter. It should be noted that in Refs. 6 and 7, the value of T_g differs slightly from the adiabatic flame temperature indicated above. Also, it is emphasized that Eq. (1.17) is valid only when there is a well-defined external gross flow pattern which is fixed except for local mechanical perturbation by interface mass transfer $\rho\,v$. For the types of fires treated in Refs. 6 and 7, the gross flow patterns are directly influenced by $\rho\,v$.

1.2 COMMENTS ON INTERFACE CONVECTIVE HEAT AND MASS EXCHANGE

In the preceding section, boundary conditions at the interface between a condensed phase and flame gases have been discussed. In general these boundary conditions consist of conservation statements of energy and chemical species, and equilibrium or kinetic statements relating temperature and concentrations, or gradients of them, at the interface. The boundary conditions provide for coupling of the processes within the condensed-phase medium and those of the gas-phase flow regime. Major subsequent sections of this book are devoted to these specific processes.

For the gas-phase regime, both radiative and convective processes are important. Convective processes in fires, however, are conveniently split into two general classes. One class is directly associated with the gross fire flow field; the extent of the flames, the convection column above them, air entrainment, etc. The other consists of those processes by which the gross flow of the fire interacts with fuel or other condensed-phase elements. The last mentioned class of convective processes is the subject of the following comments.

As has been mentioned earlier, the interactions of a convective flow and a condensed-phase object consists in general of both random (conduction or diffusion) and orderly (bulk flow) molecular processes. This is illustrated in Fig. 1.4 which depicts convective heat transfer to a solid object. Far from the object, the primary mechanism of energy transport is the bulk motion of the fluid. At the surface of the solid, there is no bulk motion normal to the surface. Thus energy transport across the solid surface is entirely by conduction. In general, the relative importance of diffusive transport decreases with distance from the solid surface. For a liquid or solid fuel element in a fire, there is an outward bulk flow because of the net mass loss from the condensed-phase fuel to the flames. Conduction counter to this outward flow is essential if heat is to be transferred from the flames to the fuel by molecular processes. On the other hand, mass transfer from the fuel element can in principle take place by bulk flow alone. In general, there is no constructive way to separately account for diffusive and bulk transport in the gas-phase regime of fires. The combined processes are denoted simply as convection, and characterized by an appropriate unit conductance α .

For boundary layer type flow configurations, as sketched in Figure 1.5, it is possible to make reasonable estimates of on the basis of general relations from engineering heat transfer.[8] Suppose that a characteristic boundary layer length ℓ_{BL} can be estimated. A dimensionless expression for α is the Nusselt number Nu ,

$$Nu = \alpha\, \ell_{BL}/\lambda^{*} \qquad (1.18)$$

where λ^{*} is a characteristic gas-phase thermal conductivity. Suppose also that a characteristic external velocity u^{*} can be estimated. Then for gases, a Reynolds number Re can be expressed approximately as,

$$Re \approx u^{*}\, \ell_{BL}/D^{*} \qquad (1.19)$$

where D^{*} is a characteristic diffusion coefficient. Equation (1.19) takes advantage of relations[1] between viscosity and transport coefficients of species and energy which are approximately valid for gases. If Re as here defined is less than about 10^5, the boundary layer is laminar. If greater than about 5×10^5 it is turbulent. In the intervening range of Re , whether the boundary layer is laminar or turbulent generally is questionable. For fires, factors which promote boundary layer turbulence are usually present and it is reasonable to assume turbulence for $Re > 10^5$. Approximate expressions for Nu are

$$\left.\begin{array}{ll} Nu \approx 0.5\ Re^{1/2} & \text{(Laminar)} \\ Nu \approx 0.03\ Re^{4/5} & \text{(Turbulent)} \end{array}\right\} \qquad (1.20)$$

Corresponding values of the boundary layer thickness are also frequently of interest. These are

$$\left.\begin{array}{ll} \delta/\ell_{BL} \approx 5\ Re^{-1/2} & \text{(Laminar)} \\ \delta/\ell_{BL} \approx 0.3\ Re^{-1/5} & \text{(Turbulent)} \end{array}\right\} \qquad (1.21)$$

The foregoing equations are set forth in an approximate spirit only. To the level of accuracy intended, no distinction between local and average values of α and δ is meaningful. However, it is essential to remember the physical meaning of ℓ_{BL} , the boundary layer length. For example, if the gross flow is perpendicular to a cylindrical object, then the diameter is a good estimate of ℓ_{BL} . If the gross flow is parallel to the cylindrical object, then its length would be an appropriate estimate of ℓ_{BL} . Because temperatures vary so widely in fires, λ and D differ substantially from wall temperature to flame temperature. In the spirit of this approximate exposition, the arithmetic means of wall and flame values are appropriate for λ^{*} and D^{*}.

The real limitation on the usefulness of Eq. (1.20) is uncertainty in the characteristic velocity u^*. In Chapter 5, some estimates of u^* are set forth. For large, destructive fires, substantial local external velocity excursions should be anticipated. The complexity of the flow geometry and uncertain effects of fire turbulence demand an experimental convective heat transfer data base.

Such a data base is only now coming into being. For larger fires, radiative, as opposed to convective, heat transfer appears to dominate, at least on the average. So far not enough measurements have been made of the split between radiative and convective heat transfer in large fires to establish a clear picture, particularly for erratic configurations typical of accidental fires.

On a smaller length scale, a reasonably clear picture has emerged for burning pools. It is remarked that, for very small pools, the essential convective energy feedback process does not fit any straightforward boundary layer model. Figure 1-6 shows typical flow configuration for pool diameter around 10 cm. There is a flame boundary layer between the central fuel gas region and the external air. But convective energy feedback to the pool either is from the "leading edge" of the flame boundary layer (where it touches the pool perimeter)[9] or as a result of recirculation in heated combustion projects in the central fuel gas region. [10]

Pool fire energy feedback is inferred from measured burning rates, interpreted via Eq. (1.15), or measured calorimetrically with horizontal gas burners which simulate a burning pool by providing a uniform, and independently controllable, flux of fuel.[10] Figure 1.7 is a sketch of typical heat transfer characteristics[3] for a simulated pool of diameter on the order of 10 cm. At very high mass flux, the convective heat flux component q_{CONV} declines with increasing mass flux because of increasingly effective blockage by outflowing cold fuel gas. At low mass flux, q_{CONV} declines with decreasing mass flux because the mixture over an increasing fraction of the burner area becomes too lean to burn. Simulated pool-fire data may differ quantitatively from the behavior of a real pool at very high and very low mass transfer rates. This is because at these limits the heat feedback distribution, as judged from fire geometry, must be spatially non-uniform whereas the spatial distribution of mass flux in the simulated pool fires is nearly uniform, and is the same for the complete mass flux range. This topic is further discussed in Chapter 5. The figure shows that, while both radiation and convection are important, the burning rate indicated by the balance of total energy feedback q and latent heat requirement $\rho v L$ is controlled primarily by convection. The split between convective and radiative feedback is quantitatively characteristic of moderately luminous fires; e.g. with such fuels as propane. Fig. 1.8 shows the general variation of burning rate with pool diameter d.

---

*In this section this symbol q is used differently than in Section 1.1. In Section 1.1, e.g. Eq. (1.2), q denotes the outward component normal to a condensed phase, of heat flow by diffusive mechanisms. Here q denotes the net heat transferred to the surface from the gas phase, by both radiative and conductive mechanism.

At very small diameters, details of burner construction become important. If the burner is carefully constructed to avoid extraneous
losses, the burning rate tends to vary inversely with d as $d \to 0$.
This variation is also characteristic of fuel droplets burning in air.[11]
This low d limit may be interpreted as a diffusion limit, bulk flow
of the flame gases then playing a negligible role in determining the
energy feedback. For d above 20 to 30 cm, the flames become turbulent at all elevations, apparently without sharply increasing the convective heat transfer. With increasing d the relative importance of
radiation apparently grows, but not enough data is available to support
concrete statements. In the turbulent range, burning rate appears to
increase relatively rapidly with d for fuels with luminous flames.
Available data suggests that for symmetrical fires with d in the 100
to 200 cm range, the convective fraction of energy feedback falls somewhere between 10-25 percent. For higher d , reliable data is simply
not available. For the intermediate range (d ≃ 10 cm) in which pool
fires have the type of structure shown in Fig. 1.6, the dependence of
energy feedback on fuel density and stoichiometric air requirement has
been systematically worked out.[9] However, differences in burning rates
among various fuels primarily reflect differences in the effective
latent heat L .

In concluding this discussion, attention is directed to the general significance of convective heat transfer to condensed-phase surfaces exposed to fire gases. Why is it important to distinguish between convective and radiative modes of transport? There are at least
two important reasons. First, the dynamic state of a fire depends on
its pertinent "operating" characteristics. Certainly local gasification rate ρv versus heating rate q (or some other measure of thermal attack) is an important operating characteristic. From the standpoint of stability, the convective and radiative heating characteristics of a fire may be quite different. The second reason is that the
many condensed-phase surfaces are designed to withstand the fire environment. Sometimes the design problem depends in a crucial way on
the split between convective and radiative heat transfer.

The significance of heat transfer characteristics has already
been hinted at in the discussion of energy feedback to real or simulated liquid fuel pools, specifically in Fig. 1.7. If all heat loss
from the fuel is accounted for in the net feedback terms q_{RAD} and
q_{CONV} , that figure indicates a stable equilibrium state at the point
of intersection between the $q = q_{RAD} + q_{CONV}$ characteristic and the
latent heat requirement line $\rho v L$. It is stable because a small
perturbation of ρv would result in an opposing shift in q , thus
driving the state of the fire back to the equilibrium point. In general, if $\rho v L$ has a more positive slope than q at the operating
point, the fire would be stable to small disturbances. But fires are
notoriously erratic. Fuel bed irregularity and atmospheric fluctuations may be expected to lead to substantial transient disturbances.
Under such circumstances, even if the q and $\rho v L$ characteristics
are stable to small disturbances, considerable instability could result if the characteristics were nearly parallel at the equilibrium
point, as sketched on the left side of Fig. 1.9. The q_{RAD} characteristic is generally monotone increasing with ρv , as indicated in

Fig. 1.7. The q_{CONV} characteristic probably decreases with increasing ρv in the range of practical interest, as also indicated in Fig. 1.7. Thus convectively controlled fire states should be relatively stable in comparison with radiatively controlled fire states. Beyond this is the possibility that a radiatively controlled equilibrium state may not exist at all for small fires. Indeed, this situation is indicated in Fig. 1.7, which is quantitatively realistic. Another possibility is the existence of multiple equilibrium states, some of which are stable and others unstable to small disturbances, but which might jump from state to state as the fire responds to larger disturbances. An imaginary characteristic which might lead to such behavior is at the right of Fig. 1.9. The situation becomes still more complicated for fuels which do not respond instantaneously to changes in the heat transfer q , as is the case with charring fuels. It is beyond the current state of knowledge to outline in any detail such possibilities for transient fire behavior. Yet transient and unstable fire behavior is plainly of great practical importance. Any modeling procedure which explains this class of behavior will be dependent on which mode of heat transfer is dominant.

To illustrate how design for the fire environment depends on the split between radiative and convective heat transfer, consider the problem of specifying an insulative coating for a tank containing a liquid near ambient temperature T_∞ . The tank is assumed to be immersed in a large fire due to a hydrocarbon fuel spill. Frequently the heat loading from such a fire is characterized merely by a so-called cold-wall heat flux. This is the measured flux from the fire to an object whose surface is maintained essentially at ambient temperature. The rationale is that the heat flux in a large fire is dominated by radiation. The radiative environment is supposed to be characterized by the incident energy flux alone. As has already been noted, the asserted dominance of radiation has not been conclusively demonstrated, particularly for fires with irregular fuel distribution and subject to atmospheric disturbance. Of particular concern is locally high convective heat flux which could result in tank failure. It is noteworthy also that the convective heat flux component depends through the unit conductance α on the size of the object immersed in the fire, in accordance with Eq. (1.20), although in the turbulent regime the size dependence is weak $(\propto \ell_{BL}^{-1/5})$.

A simple thermal characterization of the fire environment which includes potential for both radiative and convective heat transfer consists of $q_{RAD,inc}$, α and T_c . Here $q_{RAD,inc}$ denotes the incident radiation flux. The value of α takes into account the size and shape of the tank and the geometry of the postulated convective exposure. The temperature T_c is the effective fire temperature for convection and is not necessarily the same as characteristic peak flame temperature T_g . In fact T_c is an average temperature near the exposed surface and is probably substantially lower than T_g .

Many practical insulation coatings deteriorate at excessive temperatures. For simplicity the insulation is assumed to burn away if its outer temperature exceeds a critical value T_I , here assumed to be less than T_c . This limits the thickness of the insulation. Limiting its thickness places a lower limit $\alpha_{I,min}$ on the unit conductance α_I of the insulation. This in turn imposes a minimum on the amount of heat conducted through it. Denoting this minimum by q_{min} and taking the tank shell as at ambient temperature T_∞, yields

$$q_{min} = \alpha_{I,min} (T_I - T_\infty)$$
$$= q_{RAD,inc} - \sigma T_I^4 + \alpha (T_c - T_I) \qquad (1.22)$$

The insulation is assumed to be perfectly absorptive to radiation. The term σT_I^4 is the radiation emitted by the insulation; here $\sigma = 1.35 \times 10^{-12}$ cal/cm^2/sec/°K^4 is the Stefan – Boltzman constant, (Part II) and T_I is expressed in degrees Kelvin. Thus Eq. (1.22) states an energy balance on the outer surface of the insulative coating.

Now consider three example fire thermal environments each with a cold-wall heat flux of 2.5 cal/cm^2/sec with T_∞ = 20°C:

Environment	q_{RAD} (cal/cm^2/sec)	α (cal/cm^2/sec/°C)	T_c (°C)
(A)	2.5	0	--
(B)	1.25	1.6×10^{-3}	800
(C)	1.25	1.2×10^{-3}	1060

Next suppose that insulations with T_I = 500°C and 600°C are installed. The corresponding values of q_{min} and $\alpha_{I,min}$ in the above indicated units are tnen:

Environment	T_I = 500°C		600°C	
	q_{min}	$\alpha_{I,min}$	q_{min}	$\alpha_{I,min}$
(A)	2.0	4.2×10^{-3}	1.75	3×10^{-3}
(B)	1.25	2.6×10^{-3}	.8	1.3×10^{-3}
(C)	1.45	3.0×10^{-3}	1.0	1.7×10^{-3}

These results show that the split between radiative and convective heating, and the severity of the convective environment as characterized by T_c , make a major difference in the results attainable by insulative coating. This is especially true as T_I approaches the range where such coatings can do some real good.

REFERENCES

1.   J.O. Hirshfelder, C.F. Curtiss and R.B. Bird, "Molecular Theory of Gases and Liquids", Wiley, New York (1954) Ch. 1, 7, 8.

2.   W.A. Sirignano and I. Glassman, Combustion Science and Technology, 1, 4 (1970) pp. 307-312.

3.   F.A. Williams, "Combustion Theory", Addison-Wesley, Reading, Massachusetts (1965) Ch. 12.

4. D.B. Spalding, ARS Journal, 31 (1961) pp. 763-770.

5. D.B. Spalding, "Convective Mass Transfer", McGraw-Hill, New York
    (1963) Sections 2.3, 2.4.

6. J.S. Kim, J. de Ris and F.W. Kroesser, Thirteenth Symposium (Inter-
    national) on Combustion, The Combustion Institute, Pittsburgh
    (1971) pp. 949-962.

7. J. de Ris and L. Orloff, Combustion and Flame, 18 (1972) pp. 381-
    388.

8. E.R.G. Eckert and R.M. Drake, Jr., "Heat and Mass Transfer", 2nd
    Ed., McGraw-Hill, New York (1959) Part B.

9. R.C. Corlett, Combustion and Flame, 14 (1970) pp. 351-360.

10. K. Akita and T. Yumoto, Tenth Symposium (International) on Combus-
    tion, The Combustion Institute, Pittsburgh (1965) pp. 943-950.

11. F.A. Williams, op. cit., Ch. 3.

WOVEN

AGGREGATE

LAYERED

HETROGENEOUS CONSTRUCTION

HOMOGENEOUS

gas-phase pyrolysis
products

char
virgin fuel

CHAR- FORMING, INITIALLY
HOMOGENEOUS

FIGURE 1.1 Surface configurations.

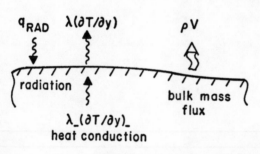

q_{RAD} $\lambda(\partial T/\partial y)$ ρV

radiation

bulk mass
flux

$\lambda_-(\partial T/\partial y)_-$
heat conduction

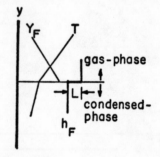

y

Y_F T

gas-phase

L

h_F condensed-
phase

FIGURE 1.2 Boundary conditions—homogeneous, uniformly
transforming model.

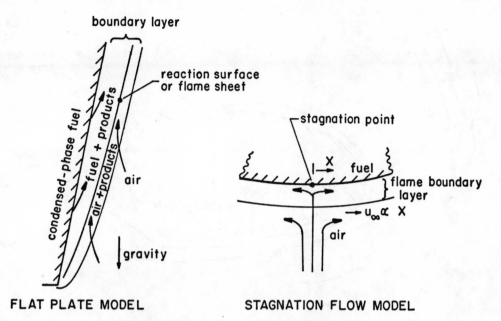

FLAT PLATE MODEL   STAGNATION FLOW MODEL

FIGURE 1.3 Boundary layer diffusion flame configurations.

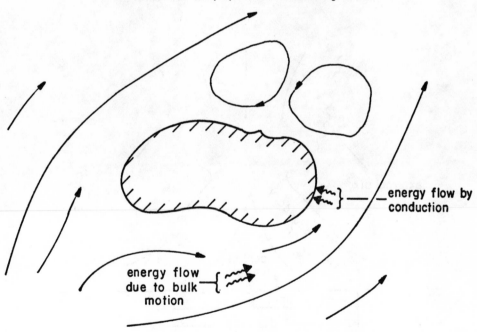

FIGURE 1.4 Interaction of gross flow with immersed object.

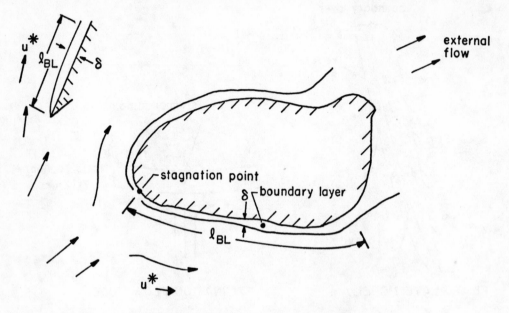

FIGURE 1.5 Boundary layer estimation parameters.

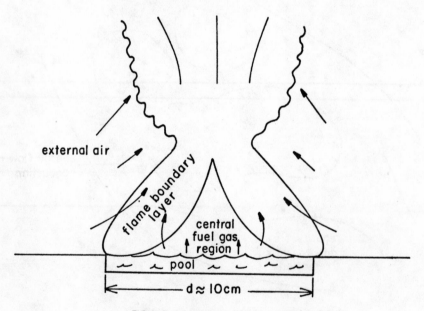

FIGURE 1.6 Pool fire flow configuration.

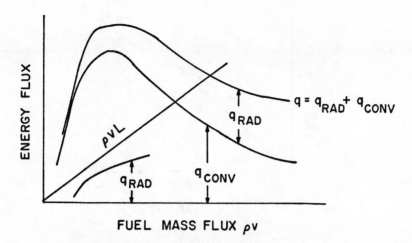

FIGURE 1.7 Typical thermal characteristics of simulated pool fire.

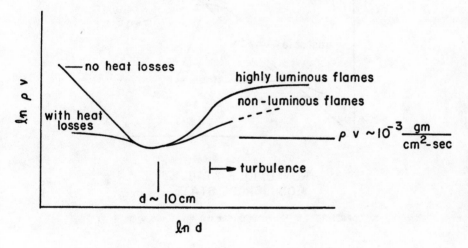

FIGURE 1.8 Variation of pool fire burning rate with diameter.

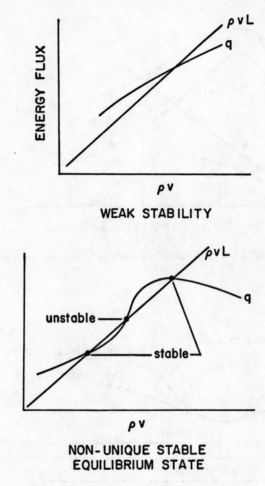

FIGURE 1.9 Hypothetical fire stability relationships.

Chapter 2

CONCENTRATION AND TEMPERATURE SIMILARITY

R. C. CORLETT
University of Washington, USA

MEASUREMENT OF SPECIES concentration and temperature distributions in fires is at best a troublesome business. Straightforward probing techniques are hampered by the hostile flame environment. Moreover, under some circumstances probing may seriously perturb the temperature and concentration fields of interest. Remote measurement techniques of considerable usefulness have been developed in recent years, but still cannot conveniently provide comprehensive field information.

The situation is no better regarding calculation of temperature and concentration fields in fires. Such calculations would be extremely difficult even if the flow fields were known in advance, because fire flow fields are geometrically complex and usually unsteady also. Because the flow in most fires is driven by buoyant forces, or at least is seriously disturbed by them, the temperature, concentration and flow fields are strongly coupled and have to be calculated together. Only in a few highly idealized cases are such solutions feasible.

In view of these difficulties, the principle of energy and concentration similarity is of special importance in fires. It enhances the significance of the limited experimental field data available. Furthermore, this principle provides valuable insights into the structure of fires, attainable otherwise only at great cost and effort.

In this chapter, the similarity principle is derived on the basis of modeling assumptions appropriate to real fuels burning at pressures and oxygen concentrations of the same general magnitude as in the normal atmosphere. Consequences of deviations from the modeling assumptions are explored.

NOTE: All figures quoted in text are at the end of the chapter.

The essential ideas were described by Zeldovich [1]. They have subsequently been generalized and exploited by many authors. The derivation may be outlined as follows. Consider two fields of interest, such
as sensible enthalpy* h_s and a species fraction Y . The partial
differential equations governing the two fields are shown to be similar
in the sense that h_s and Y can both be determined explicitly from
dependent similarity variables which satisfy the same field equations.
A weaker, but still useful, result would be that the field equations
were structurally similar, with identical values of important parameters. The next step is to show that the similarity variables satisfy
identical boundary conditions. The fields of the two similarity variables, satisfying the same differential equations and the same boundary
conditions, are thus identical. In general, the boundary conditions
are not exactly the same everywhere. So there are regions, hopefully
small, where the principle fails. Interpretation of the principle thus
requires estimates of the domain of influence of regions where boundary
condition similarity does not prevail. In some cases, useful inequality
statements can be derived.

2.1 FIELD EQUATION SIMILARITY

In this section, partial differential equations expressing energy
and species conservation are used but not derived. For derivations
and discussion of the properties of these equations, the reader is referred to works on the continuum mechanics of reacting systems [2,3].

In this development, a universal diffusion coefficient D with
unity Lewis number, defined by Eq. (1.3), and a common specific heat
c_P for each species, are assumed. The specific heat assumption means
that the specific enthalpy h can be split as follows between a sensible component h_s and a formation component,

$$h = h_s + \sum_i h_i^0 Y_i = \int_{T^0}^{T} c_P \, dT + \sum_i h_i^0 Y_i \qquad (2.1)$$

where Y_i and h_i^0 are respectively the mass fraction and the heat of
formation of the i-th species at reference temperature T^0. These
assumptions are only approximately satisfied, but are consistent with
the general accuracy level of the similarity principle.

An additional modeling assumption essential to the similarity
principle is the existence of a single overall chemical conversion
process, written formally as

$$\nu_F \, [F] + \nu_{O_2} \, [O_2] + \nu_I \, [I]$$
$$\xrightarrow{\Delta H} \nu_I \, [I] + \nu_{P,1} \, [P,1] + \nu_{P,2} \, [P,2] + \ldots \, . \qquad (2.2)$$

Here F , O_2 , I and P,j respectively denote fuel, oxygen, inert
species (e.g. nitrogen), and the j-th product species. The ν's are
molar stoichiometric coefficients. The heat of reaction ΔH is in
general a function of reaction temperature. But the variation is

---

* Sensible enthalpy h_s is defined by Eq. (2.1), below.

usually small compared to the absolute magnitude of ΔH. The uniformity among species of the mass specific heat c_p, assumed above, implies that at all temperatures ΔH is exactly equal to the change in formation energy for the reaction.

The heat of reaction per mass products Δh^* can be expressed

$$\Delta h^* = \Delta H / [\nu_I \; M_I + \nu_{P,1} \; M_{P,1} + \nu_{P,2} \; M_{P,2} \cdots] \quad (2.3)$$

where M indicates molecular weight. Note that this definition includes in the "products" the stoichiometric amount of inert material. The quantity Δh^* is the same as the rise in sensible enthalpy h_s if a stoichiometric fuel - oxygen - inert mixture reacted to completion adiabatically at constant pressure. That is, when Δh_s has increased by Δh^*, the mixture would have reached the "so-called" adiabatic flame temperature for a stoichiometric mixture. It is also convenient to define a characteristic product fraction $Y^*_{P,j}$,

$$Y^*_{P,j} = \nu_{P,j} \; M_{P,j} / [\nu_I \; M_I + \nu_{P,1} \; M_{P,1} + \cdots] \quad (2.4)$$

The assumption of a single overall conversion with constant heat of reaction implies that the rates of fuel consumption, oxygen consumption, sensible heat generation, and creation of each product species, are proportional.

Two interpretations of the single overall conversion are possible. If the reaction occurs rapidly, in a "flame sheet" of negligible thickness, then the path the reaction takes within the flame sheet is completely unrestricted, provided only that the species entering and leaving the flame sheet balance according to Eq. (2.2). The similarity principle fails within the flame sheet but could be valid outside. The second interpretation is that reaction proceeds strictly as indicated by Eq. (2.2) continuously over a significant distance. In real fire situations, neither interpretation is realistic. The principal discrepancy derives from the processes of soot formation and oxidation. These processes are much slower, for example, than the oxidation of hydrogen atoms. Nevertheless, the consumption of fuel and oxygen, the release of sensible heat, and the formation of major product species, are coupled closely enough to approximately justify the assumption.

Suppose that w_F is the volumetric mass consumption rate of fuel. Then, according to the assumption, the volumetric rate of production of the j-th product species $w_{P,j}$ may be expressed

$$w_{P,j} = \nu_{P,j} \; M_{P,j} \; w_F / (\nu_F \; M_F) \quad (2.5)$$

Analogously, the volumetric rate of sensible heat generation $\dot{q}_s$ is

$$\dot{q}_s = \Delta H \; w_F / (\nu_F \; M_F) \quad (2.6)$$

It is instructive to outline the development of this interpretation of $\dot{q}_s$ from a more fundamental energy equation,

$$\rho \; dh/dt = - \nabla \cdot \bar{q} - \dot{q}_{RAD} \quad (2.7)$$

Here ρ is the local density, h the total (sensible plus formation)
enthalpy given by Eq. (2.1), and the derivative dh/dt is a total
derivative; this derivative may be interpreted as the rate of change
of h in a frame of reference moving with the instantaneous local
macroscopic flow vector. The quantity $\bar{q}$ is a vector generalization
of the diffusive heat flux given by Eq. (1.2). The term "$\nabla\cdot$" is the
divergence operator; and $-\nabla\cdot\bar{q}$ may be interpreted physically as
the volumetric rate of energy accumulation by diffusion. The local
volumetric net loss of energy by radiation is denoted by $\dot{q}_{RAD}$. In
Eq. (2.7) the kinetic energy of the macroscopic flow and viscous dissipation are neglected; these are accurate simplifications, justified
in Chapter 7. Conservation of species, say the i-th, may be written as

$$\rho \, dY_i/dt = -\nabla\cdot\bar{J}_i + w_i \qquad (2.8)$$

Here $\bar{J}_i$ is the vector mass flux of the ith species and w_i its volumetric rate of production. The term $-\nabla\cdot\bar{J}_i$ may also be physically
interpreted as a volumetric rate of accumulation due to diffusion.
Substituting Eq. (2.1) into Eq. (2.7), and using Eq. (2.8) together
with the unity Lewis number assumption and Eq. (1.1) with uniform D,
yields

$$\rho \, dh_s/dt = \nabla\cdot(\rho D \nabla h_s) - \sum_i h_i^0 w_i - \dot{q}_{RAD} \qquad (2.9)$$

The single overall conversion assumption and Eq. (2.6) imply that

$$-\sum_i h_i^0 w_i = w_F \, \Delta H/(\nu_F M_F) = \dot{q}_s$$

Equation (2.9) has the physical interpretation of sensible enthalpy growth
along a particle path, balanced by energy accumulation rates due to
heat conduction and sensible energy release by chemical reaction minus
the radiative energy loss rate. This intuitively "obvious" statement
neglects effects of unaccounted for specific heat and transport property variations. These effects would probably be appreciable given
reasonable accurate field measurements; however, they are certainly
not decisive if the dominant species have molecular weights of the same
general magnitude.

To establish similarity between Eq. (2.8) as applied to a product
species and Eq. (2.9) the following similarity variables are introduced.

$$n_j = Y_{P,j}/Y^*_{P,j} \qquad (2.10)$$

$$\theta = (h_s - h_{s,\infty})/\Delta h^* \qquad (2.11)$$

Here $h_{s,\infty}$ is the sensible enthalpy of the ambient atmosphere. Its
introduction is not necessary for rendering the field equations similar but, rather, anticipates part of the boundary condition similarity
discussed in Section 2.2. With these similarity variables, the field
equations become

$$\rho \, dn_{P,j}/dt = \nabla\cdot(\rho D \, n_{P,j}) + \dot{q}_s/\Delta h^* \qquad (2.12)$$

$$\rho \, d\theta/dt = \nabla \cdot (\rho \, D \, \theta) + (\dot{q}_s - \dot{q}_{RAD})/\Delta h^* \qquad (2.13)$$

If $\dot{q}_{RAD}$ can be neglected, these equations for $n_{P,j}$ and θ are similar in the sense sought.

It is also of interest to define a total products fraction Y_P

$$Y_P = \frac{(Y_{P,1} + Y_{P,2} + \dots)(\nu_I M_I + \nu_{P,1} M_{P,1} + \nu_{P,2} M_{P,2} + \dots)}{(\nu_{P,1} M_{P,1} + \nu_{P,2} M_{P,2} + \dots)} \qquad (2.14)$$

As defined, Y_P is the sum of the mass fractions of the product species, scaled up by the ratio of the total mass of products indicated by Eq. (2.2) with stoichiometric inert gas included, to the total mass of true reaction product species only. In other words, Y_P is the product fraction obtained by counting as products the inert gas which accompanied the oxygen required to form the chemical reaction products present. But Y_P satisfies the same equation as the $n_{P,j}$. This can be seen by summing Eq. (2.8) over all products species, and using Eq. (2.3), Eq. (2.5) and Eq. (2.6).

The significance of this choice of Y_P lies in the fact that, to the accuracy of the present modeling assumptions, it cannot exceed unity and is exactly unity at the flame sheet in the limit of infinite forward reaction rate. This can be shown formally, by defining an air fraction $Y_A = 1 - Y_P - Y_F$, Y_F being the fuel vapor mass fraction, and noting that the resulting mathematical formulation for the concentration fields is exactly the same as if "fuel", "air", and "products" were each a chemically pure species. In the infinite forward reaction rate limit with the "air" a pure species, Y_A and Y_F must vanish at the flame sheet, so Y_P must be unity there. Because the formulation is no different for air diluted with inert gas, the same result for Y_P must hold. The argument can be extended to cases in which the fuel is a mixture or contains inert diluents.

Another way of expressing the similarity of the energy and concentration field equations, with $\dot{q}_{RAD}$ neglected, is to let

$$\phi = Y_P - \theta \quad \text{or} \quad \phi = n_{P,j} - \theta \qquad (2.15)$$

and write

$$\rho \, d\phi/dt = \nabla \cdot (\rho \, D \, \nabla \phi) \qquad (2.16)$$

Because Eq. (2.16) is homogeneous in ϕ, one could conclude immediately that $\phi \equiv 0$ if the boundary conditions on ϕ were also homogeneous. The extent to which this can be true is considered in the next section. Linear combinations of field variables constructed in this manner to satisfy homogeneous conservation equations are sometimes called Shvab-Zeldovich variables.

2.2 BOUNDARY CONDITION SIMILARITY

In general, three types of boundary conditions must be considered. One is at "infinity" in the undisturbed ambient atmosphere far from the fire. At this boundary, the concentration of all products species is zero and the sensible enthalpy $h_s = h_{s,\infty}$. Thus θ and all of the $n_{P,j}$ vanish at infinity.

The second type of boundary is an inert wall. Usually a wall would be regarded as impermeable to product species. Hence all of the concentration gradients normal to the wall would vanish there. Thus for rigorous boundary condition similarity the temperature gradient (or sensible enthalpy gradient) normal to the wall must vanish also; i.e. the wall must be insulated.

The third type of boundary is a surface where gasifying fuel is evolving. At such a surface, rigorous boundary similarity is virtually impossible to establish. For example, if a uniformly vaporizing surface is assumed, with negligible net radiative loss and negligible conduction to the interior of the condensed phase, then with y taken as the direction normal to the surface and L the effective latent heat, the boundary conditions (obtained from Eq. 1.8 and Eq. 1.9) are

$$\rho \, v \, n_{P,j} = \rho \, D \, dn_{P,j}/dy \qquad (2.17)$$

$$\rho \, v \, L/\Delta h^* = \rho \, D \, d\theta/dy \qquad (2.18)$$

with an additional equilibrium or kinetic equation functionally relating $n_{p,j}$ and θ . For Eq. (2.17) and Eq. (2.18) to be exactly similar would require the fortuitous circumstance that $n_{P,j} = L/\Delta h^*$ along the surface.

A more practical line of argument regarding boundary similarity at a gasifying surface is that almost everywhere along the fuel surface the quantity ϕ , defined in Eq. (2.15), is nearly zero. With similarity at infinity and along walls this implies that the solution of homogeneous Eq. (2.16) is zero nearly everywhere.

Consider first the argument for the smallness of ϕ . Looking ahead to the discussion in Section 2.3, the general range of θ and the $n_{p,j}$ is approximately from zero to unity. This means that the characteristic range of Δh_s is Δh^* , corresponding to adiabatic flame conditions. For ordinary non-charring fuels burning in air, surface temperatures never approach adiabatic flame temperatures. A typical value of θ might approach 0.2 at a pyrolyzing surface and could be on the order of $\pm$ 0.01 for some low-boiling liquids. Furthermore a gasifying fuel surface will probably exhibit a high fuel fraction Y_F over much of the surface. This would be especially true if the predominant reactions are occurring at more than a few diffusion lengths from the surface. Thus Y_P , and hence ϕ , would be small along the fuel surface.

A good example of a fire where ϕ is small over most of the fuel surface is a rapidly burning pool of low-boiling fuel. Note that even in this case Y_F must vanish somewhere along the surface, i.e. at the edge of the pool. This is ilustrated in Fig. 2.1. If infinite reaction rate model is strictly accepted, the flame sheet would reach all the way to the surface at some point, and there all of the $n_{P,j}$, and Y_P would be unity. Actually, the surface would probably tend to quench the reaction, thus reducing Y_P significantly below unity. The main point is that only over a very restricted region would the magnitude of ϕ be of order unity.

A variation of the foregoing arguments applies when temperature of the fuel surface and the walls are known to be near ambient, so that θ is effectively zero on all boundaries. Then on all boundaries $\phi \geq 0$. Physically interpreting Eq. (2.16) as a homogeneous diffusion equation allows the conclusion that $\phi \geq 0$, or $n_{P,j} \geq \theta$, everywhere.

Another variation of the argument applies to a particular product species which can be absorbed by the fuel. For example, some combustible liquids absorb water. If the condensed-phase concentration of this species is low enough to correspond to negligibly low vapor pressure, than an effectively zero concentration may be assumed in the gas phase just outside the interface. This approach has been used[4] to test the single overall reaction hypothesis for methanol pool fires by comparing the heat transfer as evidenced by fuel regression rate with the back-absorption rate of product water.

A more general line of argument is based on the facts that for normal fuels burning in air, the volume rate of fuel consumption is very small compared to the total throughput of the fire (because the stoichiometric air/fuel ratio is generally much greater than unity) and that the heat feedback to the fuel and heat transfer to nearby solid surfaces is a small fraction of the total heat release (for reasonably large combustion volumes). In other words the fire is a "reactor" which, as a first approximation, processes a large amount of air, converting chemical energy proportional to the oxygen consumed to increased sensible heat of the product gases. In this view, the fuel mass added and the heat transfer to boundary surface are minor perturbations. It follows that the really important boundary conditions on ϕ in Eq. (2.16) are those upstream at infinity, where similarity is well established. The final step of this physical argument is that deviations from field similarity are localized to the close vicinity of boundary surfaces where boundary similarity fails. In particular, the similarity rule

$$\theta \equiv Y_P \equiv n_{P,1} \equiv n_{P,2} \cdots \qquad (2.19)$$

becomes increasingly valid away from boundary surfaces and in regions of high air concentration.

2.3 FURTHER COMMENTS AND INTERPRETATION

The energy and concentration similarity principle, as expressed by Eq. (2.19), depends on several modeling assumptions only approximately valid. The uniform specific heat assumption can be improved in some respects and worsened in others by changing to a molar as opposed to mass formulation. The unity Lewis number assumption is not exactly

true but appears to be at least as valid as the uniform diffusion coefficient assumption. Moreover, on the total fire length scale, the
unity Lewis number assumption probably becomes less necessary with
increasing turbulence intensity. In general turbulence intensity increases with increasing fire size.

The single overall reaction assumption is one of the weaker elements of the development. In an overall sense, even this can be
strengthened if it is assumed that the amount of sensible energy release is proportional to the amount of oxygen consumed regardless of
which product species is being formed. With a suitable definition of
products, i.e., as the local concentration of combustion-formed oxides,
scaled up in analogy with Eq. (2.14), the identity of the θ and Y_p
fields can again be demonstrated.

The neglect of the radiation energy sink term $\dot{q}_{RAD}$ cannot be
cleanly justified. For all fires of size more than a meter or so, and
especially for those which burn luminously, radiation is a significant
component of the energy balance. The most obvious way to account for
radiation is to reduce the heat of reaction Δh^* in Eq. (2.11) defining θ . This procedure can only be approximate since the dominant
radiation processes are not directly coupled to the chemical reaction.
Instead the amount of radiative energy from a fire depends on temperature and concentration distributions and so must depend on fire size
and geometry, and on the fuel. Radiative losses from the sensibly hot
portions of turbulent hydrocarbon fires are probably on the order of
10 to 30 percent of the total chemical heat release. In general, the
best reduction in Δh^* can only be guessed. As further discussed in
Chapter 6, there are grounds to argue that the radiative correction
should eventually decrease as fire size increases. On the other hand,
for very small fires there is a tendency for the flames to be less
luminous, which also suggests a reduced radiation correction. Thus it
may be that the best correction of Δh^* for radiation loss may be a
maximum at some intermediate fire size. It should be reiterated that
correction for radiation by reduction of the effective heat of reaction
has at best only approximate validity.

In the foregoing paragraphs, the limitations of the principle of
energy and concentration similarity as deduced for fires have been recapitulated. In contrast with the limitations cited, the principle is
completely independent of geometric complexity and applies to fires
whose flow fields are turbulent and otherwise unsteady equally as well
as to steady fires.

The similarity principle as developed here indicates that, to the
extent that the flame sheet model applies, peak fire temperatures
should approach the adiabatic flame temperature of the fuel burning
in stoichiometric air, reduced somewhat for radiation losses. For most
hydrocarbon fuels burning in ordinary air, this indicates peak temperatures in the 1500-2000°C range. It is important to realize that the
peak temperature as predicted by the similarity principle is not the
same as the maximum time-average temperature of a turbulent fire.
This is because in a turbulent fire both the dimensionless enthalpy θ
and the products fraction Y_p (or any product similarity variable
$\eta_{P,i}$) fluctuate substantially; indeed, the magnitude of the variance

expressed as a fraction of the mean may be of order unity. What the
similarity principle does say is that θ , Y_P and the $\eta_{P,j}$ are highly correlated.

As stated in the preceding section, the similarity principle probably holds best on the air side of a fire. It also holds approximately for incompletely reacted products, suchas carbon dioxide and
water vapor with admixed carbon monoxide, hydrocarbon fragments and
soot, if an appropriate effective heat of reaction Δh^* is adopted.
Usually such a correction of Δh^* for incomplete combustion is only
a few percent, and thus not especially significant. The importance
of this result is that, in typical fire situations, the combustion product concentration on the air side should be well correlated with the
temperature. Indeed, the amount of soot and toxic gases formed are
widely held to be roughly proportional to the amount of fuel consumed.
The proportionality constant varies of course from fuel to fuel but
its general magnitude is apparently not dramatically sensitive to geometry, length scale, ventilation and such parameters. (At present
comprehensive data showing the extent of such variation is not available.) Thus there should be a rough correlation between temperature and
concentrations of smoke and toxic gases. This interpretation is consistent with the belief that carbon monoxide, and not heat, is the primary lethal factor in enclosed fire environments. The two apparently
go together, with carbon monoxide reaching a lethal level first.

REFERENCES

1.   Y.B. Zeldovich, Zhur. Tekhn. Fiz., 19, 10 (1949); English translation N.A.C.A. Tech. Mem. No. 1296 (1950).

2.   J.M. Richardson and S.R. Brinkley, "Mechanics of Reacting Continua",
Section F of "Combustion Processes" (Vol. II of "High Speed Aerodyna-
mics and Jet Propulsion"), B. Lewis, R.N. Pease and H.S. Taylor, eds.,
Princeton University Press, Princeton (1956).

3.   F.A. Williams, "Combustion Theory", Addison-Wesley, Reading, Mass.
(1965), Appendices C, D.

4.   Corlett, R.C. and Fu, T.M., Pyrodynamics, 4 (1966) pp. 253-269.

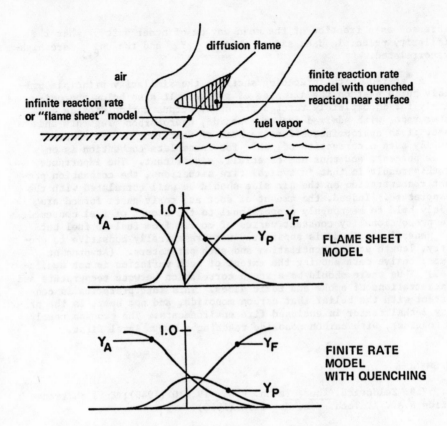

FIGURE 2.1 Total concentration properties at pool edge

Chapter 3

CONDENSED-PHASE MASS
AND ENERGY BALANCES

F. WILLIAMS
The University of California at San Diego, USA

3.1 INTRODUCTION

THE PRESENTATION IN THIS AND subsequent sections focuses on
the role of the condensed phase in the phenomenon called "fire".
Uncontrolled fires constitute a problem of greater magnitude than
usually is recognized. In the United States alone, fire losses now
average five billion dollars per year, and more than ten thousand
deaths annually are attributable to fires. Very few of these unwanted
fires involve initially gaseous fuels. In most, solid or liquid fuels
burn. Thus, fuels in the condensed phase contribute to the majority
of fires that concern society.

Superficial observation of typical fires reveals emphatically
that the vigorous combustion occurs in the gas phase. Condensed
fuels usually are gasified prior to the onset of combustion reactions.
There are exceptions to this behavior. For example, in the
glowing combustion of wood, oxygen diffuses to the surface of the
solid fuel, and combustion occurs at the solid-gas interface.
Combustion processes of this type are milder and of less concern
than gas-phase combustion. Therefore, in the fires of greatest
interest, the condensed phase can be viewed primarily as a supplier
of gaseous fuel.

Solid and liquid fuels are suited remarkably well to serve as
suppliers of gaseous fuel in support of fire. To begin with, condensed fuels store energy in an exceptionally compact manner.

NOTE: All figures quoted in text are at the end of the chapter.

Since the heat released per unit mass of fuel burned is of the same order for gaseous and condensed fuels, and since a typical gas density is one thousandth that of a liquid or solid, condensed fuels contain approximately one thousand times more energy per unit volume than gaseous fuels. Furthermore, this storage is much more compact than that of other conventional energy sources; for example, the energy of the liquid fuel in a typical filled cigarette lighter exceeds that contained in a half-dozen fully charged automobile batteries.

An additional attribute of condensed fuels, contributing to their suitability as fuel suppliers, is that upon exposure to a fire environment, these materials liberate gaseous fuel at rates matched well to the needs of the fire. Supply rates are controlled primarily by processes of heat and mass transfer; they are influenced only secondarily by rates of chemical reactions. The energy needed to convert a condensed fuel to a gas is roughly a tenth the energy liberated when the gaseous fuel burns. This ratio is small enough to favor gas-phase combustion (e.g., if the ratio were of the order of unity or larger, then not enough energy would be released to gasify the fuel, and the reaction would have to occur at the surface of the condensed phase). But the ratio also is large enough to allow the gas-phase combustion to proceed close enough to the condensed fuel for the necessary energy feedback to occur (if the ratio were very small, then gasification would occur so profusely that oxygen would be able to meet the gaseous fuel only at large distances from the condensed fuel). Thus, condensed fuels in fire environments supply fuel to quasisteadily burning gaseous flames for extended periods of time.

When it is agreed that the major role of condensed fuels in fires is to supply gaseous fuel, it clearly becomes desirable to know how to describe the rate at which gaseous fuel is liberated as a function of fire conditions, ambient conditions and any other conditions that influence the rate appreciably. Basically, this rate is governed by the principles of conservation of mass and energy. Thus, mass and energy balances for the condensed phase are useful for estimating rates at which gaseous fuel is supplied.

The present section is concerned with mass and energy balances for the condensed phase. After comparisons are made of gas-phase and condensed-phase processes, the simplest differential conservation equations for the condensed phase are discussed. Forms of these equations for both transient and quasisteady regimes will be considered. Various complicating factors, such as solid charring and liquid flow, will be treated. Overall balances will be derived, and various interface conditions will be stated and discussed. Finally, strategies will be considered for efficient use of condensed-phase mass and energy balances, in developing ideas concerning methods of fire control.

3.2 CONTRASTS BETWEEN GAS-PHASE AND CONDENSED-PHASE PROCESSES

Intuitively, condensed-phase processes seem at first to be much less complex than gas-phase processes in fires. In the interior of the solid or liquid, heat conduction — possibly transient and three-dimensional — would appear to be the only process that could be of consequence. On the other hand, in the gas, besides heat conduction, processes of convection (of both energy and chemical species), of reactant and reaction-product diffusion, of radiant energy transport, possibly finite-rate chemistry of combustion, and especially turbulent fluid flow, all seem to require consideration.

Assuredly, it is true that gas-phase phenomena in fires involve many processes. Accurate descriptions of the behavior of the gas are difficult to develop, particularly if the flow is turbulent. However, the apparent simplicity of the condensed-phase processes often can be deceiving.

Deep in the interior of condensed fuels, perhaps only heat conduction is of importance. For some fires, most of the condensed phase may experience only transient heat conduction. But more often it is true that consideration must be given to other processes occurring within the condensed phase. For all but the heaviest liquid fuels, fluid flow of the condensed fuel can be significant, and the associated convection must be considered. The flow is produced by temperature gradients and may arise either through buoyancy or through variable surface tension; it has been shown that in some cases rates of flame spread over liquid fuel surfaces are controlled by liquid flows that are driven by surface tension. For solid fuels, gases generally are produced not right at the surface, but instead by means of distributed condensed-phase pyrolysis reactions. This necessitates consideration of finite-rate chemistry along with heat conduction. During pyrolysis, solid fuels generally develop heterogeneity; sometimes they are heterogeneous initially. Eventually, a char layer forms, and gases produced in the interior must permeate through a porous solid before they can escape and feed a gas-phase flame. The description of multiphase reacting systems of this kind basically is more complex than that of a reacting gas. Thus, there are situations in which the gas-phase processes turn out to be simpler to describe than those occurring within the condensed fuel.

From a very basic viewpoint, gas-phase processes can be characterized with greater confidence than condensed-phase processes. The equations describing multicomponent reacting gas flow rest on firm ground, having been derived rigorously from the kinetic theory of multicomponent gases. The derivation is made possible by the fact that most of the time gas molecules are unperturbed entities; only during intermolecular collisons (most of which are binary) or occasional radiative transitions do energy or momenta of a molecule change. By contrast, molecules in condensed   phases continually interact with neighbors. This necessitates a less basic, more phenomenological development of conservation equations, and it greatly complicates the task of relating transport coefficients to molecular properties of the materials. Thus, the need to begin at a more empirical level in describing condensed phases contributes to the complexity of the condensed-phase problem.

Of course, the distinction between gaseous and condensed phases is not perfectly sharp. For example, gaseous flames of fires usually contain solid carbon particles, and it has been indicated that subsurface gases are produced during pyrolysis of solid fuels. The relative amount and particle size for the second phase, dispersed within the first, determines the closeness with which a pure-phase analysis can be employed for describing the existing multiphase system. For the gas phase in fires, typically the condensed particles are small enough and sufficiently well dispersed for existing equations for "dilute" systems (viz., presuming that the ratio of the size of a suspended particle to the distance between particles is small) to be applicable. For the condensed phase, correspondingly simple models seldom can be justified with confidence for describing multiphase aspects of the processes. Hence, when multiphase phenomena require consideration, it appears that multiphase aspects of gas-phase processes typically are simpler than those of condensed-phase processes.

3.3 LIQUID FUELS AND FLOW EFFECTS

Condensed fuels are conveniently divided into two classes -- liquid fuels and solid fuels. The division is an obvious one in view of the basic physical states of matter. When gasification mechanisms are considered, it will be found that the division is more useful in practice than basic physics would suggest. For present purposes, the importance of the distinction is that liquids flow while solids do not.

Consider a steadily burning pool of liquid. At the surface of the pool, liquid fuel is vaporizing. Therefore the surface temperature is near the boiling point of the liquid. Unless there is a heat source causing the liquid to boil from below, the bottom of the pool will be cooler than its surface. Since the density of a liquid is a decreasing function of temperature, the density at the surface of the liquid will be less than the density at the cool bottom. In a gravitational field this corresponds to a stably stratified system, and therefore there is no tendency for buoyancy-induced flow of the liquid to develop.

However, conditions at the sides of the pool must also be considered. As illustrated in the upper diagram of Figure 1, cool surroundings may cause heat to be extracted by conduction and/or convection. This cooling tends to cause a fuel element near the side to have a lower temperature than a fuel element at the same elevation in the interior. There is a consequent tendency for a circulatory flow pattern to develop as illustrated.

It is equally possible for the flames to transfer heat preferentially to the walls, either due to a lower reflectivity and higher absorptivity of the wall material or due to geometrical proximity of the flame to the wall. In this case the liquid fuel adjacent to the wall will be hotter than in the interior, and a buoyant circulation pattern of the type illustrated in the lower diagram of Fig. 3.1 tends to develop.

In either of these two cases, buoyant convection within the liquid is induced by the fire. In the presence of this convection, the transient heat equation alone is insufficient for describing the temperature field within the liquid. Convective transport of heat must be considered. The tendency of the convection is to steepen temperature gradients at the surface where the liquid is flowing upward and to make them more shallow where the liquid is flowing downward. In addition, the horizontal flow of liquid along most of the surface produces a generally steeper normal gradient of temperature at the surface. The enhanced heat-sink effect may cause an overall reduction in vaporization rate, although at preferred locations evaporation may be increased.

Suppose that only a small portion of the surface of a large body of liquid is afire. Then the nonburning liquid behaves qualitatively as a heat sink, and the situation resembles that in the upper diagram of Fig. 3.1, with unignited liquid comprising the regions to the side. It is seen that near the surface, the liquid tends to flow outward from the burning area. If the spread-rate of the flames through the gas is sufficiently low, then this buoyant liquid flow can constitute an important contribution to the rate of flame spread.

An additional phenomenon through which liquid motion can produce flame spread is the temperature dependence of the surface tension. The situation is illustrated in Fig. 3.2. The surface

tension of a liquid decreases as the temperature increases. This
implies that under the hot flame the surface tension is lower than
at the surface of the cold, nonburning liquid. The gradient in
surface tension, pointing from the burning surface to the unignited
surface, comprises a force which tends to move the liquid near the
surface in a direction contributing to flame spread. The resulting
flow, driven by (the variation of)surface tension, can under appropriate conditions constitute the controlling mechanism of flame
spread.

Of course, when flow contributes to energy transport within
the liquid, then momentum (and mass) conservation must be
considered in addition to energy conservation, if a quantitative
description of the phenomenon is to be obtained. This introduces,
for example, the coefficient of viscosity of the liquid as an additional
property that influences the behavior of the system. Rather
interesting problems of fluid mechanics are to be found in attempting
to describe condensed-phase effects that occur for fires above liquid
fuels.

3.4 SOLID FUEL AND CHAR EFFECTS

Except for certain synthetics which tend to soften or melt
while burning, solid fuels are not influenced significantly by liquid-
flow phenomena during burning. The fundamental modification to
the transient heat equation for solid fuels is introduced by distributed
solid-phase rate processes that generally are energetically non-
neutral, often endothermic. The polymeric nature of most combustible solids distinguishes them chemically from liquid fuels,
more strongly than does their phase difference. Rate chemistry in
the interior of the solid fuel produces various ramifications, distinct
from equilibrium fractionation of multicomponent liquid fuels.

Consider wood as the most representative example of a
solid fuel. Wood is a complex material, its major constituents
being hemicellulose (a polysaccharide producing wood sugars),
cellulose (a polymer of glucosan, $(C_6H_{10}O_5)_n$) and lignin (a multi-
ring organic compound whose molecular weight exceeds one
thousand). When wood is heated in an inert atmosphere, hemicellulose decomposes first, then cellulose, and finally lignin. The
principal decomposition product which reacts with oxygen in air is
the volatile tar levogluson, a major product of cellulose decomposition. The decomposition products of hemicellulose are in a somewhat more highly oxidized state and relatively noncombustible;
chars remaining after pyrolysis of cellulose and lignin are carbon-
rich and can burn with oxygen primarily through surface reactions.
Pure α-cellulose is a model wood whose behavior in fires qualitatively is similar in many respects to that of natural woods, although

the cellular structure of the natural material is not reproduced by the model substance.

As wood begins to be heated by a fire, first water vapor is liberated, from both absorbed water and weakly bound intramolecular groups. This liberation occurs in depth, and the vapor need not all diffuse to the surface of the wood and escape; some may diffuse inward and recondense in cooler interior portions of the solid. Thus, even before decomposition begins, the nature of the solid may be modified by the heat; a thick piece of wood may effectively have a different moisture content during steady burning than it had prior to arrival of the fire.

When hemicellulose begins to pyrolyze, its decomposition vapors may migrate like the water vapor. Soon cellulose begins to pyrolyze, liberating combustible vapor and leaving some char. The combustible vapors react with oxygen in the gas-phase flames above the wood, liberating there the heat needed to sustain further in-depth pyrolysis. As burning continues, the thickness of the char layer increases. Eventually a quasisteady situation, illustrated by the simplified model shown in Fig. 3.3, may be established. Combustible vapors liberated in the pyrolysis zone flow through the char in the direction of the arrows. Oxygen that has evaded the flame oxidizes the char at its surface, liberating some additional heat, producing CO_2 , and causing the char surface to regress at roughly the same rate as the pyrolysis zone.

The energetics of the model illustrated in Fig. 3.3 clearly are complicated. An unmodified heat equation could be applicable only in the virgin wood. In the pyrolysis zone, endothermic gas-producing decompositions modify the energy equation. In the char zone, a two-phase system must be considered, with heat transfer between gas and char, and with possible exothermic surface oxidation of the char. Development of accurate descriptions of the heat transfer is likely to be complicated by phenomena such as highly spatially variable thermal conductivities of the solid in the upper zones, irregular surface geometry to be considered in estimating radiative transfer, etc.

In view of the complexity of this simplified model of wood burning, it is desirable to investigate separately different parts of the heat-transfer process. In fact, there exist real combustible solids in which the burning process is less complex — primarily non-charring materials. For example, polymethylmethacrylate (PMMA, a plastic known commercially as plexiglas or lucite) will under many conditions in fires pyrolyze, in a distributed zone, through an unzipping process, to a nearly pure combustible monomer vapor, without leaving a char. Studies of condensed-phase processes in the combustion of such materials are simpler, yet contribute to understanding of heat transfer in the burning of woods and other substances. In the following sections, highly simplified models for condensed-phase heat transfer will be developed.

3.5 SIMPLEST MATHEMATICAL MODEL FOR CONDENSED PHASE

The simplest model for the condensed phase is that of a homogeneous, isotropic, stationary, inert, heat-conducting medium with constant properties. The equation for the temperature T in such a medium is

$$\alpha \nabla^2 T = \partial T / \partial t \qquad (3.1)$$

where t is time, ∇ is the conventional gradient operator (∇^2 being the Laplacian), and $\alpha \equiv \lambda / \rho c$ is the thermal diffusivity of the medium. In the definition of α, the quantity λ denotes the coefficient of thermal conductivity, ρ is the density of the condensed phase, and c represents the heat capacity per unit mass (i.e., the specific heat). Eq. (3.1) states that the heat flux by conduction into an element of the material is balanced by accumulation of thermal energy within that element.

Eq. (3.1) is a linear, parabolic, partial-differential equation. For this equation, it is known that a well posed, initial-value problem possesses a unique solution. Thus, if the initial temperature (at $t=0$) is specified throughout the medium, and if the temperature at every point on the boundary of the medium is specified for all $t > 0$, then T throughout the interior of the medium is determined for all positive time. There are boundary conditions other than the specification of the boundary temperature which also produce a well-posed problem. For example, the specified heat-flux condition, $\underline{n} \cdot \nabla T = b$, where $\underline{n}$ is an outward normal unit vector and b is a known function of position and time, may replace the specification of T over part of the boundary, or a mixed condition of the type $\underline{n} \cdot \nabla T = aT+b$, with a and b specified functions of position on the surface and time, may be applied. The insulated-wall condition corresponds to the special case $a = b = 0$. In the application to fire, often these latter conditions, or modifications thereof, can be of importance.

For condensed fuels in typical fires, the initial temperature is the constant ambient temperature T_∞. Qualitatively, the boundary temperatures all begin at T_∞ at $t=0$, and over at least part of the boundary (the part exposed to heat), the surface temperature increases with t, at least until the surface is burning

vigorously. Generally the increase is not specified directly but
instead is a consequence of energy interactions between the gaseous
and condensed phases. Nevertheless, such a surface-temperature
increase does occur, and Eq. (3.1) governs the induced temperature
increase in the interior of the fuel.

Numerous methods, both analytical and numerical, potentially
are available for solving Eq. (3.1). Usually it is convenient to
work with nondimensional variables. A characteristic length ℓ
of the condensed fuel is selected as the unit of distance, and the
nondimensional time is defined as $\tau \equiv \alpha t/\ell^2$, which often is termed
the Fourier number or the Fourier modulus. The nondimensional
temperature $\theta \equiv (T-T_\infty)/T_\infty$ begins at zero when $\tau = 0$, and
Eq. (3.1) becomes

$$\nabla^2 \theta = \partial\theta/\partial\tau \qquad\qquad (3.2)$$

Difficulties in solving Eq. (3.2) can arise from two major sources.
First, the three-dimensional geometry of the fuel causes four
independent variables to appear in Eq. (3.2), thereby taxing
current computing machines to the limit of their capacity if accurate
numerical solutions are sought by finite-difference methods. Second,
the boundary conditions during burning can be complicated; often
they must involve moving boundaries (to account for consumption of
the fuel).

The first of these difficulties often can be lessened by
appealing to symmetry of the fuel elements to reduce the number of
spatial dimensions. In many cases it is possible to argue that it is
sufficient to consider one spatial dimension; for example, if the
radius of curvature of the surface is large compared with the
distance that an appreciable amount of heat penetrates into the fuel,
then the distance measured normal to the surface is the only
relevant spatial coordinate.

Usually the second difficulty is handled by breaking the
history of the fire into stages, during each of which the boundary
conditions can be simplified. Thus, initially a transient stage without regression of the surface can be identified, and often this is
followed by a quasisteady stage during which the surface moves at
a constant rate. These two stages are discussed in the next two
sections.

3.6 TRANSIENT STAGE WITHOUT REGRESSION

During the transient stage, if the surface of the fuel remains
fixed, then the temperature field within the fuel is described by
Eq. (3.2) with null initial conditions and with boundary conditions

applied at fixed boundaries. The Green's function for the diffusion equation, alternatively known as the heat kernel, facilitates solution in this case. The Green's function G is defined as the solution to the nonhomogeneous heat equation, with a point source, satisfying homogeneous boundary conditions. Thus,

$$\nabla^2 G - \partial G/\partial \tau = -\delta(\underline{\rho} - \underline{\rho}_o)\delta(\tau - \tau_o) \qquad (3.3)$$

where δ denotes the Dirac delta function, $\underline{\rho}$ is the spatial coordinate vector (whose cartesian components will be denoted by ξ, η and ζ), and the subscript o identifies the source point. Since the best simple approximation to the surface boundary condition during the transient stage is $\underline{n} \cdot \nabla\theta = b$, the specified nondimensional incident heat flux (defined by $b = q\ell/\lambda T_\infty$ where q is the dimensional heat flux), the homogeneous boundary condition for Eq. (3.3) is $n \cdot \nabla G = 0$. From these results, it can be shown by Green's theorem that the solution to Eq. (3.2) may be expressed in the form

$$\theta(\underline{\rho}, \tau) = \int_0^\infty \oint \oint G(\underline{\rho}, \tau; \underline{\rho}_o, \tau_o)b(\underline{\rho}_o, \tau_o)d G_o d\tau_o \qquad (3.4)$$

where dG_o denotes an element of the surface area of the fuel.

With the nondimensional heat flux b specified at the surface, Eq. (3.4) may be used to calculate θ directly, by evaluating an integral, provided that G is known. For infinite n-dimensional space, it is known that

$$G(\underline{\rho}, \tau; \underline{\rho}_o, \tau_o) = \left[4\pi(\tau - \tau_o)\right]^{-n/2} e^{-|\underline{\rho} - \underline{\rho}_o|^2/\left[4(\tau - \tau_o)\right]} H(\tau - \tau_o) \qquad (3.5)$$

where H denotes the Heaviside unit step function. Although this particular Green's function seldom is used per se, it is helpful for constructing desired Green's functions through image methods which rely on the symmetry of the medium. For example, if the fuel occupies the three-dimensional half-space $\xi > 0$, then an image source may be placed at $(-\xi_o, \eta_o, \tau_o)$, to obtain

$$G(\underline{\rho}, \tau; \underline{\rho}_o, \tau_o) = \left[4\pi(\tau - \tau_o)\right]^{-3/2} \left\{ e^{-\left[(\xi - \xi_o)^2 + (\eta - \eta_o)^2 + (\zeta - \zeta_o)^2\right]/\left[4(\tau - \tau_o)\right]} \right.$$
$$\left. + e^{-\left[(\xi + \xi_o)^2 + (\eta - \eta_o)^2 + (\zeta - \zeta_o)^2\right]/\left[4(\tau - \tau_o)\right]} \right\} H(\tau - \tau_o) \qquad (3.6)$$

If b is independent of position on the surface $\xi = 0$, then substitution of Eq. (3.6) into Eq. (3.4) and integration over η_o and ζ_o produces

$$\theta(\xi, \tau) = \int_0^\tau b(\tau_o)\left[\pi(\tau-\tau_o)\right]^{-\frac{1}{2}} e^{-\xi^2/\left[4(\tau-\tau_o)\right]} d\tau_o \qquad (3.7)$$

Eq. (3.7) can also be derived without integration, by applying images to the case of one spatial dimension, $n = 1$. It is the type of solution that is obtained by application of a Duhamel integral. Of course, Laplace transform techniques also lead to Eq. (3.7).

As an exceedingly simple, totally explicit model for the transient stage, assume that b is constant. In this view, the fuel is at the uniform temperature T_o until exposed to the fire at $t = 0$, after which time the fire bathes the surface of the fuel with a constant heat flux. Performing the integration in Eq. (3.7). produces

$$\theta(\xi, \tau) = b\left[\sqrt{4\tau/\pi} \; e^{-\xi^2/4\tau} - \xi \, \text{erfc}(\xi/\sqrt{4\tau})\right] \qquad (3.8)$$

where erfc denotes the complementary error function. Equation (3.8) demonstrates that in this case the surface temperature of the fuel increases in proportion to the square root of time, $\theta(0, \tau) = (4b^2\tau/\pi)^{\frac{1}{2}}$.

The problem just considered contains no characteristic length. In fact, it can be verified from the solution given by Eq. (3.8) that ℓ drops out of the problem. There is an occasionally useful generalization of the preceding problem for which a characteristic length remains. Assume that the incident flux q is purely radiative and that it is absorbed not at the surface of the solid but instead in depth, according to Beer's law. If ℓ denotes the radiation absorption length for the fuel (the reciprocal of the absorption coefficient), then an additional term $B \equiv b e^{-\xi}$ appears on the left-hand side of Eq. (3.2),), and the boundary condition becomes $\underline{n} \cdot \nabla\theta = 0$. The Green's function remains useful for solving this problem; instead of an incident energy flux at the surface, there is now a distributed energy source, and Eq. (3.4) is replaced by

$$\theta(\underline{\rho}, \tau) = \int_0^\infty \iiint G(\underline{\rho}, \tau; \underline{\rho}_o, \tau_o) B(\underline{\rho}_o, \tau_o) d\mathcal{V}_o \, d\tau_o \qquad (3.9)$$

where $d\mathcal{V}_o$ is an element of volume for the fuel. The final solution is

$$\theta(\xi, \tau) = b\left\{ 2\sqrt{\tau/\pi} \; e^{-\xi^2/4\tau} - \xi \, \text{erfc}(\xi/\sqrt{4\pi}) - e^{-\xi} \right.$$
$$\left. + (e^\tau/2)\left[e^\xi \text{erfc}(\sqrt{\tau}+\xi/\sqrt{4\tau}) + e^{-\xi}\text{erfc}(\sqrt{\tau}-\xi/\sqrt{4\tau})\right] \right\} (3.10)$$

In this case, the history of the surface temperature is given by

$$\theta(0, \tau) = b\left\{\sqrt{4\tau/\pi} - \left[1 - e^{\tau} \operatorname{erfc}(\sqrt{\tau})\right]\right\}$$

which represents a lower temperature than that obtained from Eq. (3.8).

Condensed fuels of other geometries (e. g. slabs of finite thickness with cooling or insulating conditions applied at the unexposed interface, or cylindrical rods of circular or rectangular cross section) can be treated in a similar fashion. It is interesting to observe that in many fire applications, the approximation of a semi-infinite condensed phase is more widely appropriate than intuition suggests. For example, for wood typically α is of the order of $2 \times 10^{-3} \text{cm}^2/\text{sec}$, and so for a one-minute exposure time, the depth of heat penetration is only on the order of $(60 \times 2 \times 10^{-3})^{\frac{1}{2}} = 0.35 \text{ cm}$. Only for very thin fuels, or long exposure times prior to burning of the element, should the semi-infinite approximation be poor.

3.7 STAGE OF QUASISTEADY REGRESSION

As indicated earlier, the processes which occur when the fuel begins to produce gases are complex. However, there are cases in which a rapid transition occurs to a second simpler stage in which steady-state one-dimensional regression of the fuel surface occurs. Burning of the top surface of a horizontally oriented sheet of PMMA affords an example; it is found experimentally for this material that once weight loss of the fuel begins to occur, it proceeds at a nearly constant rate, with a constant rate of decrease of the thickness of the sheet. It is of interest to outline a mathematical description of this relatively simple stage.

To derive the relevant differential equation, assume that conditions remain uniform in planes normal to the ξ axis, the nondimensional spatial coordinate. At $\tau = 0$, let the fuel surface lie at $\xi = 0$, and assume now that the fuel occupies the region $\xi < 0$. Let $\nu \equiv \ell r/\alpha$ denote the nondimensional instantaneous regression rate of the fuel surface, where r is the dimensional regression rate. To keep the fuel surface at the origin of the coordinate system, introduce the transformation

$$\xi' = \xi + \int_0^\tau \nu d\tau, \quad \tau' = \tau$$

Since $\partial/\partial\xi = \partial/\partial\xi'$ and $\partial/\partial\tau = \partial/\partial\tau' + \nu\partial/\partial\xi'$, Eq. (3.2) becomes

$$\partial^2\theta/\partial\xi^2 = \nu\partial\theta/\partial\xi + \partial\theta/\partial\tau \qquad (3.11)$$

where for brevity the primes on the new variables ξ' and τ' have now been deleted.   Equation (11)   exhibits a convection term explicitly (the first term on the right).

Eq (3.11)   could   be useful for describing the transition to quasisteady regression.   After attainment of the steady state, this equation reduces to

$$d^2\theta/d\xi^2 = \nu d\theta/d\xi \qquad (3.12)$$

a simple balance between convection and conduction, with a constant regression rate ν.   The solution   to   Eq. (3.12),   satisfying   the boundary conditions $\theta = 0$ at $\xi = -\infty$ and $\theta = \theta_s$, the nondimensional surface temperature, at $\xi = 0$, is

$$\theta = \theta_s e^{\nu\xi} \qquad (3.13)$$

The dimensional version of this equation is

$$T = T_\infty + (T_s - T_\infty)e^{xr/\alpha}$$

For Eq. (3.13)   to   provide a totally explicit solution, the quantities θ_s and ν must be known.   In other words, the surface temperature T_s and regression rate r are needed.   Determination of these two quantities entails consideration of interface processes occurring at the surface of the fuel.   Both a surface energy balance and the equilibrium vs. rate character of the surface process must be considered.   These topics are considered in following sections.

It may be remarked that following the stage of quasisteady regression, various other stages sometimes occur.   For example, during combustion of cellulosic solids, after volatile fuels have been consumed a rapid transition to a stage of quasisteady glowing combustion may occur.   In addition, often there may be a terminal transient stage of cooling of noncombustible or extinguished residue. These later stages will not be discussed here.

3.8 OVERALL MASS AND ENERGY BALANCES FOR QUASISTEADY STAGE

Simple overall conservation laws can be written by phenomenological reasoning for stages during which steady one-dimensional regression of an essentially semi-infinite condensed fuel occurs.   These laws need not involve previous assumptions such as constancy of the specific heat or of the thermal conductivity. They may be developed by considering a control volume which extends from deep within the interior of the condensed phase to just outside the surface of the condensed material. The laws state that for any conserved quantity, the amount of it entering the control volume in the interior must equal that leaving the control volume at the surface.

Consider first the overall conservation of total mass. Let the subscript ∞ identify conditions deep within the interior of the condensed phase and the subscript s identify conditions in the gas phase at the surface. Mass conservation requires that the density ρ, and the component v of velocity in the direction normal to the surface, obey the equation

$$\rho_s v_s = \rho_\infty v_\infty \qquad (3.14)$$

This constant mass flux often is denoted by the symbol $\dot{m}$. Contrary to what might be guessed initially, the regression rate r of the preceding section is not the velocity v_s at the surface; it is v_∞. The reasons is that since conditions are quasisteady, the velocity at which that surface of the control volume, located deep within the interior of the condensed phase, moves into the fuel, is the same as the velocity at which the surface of the fuel (or, for example, any isotherm) regresses in the laboratory frame of reference.

Overall balances, like Eq. (3.14), apply not only for total mass but also for any other quantity that is conserved entirely. For example, chemical transformations do not destroy atomic elements such as H and C ; they merely rearrange them into different molecules. Therefore, if Y_ℓ denotes the element mass fraction of element ℓ, i.e. the total mass of that element per unit mass of the system (irrespective of the molecular forms in which the element finds itself), then the overall balance for element ℓ can be written as

$$J_{\ell s} = Y_{\ell\infty} \rho_\infty v_\infty \qquad (3.15)$$

where $J_{\ell s}$ denotes the total mass flux of element ℓ at the surface. Since diffusion processes may be important in the gas, $J_{\ell s}$ is composed of two parts, the convective part $Y_{\ell s} \rho_s v_s$ and the diffusive part $Y_{\ell s} \rho_s V_{\ell s}$. Here $V_{\ell s}$ is the effective diffusion velocity of element ℓ at the surface. Since the element may find itself in a variety of chemical species, each having a different diffusion velocity, the quantity $V_{\ell s}$ is influenced by the diffusion velocities of all species containing the element in question. If

Y_j denotes the mass fraction of chemical species j and $\mu_{\ell j}$ the mass of element ℓ , contained in the molecule, per unit mass of species j , then

$$Y_{\ell s} \rho_s V_{\ell s} = \sum_j \mu_{\ell j} Y_{js} \rho_s V_{js}$$

where V_j represents the diffusion velocity of chemical species j . The summation extends over all chemical species that contain the element ℓ .

One might be inclined to think that a formula like Eq. (3.15) is applicable also for each chemical species j . However, in general this is not true. Chemical reactions can destroy or inter-convert chemical species, and therefore in general a reaction term is needed in an equation like Eq. (3.15) for chemical species.

An overall energy balance may be obtained by reasoning which is similar to that given above, but somewhat more involved. Let h denote the thermal enthalpy per unit mass for the material, measured with respect to an arbitrarily chosen but fixed reference condition. If the reference temperature is zero and if the specific heat is constant, then $h = cT$; more generally, for a reference temperature T^O ,

$$h = \int_{T^O}^{T} c_p \, dT$$

where c_p , the specific heat at constant pressure, may depend on T . The thermal energy flux across the surface of the control volume deep within the fuel is $\dot{m} h_\infty$, where $\dot{m} = \rho_\infty v_\infty$. In the gas phase at the surface of the fuel, there is a flux of both chemical and thermal enthalpy. If L^O denotes the increase in chemical enthalpy, per unit mass, associated with forming gaseous fuel from solid fuel at the reference temperature T^O , and if h_s is the thermal enthalpy of the gaseous fuel at temperature T_s , measured with respect to the same reference temperature, then the chemical plus thermal enthalpy flux across the gas-phase surface of the control volume is $\dot{m}(h_s + L^O)$.

The overall energy balance is not simply $\dot{m} h_\infty = \dot{m}(h_s + L^O)$, because transport phenomena produce various contributions to energy transfer. Energy transport from the flame to the condensed fuel necessarily is important. The conductive energy flux from the flame is $(\lambda_g dT/dx)_s$, where λ_g is the coefficient of thermal conductivity for the gas. One should realize that although only gaseous fuel contributes to the h_s defined above, the thermal conductivity λ_g must be that of the actual gas mixture present at the surface; in many fire situations this gas is predominately nitrogen, and therefore as a first approximation λ_g may be taken as the thermal conductivity of nitrogen.

In the absence of penetration of gaseous oxidizer into the condensed phase, transport of enthalpy associated with diffusion does not modify the energy-flux contributions identified above. At the surface of the fuel, the velocity of gaseous fuels seldom ever is v ; there is a diffusive flux as well, the velocity of species j being $v + V_j$ with $V_j \neq 0$. However, for species other than fuel, the diffusive flux balances the convective flux, and for the fuel the sum of the diffusive and convective fluxes is $\dot{m}$, so it is proper to identify $\dot{m}(h_s + L^o)$ as the total enthalpy flux even in the presence of diffusion. Of course, in calculating L^o one must be sure to consider gaseous fuel in the chemical state which it actually achieves upon entering the gas. For some fuels, such as wood, this state can depend on the kinetics of decomposition; it can be influenced by the value of $\dot{m}$. For example, slow pyrolysis tends to produce gaseous fuels in a more highly oxidized state than fast pyrolysis.

There are situations, e.g. glowing combustion of wood, in which gaseous oxygen does penetrate the fuel surface. In such cases, corresponding contributions to the enthalpy flux must be included. In the extreme, no fuel vapor at all is liberated, and all fuel elements appear only in gaseous combustion products. In this case, usually it is convenient to replace the term $\dot{m}(h_s + L^o)$ by $(\dot{m} + \dot{m}_{Os})(h_{ps} - Q^o) - \dot{m}_{Os}h_{Os}$, where $\dot{m}_{Os}$ is the mass flux of oxygen crossing the fuel surface, h_O is the thermal enthalpy per unit mass for oxygen, h_p is the thermal enthalpy per unit mass for reaction products and Q^o is the heat liberated, per unit mass of product formed, when the condensed fuel and gaseous oxygen react (in the mass ratio $\dot{m}/\dot{m}_O$) at the reference temperature T^o , to form the gaseous reaction products that actually leave the surface. To be perfectly general, we should write, in place of $\dot{m}(h_s + L^o)$, the summation

$$\sum_j \dot{m}_{js} H_{js}$$

where m_{js} is the mass flux of species j at the surface (taken as positive for outward flux and negative for inward flux) and H_{js} is the thermal plus chemical enthalpy per unit mass for species j at temperature T_s . Such a term automatically includes enthalpy transport associated with diffusion. For simplicity, the most often applicable form $\dot{m}(h_s + L^o)$ will be retained, it being understood that the replacements discussed here occasionally must be made.

There often are important radiative contributions to the energy flux. Let $\dot{q}_{R\,in}$ denote the radiant energy, per unit area per second, incident on the fuel, e.g. from the flame. Let $\dot{q}_{R\,out}$ denote the radiant energy, per unit area per second, leaving the fuel (and passing across the gas-phase surface of the control volume). Then a term $\dot{q}_{R\,in} - \dot{q}_{R\,out}$ appear in the overall energy balance. Questions of reflection of radiant energy at the surface of the fuel sometimes arise. In the scheme just defined, $\dot{q}_{R\,out}$ includes both the reflected flux and the radiant energy flux emitted by the fuel. For some purposes it would be more convenient to interpret $\dot{q}_{R\,in}$ as the difference between the incident and reflected flux; then, in a gray-surface approximation, $\dot{q}_{R\,out}$ could be written simply as $\epsilon_s \sigma T_s^4$, where ϵ_s is the surface emissivity and σ is the Stefan-Boltzmann constant. In the energy-balance formula, merely $\dot{q}_{R\,in} - \dot{q}_{R\,out}$ will be written, thereby allowing either interpretation to be employed.

A convective heat loss from the side of the fuel can be included through a term $\dot{q}_c$ (energy per unit surface area of the fuel, per second). Such a term, strictly speaking, should not be present in a one-dimensional geometry. However, its inclusion does allow a heat-loss effect, which sometimes can be important in practice, to be accounted for in a phenomenological way. Since the loss occurs from the sides, the ratio δ of side area to surface area appears in any formula for $\dot{q}_c$. For a circular cyclinder of diameter d and height w, we have $\delta = 4w/d$. In terms of a mean heat-transfer coefficient $\bar{h}$ and the mean temperature $\bar{T}$ of the fuel, the expression $\dot{q}_c = \delta \bar{h}(\bar{T} - T_\infty)$ may be employed.

The overall energy balance, obtained by adding together all of the effects discussed above, is

$$\dot{m}h_\infty = \dot{m}(h_s + L^o) - (\lambda_g dT/dx)_s - (\dot{q}_{R\,in} - \dot{q}_{R\,out}) + \dot{q}_c \quad (3.16)$$

Rearrangement produces

$$\left[(\lambda_g dT/dx)_s + \dot{q}_{R\,in}\right]/\dot{m} = L^o + (h_s - h_\infty) + (\dot{q}_{R\,out} + \dot{q}_c)/\dot{m} \quad (3.17)$$

Equation (3.17) is easily interpretable physically. Its left-hand side is the energy input, conductive and radiative, to the condensed fuel, per unit mass of fuel consumed. The right-hand side shows that this energy input is used in providing the heat of gasification L^o, the thermal enthalpy increase of the fuel, $h_s - h_\infty$, and the radiative and convective energy loss from the condensed phase.

An important limitation of Eq. (3.17) is that the process must be quasisteady. This limitation also extends to Eq. (3.15) and, in principle, to Eq. (3.14). It should be clear that if the process is not quasisteady then energy and matter can accumulate with time inside the control volume that was adopted, and consequently a term involving spatially distributed accumulation must appear in the equations. Since this term is difficult to estimate (except perhaps for its sign), the overall energy and mass balances are most useful under quasisteady conditions.

3.9 INTERFACE MASS AND ENERGY BALANCES FOR TRANSIENT STAGE

Obtaining a relationship between θ_s and ν in Equation (13) entails considering an interfacial energy balance at the surface of the fuel. In the simplest view, the surface of the fuel is a mathematical plane of infinitesimal thickness. In this case, in a coordinate system fixed with respect to the surface, the interface conservation laws are the same for steady-state and transient conditions. Thus, both steady and transient stages may be considered here simultaneously. After stating simplified interface conservation equations for this case, we shall consider extensions to cases in which the "surface" has a small but finite thickness.

Let the subscripts + and - identify conditions on the gas and fuel sides of the interface, respectively. All velocities will be measured with respect to the velocity of the interface, i.e., the surface of the fuel will be taken to be stationary in the adopted frame of reference. Then conservation of total mass states that

$$\rho_- v_- = \rho_+ v_+ \tag{3.18}$$

A corresponding mass balance for chemical species j is

$$Y_{j_-} \rho_- (v_- + V_{j_-}) = Y_{j_+} \rho_+ (v_+ + V_{j_+}) - \dot{w}_j \tag{3.19}$$

where $\dot{w}_j$ is the mass per unit area per second of species j produced by chemical reactions occurring at the surface. For a simple vaporization process, there is no chemical transformation, and $\dot{w}_j = 0$. For finite-rate surface decomposition of a polymer, the quantities $\dot{w}_j$ may have different, nonzero values for different pyrolysis products. By conservation of mass in chemical reactions, $\sum_j \dot{w}_j = 0$. By definition of diffusion velocities, $\sum_j Y_j V_j = 0$. Typically it will be true in the condensed phase that $V_{j_-} = 0$ is a good approximation for all species. In the gas phase, according to Fick's law $Y_{j_+} V_{j_+} = (D_j \partial Y_j / \partial x)_+$, where D_j is the multicomponent diffusion coefficient for species j.

The interface energy balance in this simple case can be written as

$$H_- - (\lambda \partial T/\partial x)_-/(\rho_- v_-) = H_+ - (\lambda \partial T/\partial x)_+/(\rho_+ v_+)$$

$$+ \sum_j H_{j_+} Y_{j_+} (V_{j_+}/v_+) - \dot{q}_R/(\rho_+ v_+) \qquad (3.20)$$

where $\dot{q}_R$ is the net radiant energy per unit area per second absorbed at the surface. The quantity H denotes the total (thermal plus chemical) enthalpy. The penultimate term in Eq. (3.20) accounts for energy flux associated with diffusion; energy flux associated with diffusion inside the fuel (subscript$_-$) has been neglected. If L_s denotes the heat of gasification of fuel per unit mass at the surface temperature T_s, then in the absence of oxidative surface processes, Eq. (3.20) reduces to

$$(\rho_+ v_+)L_s = (\lambda \partial T/\partial x)_+ - (\lambda \partial T/\partial x)_- + \dot{q}_R \qquad (3.21)$$

which states that the radiant energy absorbed at the surface, plus the difference between the heat conducted into the fuel surface and that conducted out of the fuel surface into the interior of the fuel, is just sufficient to provide the heat of gasification of the fuel that leaves the condensed phase.

Application of Eq. (3.21) to Eq. (3.13) is facilitated by defining $\dot{q} \equiv \dot{q}_R + (\lambda \partial T/\partial x)_+$ as the total energy per unit area per second transmitted to the surface. From Eq. (3.13), one finds that $(\partial T/\partial x)_- = (T_s - T_\infty)(r/\alpha)$. In terms of the virgin solid density $\rho = \rho_\infty$, the relationship $\rho r = \rho_+ v_+$ holds. Therefore Eq. (3.21) becomes

$$\rho r L_s = \dot{q} - r\rho c(T_s - T_\infty) \qquad (3.22)$$

where the definition of α has been employed. The relationship between r and T_s which emerges is

$$L_s + c(T_s - T_\infty) = \dot{q}/(\rho r) \qquad (3.23)$$

Extension of the interface balances to a moving surface of finite thickness necessitates introduction of terms accounting for motion of the surface and for possible transient accumulation of material and energy at the surface. If $\dot{x}$ denotes the velocity of the surface in the adopted frame of reference, then the generalized overall mass balance becomes

$$\rho_-(v_- - \dot{x}) = \rho_+(v_+ - \dot{x}) + dm_s/dt \qquad (3.24)$$

where dm_s/dt is the time rate of increase of mass, per unit surface area, at the surface.

The corresponding species conservation equation is

$$Y_{j_-}\rho_-(v_-+V_{j_-}-\dot{x}) = Y_{j_+}\rho_+(v_++V_{j_+}-\dot{x}) - \dot{w}_j + dm_{js}/dt \qquad (3.25)$$

where dm_{js}/dt is the time rate of accumulation of mass of species j , per unit surface area, at the surface.

The generalized energy balance becomes

$$H_-\rho_-(v_--\dot{x}) + \sum_j H_{j_-} Y_{j_-} V_{j_-} - (\lambda\partial T/\partial x)_-$$

$$= H_+\rho_+(v_+-\dot{x}) + \sum_j H_{j_+} Y_{j_+} Y_{j_+} - (\lambda\partial T/\partial x)_+$$

$$- \dot{q}_R + dU_s/dt \qquad (3.26)$$

where dU_s/dt denotes the time rate of accumulation of energy, per unit surface area, at the surface. In Eq. (3.26), diffusion in the condensed phase has been reinstated, for generality and for symmetry. Kinetic energy, work through body forces, thermal diffusion and its inverse, are neglected because almost always they are negligibly small.

The extended interface balances given in Eq. (3.24), (3.25) and (3.26) could be useful in attempts to model systems in which transient changes occur over long time scales at the surface. For example, the formation of a char layer on the surface of a cellulosic fuel could be considered by this method. The char could be included as part of the surface, but transient accumulations within this "surface" would then be important.

3.10 PROCESSES THAT DETERMINE THE SURFACE TEMPERATURE

It has been seen that an energy balance, such as that given in Eq. (3.23), provides one relationship between the surface temperature and the regression rate. The character of the surface gasification process must be considered to obtain a second relationship between these two quantities. When the surface process is analyzed, it often turns out that in a good first approximation, study of the surface process alone results in a specification of the surface temperature during burning. Then, an energy balance like Equation (3.23) provides an explicit expression for the regression rate. In other words, the problem becomes separable.

The surface process can be of two basically different types. Either surface equilibrium may prevail, or a surface rate process may occur. In practice it is usually found that equilibrium prevails for liquid fuels and a rate process occurs for solid fuels. Nevertheless, in principle, both conditions may occur for the same fuel. To illustrate this basic property, consider a simple model for the gasification process.

3.11 DERIVATION OF AN UNOPPOSED SURFACE GASIFICATION RATE LAW

The rate of gasification of a fuel often is expressed in the form

$$\dot{m} = B_s T_s^{\alpha_s} e^{-E_s/R^o T_s} \tag{3.27}$$

Here R^o is the universal gas constant, and the constants B_s, α_s and E_s are reaction-rate constants for the gasification process. The first is the pre-exponential constant, the second is the temperature exponent, and the third is the activation energy. Often α_s is set equal to zero, so that all of the temperature dependence is included in the activation energy. The product $B_s T_s^{\alpha_s}$ is the temperature-dependent "frequency factor" for the gasification reaction. Eq. (3.27) may be derived theoretically from the viewpoint of transition-state theory (otherwise known as absolute reaction-rate theory).

For illustrative purposes, consider a one-component condensed material that sublimes in a one-step unimolecular process. One may assume that, in order to enter the gas phase, a surface molecule must first pass through a thermodynamically identifiable state located at the col of a potential energy barrier and termed an activated complex. One may further assume that the activated complex is in chemical equilibrium with the unactivated surface molecules and that the rate of gasification equals the rate at which the complex decays to gaseous molecules.

If $c_s^{\neq}$ denotes the number of moles of activated complex per unit surface area and τ is the average time for a complex to decay to gas-phase products, then the gasification rate is

$$\dot{m} = mc_s^{\neq}/\tau \tag{3.28}$$

where m is the molecular weight of the gasifying species. According to the equilibrium hypothesis,

$$c_s^{\neq} = K_s^{\neq} c_s \tag{3.29}$$

where $K_s^{\neq}$ is an equilibrium constant for the activation process and c_s is the actual number of moles of condensed material per unit surface area. As with all surface equilibrium constants, $K_s^{\neq}$ can be expressed in terms of a ratio of partition functions per unit surface area; specifically, it is the ratio of the partition function per unit surface area for the complex to that for the unexcited surface molecules. If we recall that the activated complex differs from an unexcited molecule principally in that one of the vibrational degrees of freedom is replaced by a translational degree of freedom, e.g., in the coordinate normal to the surface of the condensed material, then we may explicitly remove the extra translational degree of freedom from the complex and at the same time remove the factor representing the ground-state energy of the complex, thereby obtaining

$$K_s^{\neq} = (2\pi mKT_s/h^2)^{\frac{1}{2}} \delta (q_s^{\neq}/q_s) e^{-E_s/R^oT_s} \tag{3.30}$$

where k is Boltzmann's constant, h is Planck's constant, E_s is the ground-state energy of the complex, q_s is the partition function of the unexcited surface molecules, $q_s^{\neq}$ is the partition function for the complex (with translation in the reaction coordinate and ground-state energy factors removed) and δ is the distance in the reaction coordinate at the col over which the complex is assumed to exist. The quantity δ/τ may be interpreted as the average molecular velocity in the direction of gasification at the col and may be assumed to be given by the Maxwellian (thermodynamic equilibrium) translational velocity formula,

$$\delta/\tau = (kT_s/2\pi m)^{\frac{1}{2}} \tag{3.31}$$

Combining Eqs. (3.28) through (3.31) yields

$$\dot{m} = mc_s(kT_s/h)(q_s^{\neq}/q_s) e^{-E_s/R^oT_s} \tag{3.32}$$

as the formula for the gasification rate. Equations (3.32) and (3.27) become identical when the empirical approximation

$$mc_s(kT_s/h)(q_s^{\neq}/q_s) = B_s T_s^{\alpha_s} \tag{3.33}$$

is made.

Geometrically, $c_s^{3/2}$ can be seen to be the number of moles per unit volume in the condensed phase, whence

$$c_s = (\rho_c/m)^{2/3} \tag{3.34}$$

where ρ_c is the density of the condensed phase. In the simplest case, the complex differs from an unexcited surface molecule only in that one vibrational degree of freedom is converted to a translational mode. If q_v denotes the vibrational partition function that is converted, then

$$q_s^{\neq} q_s = 1/q_v \qquad (3.35)$$

and Equation (3.33) becomes

$$\dot{m} = m^{1/3} \rho_c^{2/3} (kT_s/h)(1/q_v) e^{-E_s/R^o T_s} \qquad (3.36)$$

Equation (3.36) provides a means for estimating the constants B_s and α_s, in terms of the molecular structure of the condensed material. However, the approximation given in Eq. (3.35) which leads to Eq. (3.36), is a gross oversimplification for many gasification processes. For example, rotational degrees of freedom of the complex may be vibrational modes (librations) of the normal surface molecule. Therefore, it is best to use the more general expression given in Eq. (3.33) for estimating B_s and α_s. In many applications, theoretical estimation of $q_s^{\neq}/q_s$ is so difficult that it becomes expedient to measure $\dot{m}$ as a function of T_s, thereby determining B_s and α_s experimentally, then use Eq. (3.33) to compute $q_s^{\neq}/q_s$, in an effort to ascertain how the structure of the activated complex differs from that of the normal surface molecule. It should be apparent from this abbreviated discussion that although it involves a host of assumptions, transition-state theory is richly versatile in its ability to describe a wide variety of rate processes.

3.12 INTERFACE EQUILIBRIUM AND LIQUID FUELS

The preceding development is more likely to be revelant for liquid fuels than for solid fuels, because in the latter case the gasification processes typically are more complex than the simple one-step process considered. However, for liquid fuels the reverse of the gasification process analyzed above typically will be fast compared with the calculated $\dot{m}$. In this case, surface equilibrium is maintained, and instead of having a formula for $\dot{m}$ as a function of T_s, we have a formula for concentrations of gas-phase constituents, c_j, as a function of T_s.

If the overall surface process is condensed material $\rightarrow \Sigma_j \nu_j'' C^j$, where C^j is the symbol for chemical species j and ν_j'' is the corresponding stoichiometric coefficient, then under the assumption that the gas is ideal, the surface equilibrium condition can be written as

$$\pi_j c_j^{\nu_j''} = K_s \tag{3.37}$$

where K_s , a function only of T_s , represents the equilibrium constant for concentrations, for the gasification process. According to the ideal gas law,

$$c_j = Y_{js} \overline{m} p / m_j R^\circ T_s \tag{3.38}$$

where Y_{js} is the gas-phase mass fraction of species j at the surface, m_j is the molecular weight of species j , $\overline{m}$ is the average molecular weight of the gas at the surface, and p is the total gas pressure.

The simplest version of Eq. (3.37) is that for which only one species enters the gas from the condensed phase, e.g. a simple vaporization process. In that case, K_s can be approximated as well as

$$(p_r / R^\circ T_r) e^{\Delta H_v / R^\circ T_r} e^{-\Delta H_v / R^\circ T_s}$$

where T_r is a constant reference temperature, p_r is a constant equilibrium vapor pressure at temperature T_r , and ΔH_v is the molar heat of vaporization of the fuel. The combination of Eqs. (3.37) and (3.38) then becomes

$$Y_{Fs} \overline{m} p / m_F = p_r e^{\Delta H_v / R^\circ T_r} e^{-\Delta H_v / R^\circ T_s} \tag{3.39}$$

where the subscript F identifies the gaseous fuel species.

There are both similarities and differences between Eq. (3.27) and Eq. (3.39). They both involve an exponential function of $1/T_s$; for the rate process, the activation energy appears in place of the heat of gasification of the equilibrium process. However, Eq. (3.27) involves the rate m but not the pressure p, while the reverse is true for Eq. (3.39). In addition, the gas-phase concentration Y_{Fs} appears in Eq. (3.39); this may be found only by considering gas-phase processes. Thus, in general, a formula like Eq. (3.39) provides a coupling between gas-phase and condensed-phase processes; it does not explicitly determine T_s .

There is a further approximation in which a value of T_s may be extracted directly in the case of surface equilibrium. Assume that $\Delta H_v / R^\circ T_s$ is large, i.e., the heat of gasification is large. In this case, a small change in T_s produces a large change in the right-hand side ofof Eq. (3.39). The precise value of Y_{Fs} then is not very important; a small change in surface temperature can produce a large change in concentration. Since we know that Y_{Fs} is less than unity and cannot be much less than 0.1 during burning, we may therefore set $Y_{Fs} \approx 1$, obtaining simply the total pressure on the left-hand side of Eq. (3.39). The result is that Eq. (3.39) states that T_s is the normal boiling point of the fuel, a known, fixed constant.

This last approximation often is not well justified because ΔH_v often is of the same order as $R^o T_s$. During burning, the true surface temperature of a liquid fuel may be 10^oC to 20^oC below the normal boiling point. However, there are examples in which an error in surface temperature of the order of 10^oC does not introduce large inaccuracies in condensed-phase analyses.

3.13 INTERMEDIATE SURFACE CONDITIONS

In principle, the surface condition, for a reversible inter-face process, always lies between the limits considered in the two preceding sections. Methods of molecular gas kinetics may be employed in discussing the intermediate condition under which finite recondensation rates occur. We shall briefly illustrate the reasoning by considering a one-component system experiencing a one-step, unimolecular gasification process.

The number of fuel vapor molecules, per unit area per second, that strikes the surface is

$$c_F \sqrt{kT_s/2\pi m_F}$$

in which the second factor is the mean molecular velocity normal to the surface. Letting α_v represent the fraction of gas-phase molecules of species F striking the surface which adhere to the surface, we find through a simple mass balance that Eq. (3.27) must be replaced by

$$\dot{m} = B_s T_s^{\alpha_s} e^{-E_s/R^o T_s} - m_F c_F \alpha_v \sqrt{kT_s/2\pi m_F} \qquad (3.40)$$

The quantity α_v is sometimes termed a surface accommodation coefficient for species F; it may depend upon T_s but is usually independent of pressure and gas-phase composition and its value can, in principle, be predicted by molecular reaction rate theories such as the transition state theory. An explicit pressure and composition dependence appears in Eq. (3.40), unlike Eq. (3.27), through the factor c_F in the last term. When α_v is independent of pressure and gas-phase composition, it is possible to relate α_v to the forward gasification rate

$$\left(B_s T_s^{\alpha_s} e^{-E_s/R^\circ T_s}\right)$$

and to the equilibrium vapor pressure at temperature T_s, which is a thermodynamic property of the system and is therefore relatively easily obtainable. The result indicates that α_v generally exhibits the approximate form

$$A T_s^a e^{-b/T_s}$$

where A, a and b are nonnegative constants. The resulting form of Eq. I3.40) is

$$\dot{m} = B_s T_s^{\alpha_s} e^{-E_s/R^\circ T_s} \left[1 - p X_{Fs}/p_F(T_s)\right] \qquad (3.41)$$

where $p_F(T_s)$ is the equilibrium vapor pressure at temperature T_s, and X_F is the mole fraction of fuel vapor in the gas. Eqs. (3.40) and (3.41) are modified somewhat when more general surface stoichiometry is considered.

Simultaneous surface rate processes in the forward and reverse directions have not been considered in studies of condensed phases of burning fuels. The reason for this is the belief that under most conditions for which models of simplicity comparable to that discussed here may be applicable, the reverse rate is believed to be large compared with the forward rate. An unusually small value of α_v is needed for the reverse rate to become slow. Systems whose gasification processes are simple most likely maintain surface equilibrium. The polymeric fuels considered in the next section have complex gasification processes and generally obey a law resembling Eq. (3.27).

3.14 INTERFACE RATE PROCESSES AND SOLID FUELS

Polymers typically gasify in a complex series of steps. Some examples will be considered later. The degree of molecular rearrangement is so great that the reverse, polymerization process occurs with practically zero probability from the gas. Therefore surface equilibrium is not maintained. The solid-fuel polymer is stable at room temperature only because its gasification rate remains extremely small until an elevated temperature is achieved.

It is possible that the rate of each of the steps in the complex gasification mechanism of a solid fuel can be described by a theory like that developed following Eq. (3.27). Thus, a rate formula analogous to Eq. (3.27) is expected to be applicable for each step in the gasification mechanism. The slowest step (or steps) determine the net rate of gasification. Therefore, inspite of the complexity of the process, it usually is found that a formula like Eq. (3.27) describes the gasification rate of a polymeric fuel, in a reasonable approximation.

With these solid fuels, the activation energy E_s usually is high, in the sense that $E_s/R^o T_s \gg 1$. The result is that the gasification rate $\dot{m}$ depends very strongly on T_s . The burning rate of a fuel usually does not vary over more than an order of magnitude in a fire. Consequently, the value of T_s , obtained from Eq. (3.27), cannot vary greatly over the burning period. It is in this sense that the kinetics of gasification of solid fuels can be said to determine a unique surface temperature T_s during combustion.

3.15 COMPLICATIONS

From the preceding extended discussion of interface conditions, it would be clear that at least in some approximation the burning surface temperature can be specified. With T_s specified, a simple energy balance relates the regression rate r to the energy flux $\dot{q}$ incident on the condensed fuel from the fire (e.g., Eq. (3.23)). A simple model, such as that whose solution is given in Eq. (3.13) then completely determines condensed-phase behavior for a fuel in a fire. Although this simple view often can provide a first approximation to a description of the condensed fuel in a fire, there are various complications that should be considered in reality.

First, Eq. (3.13) is based on the assumption of constant properties (λ, ρ and c). As the temperature of the condensed material increases, these properties will change. Good data often is lacking on temperature dependences of condensed-phase properties. If the data were available, then it is possible that some of the variations could be included in theoretical analyses in which transformations of variables are used to generalize the analysis that leads to Eq. (3.13). Very little research has been persued along these lines.

It is possible that property variations associated only with temperature changes will not be very important. However, for polymeric fuels, property variations associated with chemical decomposition almost certainly are substantial. When degradation of the solid begins to occur, and when gases begin to be produced, large changes in properties are expected to occur and to influence the temperature profile. Little theoretical work has been done in attempting to account for these changes. Very little experimental work has been completed on measurement of the magnitudes of property changes associated with decomposition.

The experimental difficulties are severe. For example, for cellulosic materials that char, consideration should be given to heat capacities and thermal conductivities of gaseous and residue materials separately, and also to ways of combining these to obtain suitable mean properties for the system. Thermal conductivities of gases in the char layer will be lower than or comparable with that of the undecomposed fuel, and although the thermal conductivity of carbon is high, the structure of the carbonaceous char may be such that the conductivity of the matrix is appreciably less than that of the virgin cellulose. Neither its magnitude nor its direction dependence, if any, has been determined; undecomposed woods are known to have noticably different conductivities in different directions. To correctly account for property variations associated with decomposition poses a formidable problem which deserves further study. It has been speculated that char layers with gases flowing through them are highly effective insulators of the material beneath.

A second phenomenon which can invalidate the model that gives Eq. (3.13) is natural heterogeneity of the condensed fuel. Except for some synthetic plastics, solid fuels usually are not homogeneous. For example, natural woods consist of cells and fibres, typically larger than they are wide, with characteristic dimensions ranging from 20μ to 2 mm. Phenomena such as preferential heat flow along paths of enhanced conductivity or preferential gasification at internal cites of relative unstabe fuel may occur. Research has not progressed far enough to allow us to assess the overall importance of solid heterogeneity.

It is known, however, that heterogeneity (e.g., char) produced by fire is important in the burning of many solid fuels. The effect on conductivity has been noted. Effects also occur directly in energy and mass conservation equations. For example, the existence of a char layer implies that gas production has occurred in depth through distributed pyrolysis reactions. Thus, Eq. (3.11) now needs an associated distributed energy source term; it becomes

$$\partial^2 \theta / \partial \xi^2 = \nu \partial \theta / \partial \xi + \partial \theta / \partial \tau + \gamma \omega \qquad (3.42)$$

where γ is a nondimensional heat absorbed in the condensed-phase pyrolysis reaction, and ω is a nondimensional reaction rate. Usually diffusion of reactants within the solid is neglected, so that the reactant mass conservation equation, coupled with Eq. (3.42) becomes

$$\partial \epsilon / \partial \tau + \nu \partial \epsilon / \partial \xi = -\omega \equiv -\delta \epsilon^n e^{-\beta/(1+\theta)} \qquad (3.43)$$

where ϵ is a normalized reactant content (going from unity initially to zero when pyrolysis is complete), δ is a nondimensional pre-exponential factor for the reaction rate, n is the reaction order and β is the nondimensional activation energy for the distributed condensed-phase reaction. Eqs. (3.42) and (3.43) have been considered by a number of investigators as possible models for fire-induced decomposition of both charring and noncharring polymeric solids. Of course, there might well be more than one kinetic process distributed in the condensed phase.

When gases are produced in depth, they must flow through the residual char to leave the solid. Therefore, different velocities for gaseous and condensed constituents within the fuel should be allowed. This further complicates the conservation equations and adds an additional convective contribution to the energy flux. It has been estimated that this new term would not be negligible in Eq. (3.42).

A number of other possible associated effects deserve further thought. Char may provide a barrier to escape of gases, so that through momentum conservation, a possible increased pressure in the interior of the solid might be considered. The gases and char locally may have different temperatures, and heat transfer between the two phases may be considered. Further pyrolysis of gases may occur catalytically, preferentially on char surfaces; thus, the possibility of internally distributed heterogeneous reactions needs consideration. Clearly, the presence of a char layer can introduce many complexities.

A variety of other types of complications also may arise. For example, for liquid fuels such as kerosine, fractional distillation may occur; i.e., the lighter fuel fractions may vaporize first, leaving heavier residuals, with higher boiling points, to vaporize later. The result is not only a time-dependent change in liquid fuel composition but also an induced spatial distribution of composition within the fuel. Liquid flow effects, non-one-dimensional heat losses from solids, etc., may introduce further complications. Since the problem rapidly can become too complicated to be understood, it is best to introduce complicating factors one at a time and to consider each carefully in the absence of others. Much more research along this line needs to be performed.

3.16 STRATEGIES FOR DESCRIPTION OF CONDENSED PHASE

Faced with a problem as complicated as that indicated above, many theoreticians are tempted to write large computer programs which try to include everything. The difficulty with this approach is that first, not everything can be included, and second when too much is included, it becomes less and less clear which phenomenon is producing which effect. A more efficient approach therefore would appear to be to add each new phenomenon separately, attempting at each stage to explore thoroughly the theoretical implications and to compare carefully with experiment.

It would seem that this particular approach is especially desirable in connection with efforts directed toward understanding mechanisms of fire suppression. Does water really act as an extinguishant primarily through cooling of the condensed phase? The way to approach this question is through detailed modeling of cooling effects of water applied to a burning fuel.

REFERENCES

1. Fire Research Abstracts and Reviews, Committee on Fire Research, Division of Engineering, National Research Council, National Academy of Sciences, Washington, D.C., Vol. 1, (1968) through Vol. 15, (1973).

2. Symposium (International) on Combustion, The Combustion Institute, Pittsburgh, Pa., Ninth, (1963) through Fourteenth, (1973).

3. Combustion Science and Technology, Gordon and Breach Publ., New York, Vol. 1, (1969) through Vol. 6, (1973).

4. Browne, F.L., "Theories of Combustion of Wood and Its Control - A Survey of Literature," U.S. Forest Products Laboratory, Madison, Wis., Report No. 2136, (1958).

5. Krishnamurthy, L. and Williams, F.A., "On the Temperature of Regressing PMMA Surfaces", Combustion and Flame, Vol. 20, 1973, pp. 163-169.

6. Lengelle, G., "Thermal Degradation Kinetics and Surface Pyrolysis of Vinyl Polymers," AIAA Journal, Vol. 8, 1970, pp. 1989-1996.

7. Murty Kanury, A., "An Evaluation of Physico-Chemical Factors Influencing the Burning Rate of Cellulosic Fuels and a Comprehensive Model for Solid Fuel Pyrolysis and Combustion," Ph.D. Thesis, University of Minnesota, Minneapolis, Minn., 1969.

8. Williams, F.A., "Combustion Theory", Addison-Wesley Publishing Company, Reading, Mass., 1965.

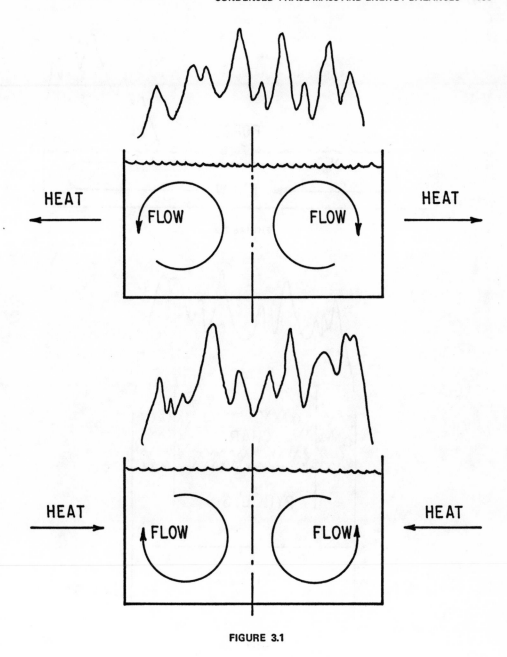

FIGURE 3.1

NOTE: Figure captions for this chapter are listed on page 195.

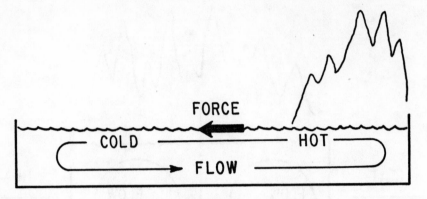

FORCE

COLD ——————— HOT

FLOW

FIGURE 3.2

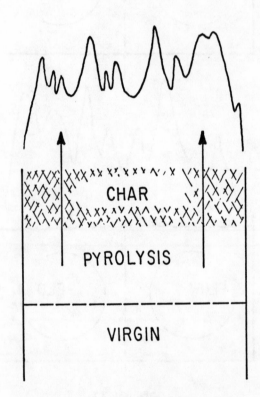

CHAR

PYROLYSIS

VIRGIN

FIGURE 3.3

FIGURE CAPTIONS

Figure 1   Diagram of Convection Patterns in Liquid Fuels.

Figure 2   Diagram of Flame Spread Induced by Surface Tension.

Figure 3   Mechanism of Flaming Combustion for Solid Fuel.

Chapter 4

CHEMICAL KINETICS OF PYROLYSIS

F. WILLIAMS
The University of California at San Diego, USA

4.1 INTRODUCTION

IT HAS BEEN INDICATED THAT condensed fuels in fires can be
viewed as suppliers of gaseous fuels. From this viewpoint, pyrolysis
of the condensed fuel, viz. the chemical breakdown of the fuel constituents under the influence of heat, plays a central role in the
phenomenon of fire. At the molecular scale, it is the mechanism
whereby the gaseous fuel is liberated to support the fire.

Pyrolysis, in the conventional sense of the term, is relatively
unimportant in the condensed phase for liquid-fueled fires. Usually
temperatures within the liquid phase are too low for a significant
amount of chemical breakdown to occur there in the time available
during a typical fire. This may not be true for high-boiling liquids
or for the higher-boiling residues that remain toward the end of a
fire in liquid-fuel mixtures. In a broader sense, pyrolysis may be
interpreted as encompassing the liquid-gas transition (vaporization)
which occurs (usually maintaining equilibrium) at the surface of a
burning liquid. This process has been considered in the previous
section. In general, pyrolysis processes are of lesser significance
within liquid fuels than they are within solid fuels. Therefore, the
discussion in the present section is directed primarily toward solid
fuels.

For solid fuels, condensed-phase chemical kinetics are of
relatively great importance in fires. There can be exceptions to

NOTE: All figures quoted in text are at the end of the chapter.

this rule. For example, if a metal such as aluminum burns in air
with a gas-phase flame, then an equilibrium or sublimation process
occurs at the surface of the metal, there being no appreciable
chemical transformations occurring within the interior of the metal.
However, this type of solid-fuel combustion is not what one normally
thinks of in terms of solid-fuel fires; the ignition energies or
ignition temperatures required are too high for such fires to occur
often. Carbon combustion is another example in which condensed-
phase pyrolysis is unimportant, but combustion of pure carbon is not
encountered very often in practice, and even for charcoal and common
anthracites, there are enough hydrogen-containing large-size molecular constituents within the solid for chemical kinetics of condensed-
phase pyrolysis to contribute appreciably to the burning mechanisms.
Thus, pyrolysis kinetics must be considered in describing the combustion of commonly encountered solid fuels.

A researcher's job would be easy if a single pyrolysis
mechanism described the chemical breakdown of all solid fuels of
practical interest. But this is not the case. Most solid fuels, both
natural and synthetic, are similar in that their structure is primarily
polymeric. However, the polymers differ, and kinetic schemes for
pyrolysis possess a high degree of specificity (as well as, often, a
high degree of complexity). Synthetic plastics differ from natural
cellulosics and also from one another. Even cellulosics differ somewhat, linen differing from cotton and birchwood from oak. Moreover, even for a given polymer, e.g., cellulose, the pyrolysis
mechanism appears to depend strongly on the contents of impurities
such as inorganic ash-producing compounds and fire retardants.
Thus, each polymer must be considered separately in developing
descriptions of pyrolysis mechanisms.

The complexity of the breakdown schemes of most solid fuels
is so great that appreciable uncertainties always remain in description of their pyrolysis mechanisms. A newcomer therefore should
realize in advance that speculative elements exist in all proposed
kinetic schemes for pyrolysis. Various steps that are known to
occur often are ignored. The objective is a good one, viz. to obtain
a mechanism for pyrolysis which is as simple as possible and yet
encompasses most of the phenomena of interest with respect to
behavior in fires. Without simplification, the mechanisms would
be too complex for useful conclusions to be drawn. It should be
realized that pyrolysis kinetics described below represent
idealizations.

Since pyrolysis mechanisms differ for each solid fuel, we
must consider different fuels separately in the following discussion.
A few examples will be considered to illustrate the types of processes that may occur.

4.2 POLYMETHYLMETHACRYLATE AS A RELATIVELY SIMPLE EXAMPLE

The acrylyl radical is

$$
\begin{array}{ccc}
\text{H} & \text{H} & \\
| & | & \\
\text{C} = \text{C} - \text{C} - \\
| & & \| \\
\text{H} & & \text{O}
\end{array}
$$

and the methacrylyl radical has the hydrogen on the central carbon replaced by a methyl group, viz.,

$$
\begin{array}{ccc}
 & \text{H} & \\
 & | & \\
\text{H} & \text{H} - \text{C} - \text{H} & \\
| & | & \\
\text{C} = = \text{C} - - \text{C} - \\
| & & \| \\
\text{H} & & \text{O}
\end{array}
$$

In methacrylic acid, an OH group is attached to the free bond. The methyl methacrylate molecule then is seen to be

$$
\begin{array}{ccc}
 & \text{H} & \\
 & | & \\
 & \text{H} - \text{C} - \text{H} & \\
\text{H} & | & \text{H} \\
| & | & | \\
\text{C} = \text{C} - - \text{C} - \text{O} - \text{C} - \text{H} \\
| & \| & | \\
\text{H} & \text{O} & \text{H}
\end{array}
$$

and it constitutes the monomer of PMMA. To form the polymer, the double C—C bond is broken and reattached intermolecularly. The polymer then can be represented as

$$
\left[
\begin{array}{ccc}
 & | & \\
 & \text{H} - \text{C} - \text{H} & \\
\text{H} & | & \text{H} \\
| & | & | \\
\text{H} - \text{C} - - \text{C} - - \text{C} - \text{O} - \text{C} - \text{H} \\
| & | & \| & | \\
\text{H} & | & \text{O} & \text{H}
\end{array}
\right]_n
$$

with n of the indicated units attached by the bonds drawn long. The chain is terminated at one end by the radical that served to initiate the polymerization and at the other by either an extra proton or a double $C = CH_2$ bond with a one-proton deficiency. In commercial products, chains contain between 10^3 and 10^6 of the indicated monomer units. This number is termed the degree of polymerization.

Before stating explicitly what the pyrolysis kinetic scheme is for PMMA, it is of interest to consider various simple possibilities. Almost invariably, the degradation mechanisms are chain processes. Therefore, they have initiation steps, propagation steps and termination steps.

There are three main types of chain initiation steps. In end initiation, the monomer at the end of the chain splits off, leaving a radical at the chain end. In a random-scission initiation process, thermal fluctuations break the polymer at random points along the chain, producing radicals on each side of the scission. In a weak-link initiation process, the polymer is broken internally at preferred high-strain spots, again producing radical-ended chains.

The most popular chain-propagation step is unzipping, whereby a single monomer unit is formed and detached at the radical end of the chain. For PMMA this process can be indicated as

$$
\text{[chemical structure diagram]}
$$

```
            H                 H                         H                 H
            |                 |                         |                 |
         H—C—H            H—C—H                      H—C—H            H—C—H
            H                 H                         H                 H
            |                 |                         |                 |
 R———————C———C————————C——C——         R———————C——C——————      +    C===C
            |         ‖         |     ‖                 |   ‖              ‖   ‖
            H     C==O      H  C==O                  H  C==O          H  C==O
                  |                   |                  |                 |
                  O                   |                  O                 O
                  |                                      |                 |
               H—C—H            H—C—H                  H—C—H            H—C—H
                  |                 |                     |                 |
                  H                 H                     H                 H
                               n                                       n
```

where R is the radical that initiated polymerization. Alternatives to unzipping are breakoff of higher units from the end (dimers, trimers, etc), termed intramolecular transfer, and interchain reactions. An example of an interchain process is the intermolecular transfer process, in which the radical at the end of an active chain breaks another, stable polymer somewhere in the middle, forming a new stable polymer with half of the attacked chain and leaving as active the other half of the attacked chain.

Termination steps include the direct combination of the radicals at the ends of two active chains, to form a single stable polymer. Another termination step is disproportionation, whereby two active radicals essentially deactivate each other, forming two stable polymers. Disproportionation for PMMA may be diagramed as

Expressions for the overall rate of volatile production can be derived from mechanisms including the steps defined above. Examples of derivations of rate expressions may be found in Refs. 1 to 6. Let us turn our attention specifically to PMMA.

In Table 4.1, taken from Ref. 2, it is seen that the vapors produced when PMMA is pyrolyzed in vacuum are almost all monomer. This continues to occur at higher temperatures, at least up to $800^{\circ}C^{2}$. It suggests that unzipping is the dominant propagation step for PMMA.   Coupled with observations of the residue molecular weight as a function of the extent of volatilization, and of the initial rate of volatile formation as a function of the initial molecular weight, it has been inferred that thermal degradation of PMMA involves chain-end initiation, unzipping, and bimolecular termination [1].   The kinetic chain length (or zip length), i. e., the number of chain-propagation steps that occur prior to termination, has been inferred to be at least 2000 at a temperature of $220^{\circ}C$ (a depolymerization rate of 1%/min) [1]. At higher temperatures, one might expect even longer kinetic chain lengths. Therefore, the termination step is expected to be relatively unimportant in a fire environment, and since propagation appears to be fast compared with initiation, the pyrolysis rate should be controlled mainly by the rate of the initiation step. There results a first-order rate expression for the dependence of the condensed-phase mass m on time t in a homogeneous system, viz.,

$$dm/dt = -k_i m \qquad (4.1)$$

where k_i is the rate constant for the initiation step.

Expressions for k_i have been obtained from the experimental data [1,6]. In terms of the preexponential factor B and the activation energy E, the rate constant k_i is given by

$$k_i = B e^{-E/R^o T} \qquad (4.2)$$

Reported values are [1] $B = 2.82 \times 10^9 sec^{-1}$, E = 31 kcal/mole and, as bounds [6], $B = 10^{9.0 \pm 1.5} sec^{-1}$, E = 29.6 ± 4.4 kcal/mole. The resulting activation energies agree well with the corresponding approximate value appearing in Table 4.2, which has been taken from Ref. 2.

The conclusion that PMMA pyrolysis is basically a first-order process, yielding essentially pure monomer at a rate determined by the rate of the initiation step, is an exceedingly simple result. Extrapolation of this result from laboratory conditions to the more severe conditions prevailing in a fire environment is hazardous in the absence of supporting experimental verification. Preliminary verification along these lines has been obtained recently [7, 8]. Therefore,   there appears to be some justification for using the results described here in studying the burning of PMMA.

4.3 POLYSTYRENE AND OTHER SYNTHETICS

Styrene is a vinyl molecule containing a benzene-type group, as shown

For brevity of display, the ring will be written simply as ⬡
In polystyrene, again linking occurs through the double bond, so that the polymer can be represented as

Polystyrene affords a good example of a polymer whose thermal degradation mechanism is appreciably more complicated than that of PMMA. An indication of this is provided by Table 4.1, in which it is seen that only 40% of the volatiles are found as monomer units. Controversy still exists concerning the proper pyrolysis mechanism   for polystyrene. A discussion of various possibilities should prove instructive. Here we follow Ref. 7.

Two observations in isothermal bulk degradation experiments are that the products are mainly monomer, dimer, trimer and tetramer, up to 500°C,[2,9] and that the molecular weight of the residue drops very sharply at first and then more gradually as volatilization progresses [9,10] Jellinek[1] explains the latter observation by rapid breaking of a finite number of weak links distributed at random over the backbone of the polymer chain. This is presumed to be followed by the true depolymerization reaction, which consists of initiation at chain ends and unzipping with a short kinetic chain length P_k. Let the effective degree of polymerization, after all weak links have been broken, be $P_f = P_0/(1+w_0)$, where w_0 is the number of weak links in a polymer chain of length P_0 . Then there are $m_0/P_f M_m$ chains in the sample after initial breakdown, where m_0 is the initial mass of the sample and M_m is the monomer molecular mass. Since P_f is large compared with P_k , the number of chains remains constant during most of the reaction. Therefore the rate at which chain ends are activated remains constant at the value $k_i m_0/P_f M_m$. Since each activation produces P_k monomers, the mass loss rate is predicted to be

$$dm/dt = -(P_k/P_f)k_i m_0 \qquad (4.3)$$

a zero-order reaction. Jellinek finds experimentally [1] that the mass loss rate is constant over much of the reaction history, in agreement with Eq. (4.3). From the results, he deduces that

$$P_k k_i = 3.54 \times 10^{13} \exp(-44,700/R^\circ T) \text{ sec}^{-1} \qquad (4.4)$$

for the proposed model. Although various observations, such as the occurrence of dimers and larger fragments,   are not explained by this mechanism, it might provide a useful first approximation, with minimum complexity.

Simha and Wall [5] observe additionally that in fact the rate of production of volatiles, as a function of the extent of volatilization, shows a pronounced maximum at about 30% to 40% volatilization [9,10]. Believing this to be a significant feature to describe by the kinetic mechanism, they abandon the weak-links idea and assume random initiation along the chains, unzipping with a very short kinetic chain length, and an important contribution from intermolecular transfer, resulting in further scission of the chains. An added aspect [11] is the postulate of a critical chain length P_c, at or below which the chain will evaporate rather than degrade further. If k_r is the rate constant for random splitting of any one bond, combining random initiation and intermolecular transfer, then it was found that under isothermal conditions

$$dm/dt = -k_r m_o (1+P_c) P_c e^{-P_c k_r t} (1-e^{-k_r t}) \qquad (4.5)$$

which possess a maximum value of $m_o k_r P_c \left[P_c/(1+P_c) \right]^{P_c}$ at

$$\frac{m_o - m}{m_o} = 1 - \frac{1+2P_c}{1+P_c} \left(\frac{P_c}{1+P_c} \right)^{P_c}$$

Identifying P_c approximately with the ratio of the energy required to break one $C-C$ bond in the polymer backbone to the energy required to evaporate a monomer[12], it is found[5] that the maximum mass-loss rate occurs at approximately $(m_o - m)/m_o = 25\%$. A sharp decrease in degree of polymerization, roughly consistent with experiment, is also predicted by this mechanism [11]. Therefore, all observations are explained qualitatively. Using $P_c \approx 7$ (Ref. 1, p. 100) and the maxima observed at several temperatures, [9] it is found from Eq. (4.5) that

$$k_r = 1.54 \times 10^{14} \exp(-55,000/R^o T) sec^{-1} \qquad (4.6)$$

which of course should be used only in connection with the model which produced Eq. (4.5). A result like Eq. (4.5) also is obtained in Ref. 13.

Grassie and co-workers [10,14,15] hold a third view. They revive the weak-links explanation for the initial decrease in molecular weight, presuming that breaking of weak links results in stable molecules by disproportionation. Volatile production is considered to arise from end-initiated unzipping, with the larger volatiles coming from intramolecular transfer. The rate-curve maximum is ascribed to continual chain-end creation through finite-rate weak-link scission. Random scission by intermolecular transfer is excluded as a cause of the initial sharp drop in molecular weight, by showing [15] that the drop can be   separated experimentally from volatile production, which presumably promotes intermolecular transfer as a by-product of unzipping. An analysis consistent with these conclusions has been implied by Gordon[16,17]and developed by Lengelle [7].

If w denotes the number of weak links per original chain, then the rate at which these bonds are broken may be assumed to be of first order, viz.,

$$dw/dt \equiv -k_w w \qquad (4.7)$$

where k_w is the rate constant for breaking weak links. It is desirable to relate w to the degree of polymerization P and the degree of volatilization. This can be done through the total number of bonds broken per original chain, which is $w_o - w + n_o - n$, where n is the number of normal bonds per original chain. It is presumed that breaking of a normal bond occurs only through unzipping and leads to production and volatilization of a monomer; higher fragments are assumed not to volatilize. It follows that in the residue the number of bonds that have been broken per original chain is $w_o - w$. Hence the number of segments remaining in the residue per original chain is $1 + w_o - w$. Since the maximum degree of polymerization that could occur in the residue, after volatilization   mass m_o to mass m via breaking of normal bonds (without breaking any weak links) is $P_o(m/m_o)$, we see that the number-average   degree of polymerization of the residue (including the effect of weak-link breaking) is $P = P_o(m/m_o)/(1+w_o-w)$. Therefore $w = 1+w_o - P_o m/Pm_o$, and Eq. (4.7) can be written as

$$d(m/P)dt = k_w m_o (1/P_o +N_w -m/Pm_o) \qquad (4.8)$$

where $N_w \equiv w_o /P_o$ is the number of weak links per monomer unit in the original chain.

Since both m and P vary with time, an additional equation is needed to obtain dm/dt. This equation can be developed by considering directly the kinetics of volatilization. Assuming that monomer (and volatile) production occurs only by unzipping, we have

$$dm/dt = -M_m k_p n_A m \qquad (4.9)$$

where n_A is the number of active chains per unit mass and k_p is the rate constant for propagation (unzipping). Let k_i be the rate constant for end initiation of a chain (assumed to be the only process leading to unzipping) and k_t be the rate constant for a second-order termination process (viewed as disproportionation). Then the steady-state hypothesis states that $k_i n_I = k_t n_A^2$, where n_I is the number of inactive chains per unit mass. It is assumed that $n_A \ll n_I$, so that n_I approximately equals the total number of chains per unit mass. Since the definition of P implies $M_m P$ is the mass per chain, it follows that $n_I \approx 1/M_m P$. With this approximation, we obtain $n_A = [k_i/k_t M_m P]^{1/2}$, and Equation (4.9) becomes

$$dm/dt = -Km/\sqrt{P} \qquad (4.10)$$

where $K \equiv k_p \sqrt{k_i M_m/k_t}$ is a combination of rate constants. Equations (4.8) and (4.10) can be solved simultaneously for m and P as functions of t.

To obtain the constants appearing in Eqs. (4.8) and (4.10) from experimental data, note first that if no normal bonds are broken, then $m = m_o$, and Eq. (4.8) becomes $d(1/P)dt + k_w(1/P_o+N_w-1/P)$. From this it is seen that P decreases to $P_o/(1+N_w P_o)$ when all weak links are broken. Since the initial drop in molecular weight is interpreted as breaking of weak links, the number of weak links per monomer can be estimated from P_o and P_f; the result, $N_w = 1/P_f - 1/P_o$, varies between about 10^{-4} and 8×10^{-4}, depending on the method of polymerization and on P_o [14] After all weak links have been broken, Eq. (4.8) predicts that m is proportional to P, in approximate agreement with experiment.[15] By applying Eq. (4.10) at at t = 0 and knowing P_o, the data of Modorsky [9] can be used to obtain

$$K = 4.6 \times 10^{17} \exp(-58,000/R^o T)\sec^{-1} \qquad (4.11)$$

The value of the remaining constant, k_w, can be obtained from experimental curves of P as a function of t for different temperatures [10]. At very small times, Eq. (4.8) reduces approximately to $1/P - 1/P_o = N_w k_w t$, and comparison with the data produces [7,10,14]

$$k_w = 1.8 \times 10^{19} \exp(-65,000/R^o T)\sec^{-1} \qquad (4.12)$$

A fourth possible model is to adopt the reasoning leading to Eq. (4.10) but attribute the initial drop in molecular weight to

random scission instead of weak links. In this case, it can be
shown that [3,16]

$$d(1/P)/dt = k_s(1-1/P) \qquad (4.13)$$

where k_s is the rate constant for random scission. Equations
(4.10) and (4.13) are now to be employed simulaneously and produce
the observed molecular weight drop and peak volatilization rate.
The small-time version of Eq.   (4.13), viz., $1/P-1/P_o = k_s t$,
produces, from the data, [7,10,14]

$$k_s = 7.8 \times 10^{15} \exp(-65,000/R^o T) \sec^{-1} \qquad (4.14)$$

The activation energies appearing in Eqs. (4.11) and (4.14) are close
enough for the dependence of residue molecular weight (after the
initial rapid drop) on the extent of volatilization to be fairly insensitive to temperature, as observed experimentally [10] (and as also
is predicted by the weak-links model). It may be noted that the
random-scission rate constants appearing in Eqs. (4.6) and (4.14)
differ, because the models to which they correspond differ.

The preceding discussion should give some idea of the kinds
of controversies that exist in studies of the kinetics of polymer
degradation. Disputes are difficult to resolve, and it often may
occur that more schemes than one are consistent with the observations. It is possible that in application to fires, any one of the
views would be sufficiently accurate. But one could not be sure of
this without checking.

Polymers appearing as the later entries in Table 4.1 pyrolyze
by mechanisms of the general character discussed in connection
with PMMA and polystyrene. Those with low monomer yield, such
as polyethylene, the simple polymer whose monomer unit is

$$
\begin{array}{c}
\text{H} \quad \text{H} \\
| \qquad | \\
-\!\!-\!\!-\!\!\text{C}-\!\text{C}-\!\!-\!\!-\!\!- \\
| \qquad | \\
\text{H} \quad \text{H}
\end{array}
$$

do not exhibit any contribution from unzipping. In this last case,
random scission is believed to be the dominant mode of volatile
production [18]. It is possible to consider the random decomposition
to occur by a chain process with transfer steps carrying the
propagation[18].

There are synthetics that exhibit special features in their
degradation mechanisms [18].   As an example, consider polyvinyl-
chloride, the monomer unit of which is

$$\begin{array}{c} \text{H} \quad \text{H} \\ | \quad\;\; | \\ -\text{C}-\text{C}- \\ | \quad\;\; | \\ \text{H} \quad \text{C}\ell \end{array}$$

Pyrolysis of this material produces volatile HCℓ , leaving a carbon-rich residue behind.   Benzene is also observed as a volatile, presumably formed in the carbon-rich residue.   A carbon char remains after volatilization.   The general character of the process, viz. side-group splitting to produce a charred residue, is shared by a number of char-producing polymers.   A second example is acrylonitrile, whose monomer unit is

$$\begin{array}{c} \text{H} \quad \text{H} \\ | \quad\;\; | \\ -\text{C}-\text{C}- \\ | \quad\;\; | \\ \text{H} \quad \text{C}\equiv\text{N} \end{array}$$

and which produces small amounts of HCN in pyrolysis.

Polymers containing aromatic links in their chains, e.g., polyphenyl whose monomer unit is

tend to cross-link during pyrolysis [18], producing a high-carbon residue and light volatiles, such as H_2 .

Oxidative degradation has been shown to occur for various polymers [18]. This means that the presence of gaseous oxygen enhances the rate of degradation.   The enhancement is viewed in one model as oxidative chain-end initiation of unzipping.   In one series of tests [18], under oxygen pressures   from 10^{-3} to 1/5 atm, the reaction order with respect to O_2 was observed to be roughly 0.5, which is consistent with the k_i , appearing in the definition of K in Eq. (4.10), being proportional to the oxygen concentration. It is tempting to infer that degradation mechanisms in fire environments, where oxidation obviously is occurring at least in the gas phase, may be quite different from those in laboratory vacuum-pyrolysis experiments.   For a number of reasons, however, this seems unlikely.   First, the extent of enhancement by atmospheric oxygen is likely to depend strongly on the preconditioning of the polymer, and materials aged in air could be less sensitive to the oxygen content of the atmosphere in which they are pyrolyzed [18]. Second, the amount cf oxygen that penetrates typical gaseous

diffusion flames is quite low, and therefore under many conditions
pyrolyzing polymers in fires will not be exposed to very much
oxygen. In the absence of further evidence, the most logical course
would seem to be to assume that in fire environments the degradation
mechanisms are the same as those that the polymer chemists have
defined.

4.4 CELLULOSICS

The monomer unit of cellulose is a glucosan, as illustrated [19]

It should be possible to visualize how the monomer is related to
α-glucosan (1,2 anhydroglucose) and how this, in turn, is related to
glucose, viz.,

α-glucosan   glucose

We bypass a thorough discussion of the structure of the molecule,
since it entails an appreciable amount of organic chemistry.

Pyrolysis of cellulose has been and continues to be studied
very extensively. Numerous techniques have been employed, and
a multitude of facts have been established. The current situation is
quite complex. However, a few simplified, unifying principles
appear to have been established. Since these principles provide
useful insights into the pyrolysis mechanism, they will be outlined
here. Much of the discussion is taken from investigations of Broido
and co-workers [20,21]

The first principle is that pyrolysis of cellulose occurs by two distinct paths, as shown

$(200-280^\circ C)$

Cellulose

$(280-340^\circ C)$

"Dehydrocellulose" $\xrightarrow{\text{(exothermic)}}$ CHAR + H_2O + CO_2 + CO + $\cdots$
+ H_2O

(slightly endothermic)

(endothermic)

TAR (primarily Levoglucosan)

The tar is volatile and constitutes the major gaseous fuel to support a gas-phase flame. The gases evolved in the "dehydrocellulose" route are primarily noncombustible, and the char which remains can oxidize only through a surface ('glowing") reaction. That the two paths compete with each other possesses implications concerning fire retardation (discussed later).

Without delving further into the details of the mechanism, already one can use the two-path principle to write a rough phenomenological expression for the degradation rate, viz., [19]

$$dm/dt = -m^{n_1}B_1 e^{-E_1/R^\circ T} - m^{n_2}B_2 e^{-E_2/R^\circ T} \qquad (4.15)$$

where the subscripts 1 and 2 refer to the dehydration and tar-production paths, respectively. The kinetic parameters appearing here must be obtained from pyrolysis data. Rate parameters that have been derived from pyrolysis experiments are summarized in Table 4.3, which was taken from Ref. 22. In spite of questionable interpretations of some of the data, the most logical approach is to identify the first stage with dehydration and the second stage with tar production. This leads to reaction orders $n_1=0$ and $n_2=1$, which must be used in connection with the listed data regardless of whether they are the true orders of the indicated paths; (in general, the data has not been treated properly, and probably is not sufficiently accurate, for extracting true reaction orders). For pure cellulose, reasonable estimates, obtained from the table, appear to be $B_1 = 5 \times 10^9 g/cm^3 sec$, $E_1 = 35\,kcal/mole$, $B_2 = 3 \times 10^{17} sec^{-1}$ and $E_2 = 55$ kcal/mole.

A reasonable mechansim can be suggested for the tar-production path[20]. The yield of levoglucosan is so high that probably some sort of an unzipping process is indicated. It has been proposed that the chain may be initiated either by random scission or by end-initiation, through attacks by a hydroxyl group (one of which is attached to the C atom at the end of each chain). After the monomer breaks off, propagation could be sustained by the free oxygen bond. It is the reason for the monomer appearing as levoglucosan which requires explanation. A proposed model for this process is a two-step attack[20],

first by the free oxygen and next by the hydroxyl. The final molecule shown is levoglucosan (β-glucosan or 1,6 anhydroglucose). The first step is endothermic and the second exothermic.

For the dehydration process, it has been reasoned [20] that an out-of-plane, intermolecular interaction must be the cause. The hydroxyl in an H_2COH group of one chain can attack the carbon-oxygen linkage of an adjacent chain, breaking that chain in such a way that half of it is linked to the attacking chain while the other half gives up H_2O in forming a stable end-group. Hypotheses for the mechanism of the further decomposition toward char through production of H_2O and CO have also been developed [20]. Thus, the dominant features of the uncatalyzed pyrolysis of pure cellulose can be understood self-consistently.

4.5 EFFECT OF FIRE SUPPRESSANTS ON CELLULOSICS

Fire suppressants act in various ways. Some are believed to influence primarily gas-phase processes. Many investigators believe that the main effect of water as a suppressant is to remove the source of gaseous fuel, by cooling the condensed phase to a temperature at which it no longer supplies gaseous combustibles. In contrast to this physical mechanism, a number of substances are believed to influence condensed-phase processes chemically.

These substances include fire-retardant materials for plastics and natural cellulosics.

Many different materials exhibit some effect of an apparently chemical character on the pyrolysis of cellulosic substances. If catalysis occurs in such a way as to reduce the yeild of combustible gases, then the additive qualifies as a fire retardant, at least by some standards. Materials that have been considered for use as fire retardants include [21,23,24] potassium bicarbonate ($KHCO_3$), potassium carbonate (K_2CO_3, hydrated), monoammonium phosphate ($NH_4H_2PO_4$), diammonium phosphate (($NH_4)_2HPO_4$), ammonium sulfate (($NH_4)_2SO_4$), ammonium sulfamate ($NH_4SO_3NH_2$), ammonium chloride (NH_4Cl), sodium tetraborate ($Na_2B_4O_7$, hydrated), monosodium sulfate ($NaHSO_4$), monosodium phosphate (NaH_2PO_4), trisodium phosphate (Na_3PO_4, hydrated), sodium carbonate (Na_2CO_3) and sodium chloride ($NaCl$). Some of these(e.g., $KHCO_3$, $NH_4H_2PO_4$, ($NH_4)_2HPO_4$, $NaHSO_4$ and NaH_2PO_4) obviously are acidic in character and may be classified as acidic fire retardants. Others are less acidic, some basic in character. There appear to be significant differences between catalytic effects of acidic and basic retardants. Before discussing these differences, it is of interest to define a unifying principle of fire retardation of cellulosic materials by chemical means.

Considering the two-path mechanism for cellulose pyrolysis and realizing that the major gaseous combustible is the volatile tar, one immediately concludes that the way to prevent cellulose from supporting a gas-phase flame is to inhibit the tar path in some way. It is very difficult to invent a means for inhibiting one depolymerization process without at the same time promoting another degradation mechanism. For cellulose, the existing two-path mechanism immediately suggests that catalysis of the competing dehydration path could be a simple and convenient way to inhibit the tar path. In fact, most researchers believe that existing fire retardants work through the principle of catalysis of dehydration. The principle of retardation therefore instructs one to find materials that are most effective in promoting the formation of "dehydrocellulose".

It should be observed that this suggests that wood treated with a fire retardant in fact will lose its structural integrity more rapidly, when exposed to the heat of a fire, than will untreated wood. Although in general this aspect of fire retardation is undesirable, no better means has been found to mitigate the tendency of cellulosics to support flaming combustion. Offsetting this aspect somewhat is the fact that through enhancement of char formation, good retardants usually increase the weight of the wood-derived residue that remains after passage of the fire. In fact, there appears to be a direct relationship between the percentage of wood remaining as residue after pyrolysis to a high temperature   and the effectiveness of the additive as a fire retardant[23].

Certain initially surprising observations perhaps can be accepted somewhat more readily on the basis of the principle of the catalytic effect of retardants. For example, good fire retardants also catalyze the combustion of sugar cubes [25], which otherwise are very difficult to burn. This behavior is not inconsistent with the adopted principle.

When cellulosic materials, mixed with fire retardants, are heated gradually, recorded weight-loss histories depend strongly on whether the retardant is acidic or basic [23]. Acidic retardants clearly increase the early rate of weight loss, decreasing the temperature at which observable weight loss begins. On the other hand, basic retardants, notably sodium tetraborate, do not modify at all the early history of the weight loss of the cellulosic; (enhanced early weight loss may be observed, but it is all attributable to decomposition of the additive). The only definite influence of the basic retardant on the weight-loss curve is the enhancement of residue formation after onset of the tar-production endotherm.

The enhanced dehydration rate at low temperatures, produced by acidic retardants, is reflected in the ignition behavior of cellulose exposed to a radiant energy source[24]. Radiant flux in excess of a minimum value is needed to produce sustained flaming combustion of cellulose. A flux slightly below this minimum produces sustained glowing combustion of the cellulose. There is a somewhat lower critical radiant flux, about 0.5 cal/cm^{2}sec, below which pure cellulose exhibits no combustion whatever. Although intuitively one might expect a fire retardant to increase this minimum critical irradiance, experimentally it is found that addition of potassium bicarbonate decreases the flux required to produce sustained glowing. The critical irradiance drops to about 0.4 cal/cm^{2}sec with 1.5% KHCO$_3$ and to 0.25 cal/cm^{2}sec with 15% KHCO$_3$. The greater ease with which glowing combustion can occur in the presence of the retardant is entirely consistent with enhanced low-temperature char production through the dehydration mechanism. Also consistent with this interpretation is the observation that the energy flux needed to produce flaming combustion is increased by addition of the retardant; sustained flaming is prevented completely by addition of 1.5% of KHCO$_3$ [24]. There exist retardants (mainly phosphates such as diammonium phosphate) which for reasons that are not well understood, appear to inhibit both flaming and glowing [19].

Tar yields in the early stages of tar production appear not to be reduced appreciably by acidic retardants. In other words, retardants do not prevent a transient flaming process from occurring.

It is curious to observe that in pyrolysis tests of pure α-cellulose with acidic retardants added, the principal tar component is not levoglucosan. The occurrence of a different tar has been inferred a number of times in the past, and recently its structure has been determined[21]. The form of the molecule can be diagrammed as

```
              H
              |
          H—C————————————O
              |                    \
              C————O               \
              |                       \
              H                        \
        H—C                            C—H
            ‖                          ‖
            C              C
            |              ‖
            H              O
```

It has been called levoglucosenone, and a simple mechanism for producing it from levoglucosan by a pair of sequential acid (H^+)-catalyzed dehydrations has been suggested[21].

It is clear that acidic fire retardants have drawbacks (e.g., enhanced low-temperature weight loss and continued production of some combustible tars). Basic retardants seem to be somewhat less well characterized, but at least appear to possess the advantage of reduced low-temperature weight loss. One might therefore think that the "borate bombers" which drop retardants from the air in efforts to control California forest fires would employ the basic retardant sodium tetraborate. This is not true. At present the "borate bombers" deliver mainly the acidic retardant diammonium phosphate. The reason is that the chemical effectiveness of the retardant per se is not the only criterion for overall effectiveness in fire control. Disadvantages of borates are certain tendencies toward corrosion and toward sterilization of soils. Other aspects that must be considered are solubility in water, influences on penetration of water droplets to the cellulosic fuels, adhesion of the retardant to the woods requiring protection, amount of retardant needed for a specified amount of flameproofing, etc. The result is that the most suitable retardant depends on the intended use; for example, boron compounds often are employed by the construction industry in treatment of wood to produce fire-retarding effects.

4.6 USE OF PYROLYSIS KINETICS IN REGRESSION MODELS

In the preceding section, it has been seen that studies of pyrolysis kinetics are useful as aids to understanding modes of action of fire retardants. Pyrolysis kinetics also underlie models for transient and steady-state combustion of solid fuels. For example, Equation 4.15 could form the basis for studying the rate of growth of the char layer in wood burning. The analyses, although tractable, would be complicated, and few studies with nonuniform temperature fields have been completed. Models are easier to develop for noncharring polymers. Recently, a general approach has been developed for describing the steady linear regression of the surface of a noncharring polymer, including realistic depolymerization schemes[7]. Because of its potential usefulness in analyzing condensed-phase aspects of fires, the approach will be outlined here.

During steady linear regression, there exists a coordinate system in which time derivatives vanish. Let y denote the spatial coordinate normal to the sample surface in this system, with the gas-phase occupying the region $y > 0$. The conservation equations for mass and energy in the region $y < 0$ are

$$r d\rho_i/dy = -d(\rho_i V_i)/dy + w_i \qquad (4.16)$$

and

$$(dT/dy)\sum_i \rho_i c_i (r+V_i) = d(\lambda dT/dy)/dy - \sum_i H_i w_i \qquad (4.17)$$

Here r is the regression rate, ρ_i the mass of any species i per unit volume, V_i the velocity of species i with respect to that of the nondegraded polymer, c_i the specific heat at constant pressure for species i , H_i the enthalpy per unit mass (including thermal and chemical contributions) for species i , λ the thermal conductivity and w_i the mass of species i produced per unit volume per second by chemical reactions. The principle species considered will be the polymer (subscript p) and the monomer (the principle volatile, subscript m) .

A suitable nondimensional form for Equations (4.16) and (4.17) is

$$d\kappa_i/dY = -d(\kappa_i v_i)/dY + \omega_i \qquad (4.18)$$

and

$$\Gamma d\theta/dY = d(\Lambda d\theta/dY)/dY - \sum_i h_i \omega_i \qquad (4.19)$$

where $\kappa_i \equiv \rho_i/\rho_{p\infty}$, $v_i \equiv V_i/r$, $\omega_i \equiv w_i \lambda_{p\infty}/c_{p\infty} \rho_{p\infty}^2 r^2$, $\theta \equiv T/T_s$, $\Gamma \equiv \sum_i \kappa_i (c_i/c_{p\infty})(1+v_i)$, $\Lambda \equiv \lambda/\lambda_{p\infty}$, $h_i \equiv H_i/c_{p\infty} T_s$ and $Y \equiv yr(\rho_p c_p/\lambda_{p\infty})$. The subscript ∞ identifies conditions deep within the polymer. For convenience, the surface temperature T_s will be considered to be specified; from the results it is easy to calculate the incident energy flux at the surface and to convert this condition to a specification of the incident energy flux.

Depolymerization kinetics enter in the terms ω_i which appear in Equations (4.18) and (4.19). These reaction rates will depend on various ρ_i and T. The major temperature dependence is described by one or a number of activation energies E. From the values of E quoted for Eqs. (4.2), (4.4), (4.6), (4.11), (4.12), (4.14) and (4.15), it is seen that even at the highest temperature T_s, the nondimensional parameter $E/R^o T_s$ is large $(\gtrsim 20)$. Under these conditions, in most cases, almost all of the reaction will occur very near the peak temperature T_s. The only cases in which this deduction would not follow are those in which certain reactions have high enough rates at lower temperatures to consume essentially all of their reactants before reaching temperature T_s; this could occur for charring materials and could arise from the first term in Equation (4.15), for example. However, even in these cases it is likely that narrow reaction zones are identifiable and can be treated by methods like that indicated below. Thus the result of the high-E observation is that the equations simplify, exhibiting broad, inert, convective-diffusive zones and narrow reaction zones. In the simple case considered below, the reaction zone is at the surface of the polymer.

We assume that the result of the degradation process is conversion of polymer to monomer, and we consider Eq. (4.18) only for the polymer p. In this equation, it will be assumed that $v_p = 0$. Although thermal expansion of the solid in the heated reaction zone will cause $v_p > 0$, this effect should be negligibly small since one expects small fractional changes in specific gravity, giving $V_p \ll r$ and therefore $v_p \ll 1$. On the other hand, in the hotter parts of the reaction zone, the abundance of either high-pressure or high-velocity gaseous products may physically detach polymer particles from the undergraded substrate and rapidly accelerate them to $v_p \gg 1$. For this reason, Eq. (4.18) with $v_p = 0$ is best viewed as describing polymer regression under conditions such that the condensed phase remains intact until vaporizing. High tensile strength of condensed polymer and the absence of bubbling zones would favor small v_p. Of course, the results should be qualitatively correct if $v_p \lesssim 1$. Thus, we replace Eq. (4.18) by

$$d\kappa_p/dY = \omega_p \qquad (4.20)$$

In the subsurface convective-diffusive zone, the quantities Γ and Λ are expected to be approximately unity. Since $\omega_i = 0$

in that zone, the approximate solutions to Eqs. (4.19) and (4.20) there become

$$\kappa_p = \kappa_{p\infty} \quad , \quad \theta = \theta_\infty + (1-\theta_\infty)e^Y \qquad (4.21)$$

Strictly speaking, the constant factor $(1-\theta_\infty)$ multiplying e^Y in Eq. (4.21) should be determined by matching to the reaction zone, but to lowest order the matching produces this intuitively expected result.

To analyze the reaction zone, it is suitable to work with the stretched "inner" variables $\eta \equiv -\beta Y$ and $F \equiv \beta(1-\theta)$, where $\beta \equiv E/R^o T_s$, for a representative activation energy E , is a large parameter. If $\Lambda = 1$, then to lowest order in the stretched variables, Eqs. (4.19) and (4.20) become

$$d^2 F/d\eta^2 = -\sum_i h_i \omega_i / \beta \qquad (4.22)$$

and

$$d\kappa_p/d\eta = -\omega_p/\beta \qquad (4.23)$$

Matching conditions will be, to lowest order, $dF/d\eta \to 1-\theta_\infty$ and $\kappa_p \to 1$ as $\eta \to \infty$, while $F(0) = 0$ and $\kappa_p(0) = \kappa_{ps}$, typically zero. The depolymerization kinetics enters on the right-hand sides of Eqs. (4.22) and (4.23).

Consider first a one-step reaction of order n , a good approximation for PMMA if $n = 1$ (see Eq. (1)). Then

$$w_p = -w_m = -\rho_p^n B e^{-E/R^o T} \qquad (4.24)$$

and with the nondimensional regression rate defined as

$$\nu = r\left[\beta c_{p\infty}\rho_{p\infty}^2/\lambda_{p\infty}\rho_{p\infty}^n B e^{-\beta}\right]^{\frac{1}{2}} \qquad (4.25)$$

and the nondimensional heat of depolymerization defined as $h = h_m - h_p$, Eqs. (4.22) and (4.23) become, in lowest order,

$$d^2 F/d\eta^2 = -\kappa_p^n e^{-F}(h/\nu^2) \qquad (4.26)$$

and

$$d\kappa_p/d\eta = \kappa_p^n e^{-F}/\nu^2 \qquad (4.27)$$

Combining Eqs. (4.26) and (4.27) and employing boundary conditions gives

$$\kappa_p = 1 - (dF/d\eta - 1 + \theta_\infty)/h \qquad (4.28)$$

and the single equation

$$(g-C)dg/g^n = -D\,e^{-F}\,dF \qquad (4.29)$$

where $g \equiv C - dF/d\eta = h\kappa_p$, $C \equiv h + 1 - \theta_\infty$ and $D \equiv h^{1-n}/\nu^2$. In these equations h is evaluated at temperature T_s . Integration of Eq. (4.29) yields

$$\frac{g_\infty^{2-n} - g_s^{2-n}}{2-n} - \frac{C(g_\infty^{1-n} - g_s^{1-n})}{1-n} = -D \qquad (4.30)$$

which in view of the fact that $g_\infty = h$ and $g_s = h\kappa_{ps}$, requires that

$$\nu^2 = (2-n(1-n)\left[h(1-\kappa_{ps}^{2-n})(n-1) + (h+1-\theta_\infty)(2-n)(1-\kappa_{ps}^{1-n}) \right]^{-1} \qquad (4.31)$$

for $0 \le n < 1$. In the special case $n = 1$, the C term in Equation (4.30) becomes $C \ln(g_\infty/g_s)$, and Eq. (4.31) becomes

$$\nu^2 = \left[(h+1-\theta_\infty)\ln(1/\kappa_{ps}) - h(1-\kappa_{ps}) \right]^{-1} \qquad (4.32)$$

In view of Eq. (4.25), it is seen that Eqs. (4.31) and (4.32) provide the regression rate as a function of surface temperature, when κ_{ps} is specified. The selection $\kappa_{ps} = 0$ is acceptable unless $n = 1$, in which case a small, nonzero value must be used [7].

Since the values for ν obtained in Eqs. (4.31) and (4.32) are independent of the reaction rate, it is seen from Eq. (4.25) that the regression rate is proportional to the square root of the reaction-rate constant for the depolymerization reaction. In particular, the activation energy for r will be half that for the bulk process. This type of result is quite common in relating steady-regression results to homogeneous kinetics; it also occurs, for example, for premixed laminar flames in gases.

To illustrate that similar methods work with kinetics that are more complex, consider the weak-link model for polystyrene degradation. The kinetics are summarized by Eqs. (4.8) and (4.10), which transcribe to

$$r d\kappa_p/dy = -B_p e^{-E_p/R^oT} \rho_p/\sqrt{P} \qquad (4.33)$$

and

$$r d(\kappa_p/P)/dy = B_w e^{-E_w/R^oT} (1/P_o + N_w - \kappa_p/P) \qquad (4.34)$$

where B_p and E_p are the constants appearing in Eq. (4.11), while B_w and E_w are those appearing in Eq. (4.12), respectively. After nondimensionalization, Eqs. (4.22), (4.23) and the corresponding inner form of Eq. (34) become

$$d^2F/d\eta^2 = -e^{-F}\sqrt{\kappa_p}\sqrt{q}\,A\sqrt{h}/\gamma\nu^2 \qquad (4.35)$$

$$d\kappa_p/d\eta = e^{-F}\sqrt{\kappa_p}\sqrt{q}\,A/\gamma\nu^2\sqrt{h} \qquad (4.36)$$

and

$$dq/d\eta = -e^{-F/\gamma}(1-q)/\gamma\nu^2 \qquad (4.37)$$

Here we have used the activation energy E_p for stretching; i.e., $\beta \equiv E_p/R^oT_s$ appears in the definitions of F, η and ν (Eq. (4.25)). The new variable q is defined as $(\kappa_p/P)/(1/P_o+N_w)$. The parameter h is the same as before, while $\gamma \equiv E_p/E_w$, and

$$A \equiv \left[h(1/P_o+N_w)\right]^{\frac{1}{2}}(B_p/B_w)e^{E_w/R^oT_s - E_p/R^oT_s}$$

Eqs. (4.35) and (4.36) again produce Eq. (4.28) relating κ_p to F. Dividing Eq. (4.36) by Eq. (4.37), ignoring the factor $e^{F(1-\gamma)/\gamma}$ on the grounds that γ is near unity, and integrating from η to ∞, one obtains

$$2(\sqrt{g} - \sqrt{g_\infty}) = -A\left[G(q)-G(q_\infty)\right] \qquad (4.38)$$

where

$$G(q) \equiv -2\sqrt{q} + \ell n\left[(1+\sqrt{q})/(1-\sqrt{q})\right]$$

This solution allows Eq. (4.37) to be written as

$$(g-C)dq/dF = e^{-F/\gamma}(1-q)/\gamma\nu^2$$

and to be integrated formally to produce the regression-rate formula

$$\nu^2 = \left\{ (h+1-\theta_\infty)\ln\left[(1-q_\infty)/(1-q_s)\right] - I \right\}^{-1} \qquad (4.39)$$

where

$$I \equiv \int_{q_\infty}^{q_s} \left[g(q)/(1-q) \right] dq$$

To use Eq. (4.39), the integral defined above is evaluated numerically (an upper bound for it can be shown to be $(2/3)h^{3/2}/(A\sqrt{q_\infty})$)[7]. Also, the values q_∞ and q_s must be known. We have $q_\infty = (1+P_0N_w)^{-1}$, and q_s can be related to a critical degree of polymerization P_c for vaporization at the surface, viz., $q_s = \kappa_{ps}(P_0/P_c)/(1+P_0N_w)$, in which κ_{ps} is eliminated by evaluating Eq. (4.38) at the surface and using $g_s = h\kappa_{ps}$. In fact, one expects q_s to be near unity, and the expansion

$$q_s = 1-\exp\left\{ -G(q_\infty)-2\sqrt{h}/A+\left[h(P_c/P_0)(1+P_0N_w)\right]^{\frac{1}{2}}/A+\ln 4-2 \right\}$$

can be employed, in which $2\sqrt{h}/A$ dominates and makes q_s near one. A similar development can be completed for the random-scission kinetic model, corresponding to Eqs. (4.10) and (4.13) [7].

One might feel that a somewhat esoteric analysis such as that outlined above is unlikely to produce results that can be compared directly with experiment. There are even basic uncertainties as to whether homogeneous degradation results can be used in principle (even when built into a theory such as that just described) to describe the degradation producing linear regression in a fire (i.e., it is conceivable that the kinetic scheme changes). To test the quoted results experimentally, simultaneous measurements of the regression rate r and the surface temperature T_s are needed. It is difficult to find such measurements in experiments involving fires. But a confrontation with experiment can be achieved on intermediate ground, viz. in linear pyrolysis experiments under controlled laboratory conditions.

Surface temperatures and regression rates have been measured for polystyrene by pressing a sample of the material against a heated plate [26] and by directing the hot exhaust of a laboratory rocket motor onto the surface of the material [27]. In both cases, regression rates comparable to those occurring in fires were obtained. Fig. 4.1 shows a comparison of the theoretical and experimental results. The data of Ref. 26 do not agree with the theory; they exhibit much higher regression rates than predicted. The reason is that in the hot-plate test the polystyrene softens and spreads over the plate, so that the measured regression rate increases as the viscosity of the liquid layer decreases (consistent with the experimentally observed dependence on degree of polymerization). The hot-plate technique is expected to produce results applicable under fire conditions only if the polystyrene has a higher degree of polymerization ($P_0 \gtrsim 10^4$). On the other hand, experimental results for the hot-jet technique agree well with theory. Moreover, agreement is better for the weak-links model, (Eq. (4.39)) than for the random-scission model (for which practically no dependence on molecular weight was obtained). The comparison therefore tends to support the weak-links model as well as the general applicability of the theoretical approach.

A similar comparison has been made for PMMA [8]. In this case a simple first-order reaction is employed, using Eq. (4.32) with $\kappa_{ps} = 0.01$ in Eq. (4.25) to calculate r from the kinetic parameters given in Eq. (4.2). The results are shown in Fig. (4.2), where the three straight theoretical lines correspond to calculations made from Eqs. (4.25) and (4.32), employing the three different sets of rate parameters for homogeneous degradation cited after Equation (2). The horizontal bars represent upper and lower bounds on the surface temperature determined experimentally in Ref. 8. The situation is rather complicated because each set of experimental data suffers from its own idiosyncrasies that introduce systematic bias. However, the general conclusion which can be drawn is that, within the accuracy of the experimental data on regression rate as a function of temperature, the calculated and experimental results are mutually consistent [8]. Therefore, for PMMA too, it seems likely that the depolymerization mechanism is the same in linear regression as it is in homogeneous pyrolysis. Since this type of result appears to be rather widely applicable, there seems to be justification for persuing analyses of the general type discussed here, with the objective of developing models for describing more complex systems, such as charring cellulosics.

4.7 RELATIONSHIPS TO OBSERVED BURNING HISTORIES

Processes occurring in real fires are more complex, both spatially and temporally, than the steady linear regression just discussed. To see how to use ideas related to pyrolysis kinetics in developing an understanding of the processes that occur in real-world situations, let us consider a particular example in some detail.

Experiments have been performed on the burning of vertically oriented cylinders of α-cellulose in air[31] The cylinders were constructed specially by pressing and drying chopped filter papers, in the manner of Ref. 32. Diameters ranged from 1/8 to 1 inch and lengths from 1 to 6 inches. The cylinders simulate burning sticks reasonably well in many respects, but they provide more reproducible experimental results by virtue of the fact that they are less non-homogeneous and therefore support a smoother, more uniform flame. The cylinders may be considered to be full-scale models of free-burning sticks of cellulosic fuels.

The cylinders are mounted vertically on a pedestal placed in a balance and ignited with a bunsen burner. The weight is recorded as a function of time, thermocouple measurements of surface temperatures are made, and the combustion process is observed photographically. A representative weight-loss history and surface-temperature histories are shown in Fig. 4.3. Two thermocouples were employed, one located 0.6 inch above the lower edge of a cylinder 3.12 inches long and the other located 2.62 inches above its lower edge.

It can be seen that there is an initial heatup period, lasting about 2 minutes, during which the surface temperature is independent of vertical position but increases slowly with time. For some period of time thereafter (3 minutes duration in Fig. 4.3), the temperature of each thermocouple remains constant at a value which is a slowly increasing function of height above the lower edge. This is the steady-flamming period during which an envelope flame surrounds the entire cylinder. A generalization of the analysis of the preceding section might be applicable during this period. As the foot of the envelope flame begins to rise above the base of the cylinder, the temperature of the lower thermocouple begins to increase again while the temperature on the upper part of the surface of the cylinder changes very little (at 5 to 7 minutes in Fig. 4.3). Later, the upper thermocouple also begins to spend some time unenveloped in flame, and its temperature then also rises approaching that of the lower thermocouple. When the flame is finally completely extinguished (at 9 minutes) and only glowing combustion occurs, both thermocouples again show essentially the same temperature, 600°C. This temperature level is maintained for a few minutes as glowing combustion proceeds; as extinguishment of glowing begins to occur, the surface temperature decreases.

The range of surface temperature measured prior to flame-out, roughly 150°C to 450°C , corresponds closely to the temperature range employed in ordinary studies of bulk pyrolysis for cellulose. However, the time rate of change of temperature differs in the present tests. The nearly linear temperature rise from 150°C to about 350°C in 2 minutes represents a much higher heating rate than is usual for laboratory testing of polymers, and the temperature remains in the range 350°C to 375°C slightly longer than in a typical differential thermal analysis. Nevertheless, conclusions that have been drawn from bulk degradation experiments can be used to discuss implications of the present results.

First, in view of the temperature ranges, indicated in Section 4, as favoring dehydration and unzipping, it is seen that the temperature-time history observed in the present experiment appears strongly to favor tar formation; the temperature range of the dehydration reaction is traversed very quickly, and the surface temperature lingers in the optimum range for tar formation. Of course, temperature measurements in the interior are not shown here, and conditions may be less disadvantageous for the dehydration reaction in the interior. But in an overall sense, the surface temperature history is entirely consistent with the fact that with these cellulose cylinders, one observes an extended period of steady flaming during which most of the weight loss of the sample occurs.

The increase in the surface temperature to 600°C after flameout is consistent with the occurrence of glowing combustion; it is at this temperature that glowing combustion of cellulose char is usually observed to occur in air. When the flame is flickering, the reading of the bottom thermocouple is much higher than that of the top one, suggesting that as the flame lifts up, glowing can occur in the vicinity of the bottom thermocouple while flame is occurring around the top one.

It is clear that the heat-up and flickering-flame processes are transient. A quasisteady stage of glowing combustion may occur prior to transient burnout. An interesting question is whether the processes in the interior of the cellulose are quasisteady during the steady-flaming period.

Note that during steady flaming the top thermocouple records a temperature about 20°C below that of the bottom thermocouple. The dependence of the weight-loss rate on cylinder height strongly suggests that the rate of production of volatiles (per unit surface area) decreases with height, and a gas-phase combustion analysis for calculating the change with height is available [33]. This analysis suggests that the rate of weight loss per unit area, $\dot{m}$, varies as $z^{-1/4}$, , where z is the height above the lower edge of the cylinder. Use of this result in the simplest of possible surface-pyrolysis expressions, viz., $\dot{m} \sim \exp(-E_{eff}/R^{o}T_{s})$, produces an estimate of the corresponding effective activation energy E_{eff} from the measured temperatures T_{s1} and T_{s2} at the two different heights z_1 and z_2. The formula for E_{eff} is readily seen to be

$$E_{eff} = (R^{o}/4)\left[\ell n(z_2/z_1)\right]/(T_{s2}^{-1} - T_{s1}^{-1})$$

$$= \left[\ell n(z_2/z_1)\right] R^{o} T_{s1} T_{s2}/4(T_{s1}-T_{s2}) . \quad (4.40)$$

Using, in Eq. (4.40), surface temperature data such as that illustrated in Fig. 4.3, we find E_{eff} = 18 ii 9kcal/mole. The uncertainty of ± 9 kcal/mole reflects the experimental uncertainty in the difference of surface temperatures; the technique was designed only to provide indications of magnitudes rather than definitvely accurate results.

To interpret this result on the basis of a model of quasi-steady regression, we may return to Eq. (4.25), where $\dot{m} = r\rho_{p\infty}$, to find that E_{eff} = 18 ± 9 kcal/mole implies that the corresponding activation energy for bulk degradation, viz., $E = \beta R T_s$, is 36 ± 18 kcal/mole. To see how this value compares with typical activation energies for cellulose pyrolysis, we return to Eq. (4.15) of Section 4.4 and to Table 3. The mean value 36 kcal/mole is close to the value 35 kcal/mole cited as a representative activation energy for the dehydration process. However, the fact that tars are being produced, and the fact that the value of the surface temperature is more nearly optimum for the unzipping process, suggest that E should be identified with E_2 of Equation (4.15). The mean value cited for E_2, viz. 55 kcal/mole, coincides with the upper limit of uncertainty of the E calculated here. Thus, the current status is uncertain; quasisteady regression may or may not describe properly the behavior of the condensed phase during steady flaming in this experiment.

There are alternative explanations for the observed value of E_{eff}. Assuming that the average molecular weight of gaseous pyrolysis products is 30 g/mole, and using a heat of gasification of 710 cal/g [31], we find a molar heat of gasification of 21 kcal/mole, which is close to E_{eff}. Thus, a possible (although unlikely) interpretation of the observation is that an equilibrium dissociative vaporization process occurs at the surface of the cellulose. Another possibility is that quasisteady regression is not achieved, that highly distributed in-depth pyrolysis reactions occur, and that therefore E_{eff} should be interpreted as a homogeneous activation energy (i.e., not multiplied by two). Although this value may seem to represent a low activation energy for homogeneous degradation, it is consistent with values (14 to 22 kcal/mole) obtained from measurements of interior mass-loss histories of pyrolyzing samples [34].

From this discussion one sees that although there is an important interplay between pryolysis kinetics and burning histories, interpretations are not entirely clear-cut. In particular, it is unclear how relevant a quasisteady regression model may be for cellulose combustion. In this respect, cellulosics and other char-forming polymers are more difficult to model than noncharring materials with relatively straightforward depolymerization mechanisms, such as PMMA. Further research is needed in modeling the condensed-phase behavior of cellulosics in fires.

REFERENCES

1.   Jellinek, H.H.G, Degradation of Vinyl Polymers, Academic Press, New York, 1955.

2.   Madorsky, S.L., Thermal Degradation of Organic Polymers, Interscience, New York, 1964.

3.   Boyd, R.H., "Theoretical Depolymerization Kinetics in Polymers Having an Initial 'Most Probable' Molecular Weight Distribution," Journal of Chemical Physics, Vol. 31, No. 2, Aug. 1959, pp. 321-328.

4.   Simha, R., Wall, L.A., and Blatz, P.J., "Depolymerization as a Chain Reaction," Journal of Polymer Science, Vol. 5, No. 5, 1959, pp. 615-632.

5.   Simha, R. and Wall, L.A., "Kinetics of Chain Depolymerization," Journal of Physical Chemistry, Vol. 56, June 1952, pp. 707-715.

6.   Gordon, M., "Probability Model Theory of Chain-End Activated Polymer Degradation. III. Statistical Kinetics of the Degradation of Polymethyl Methacrylate," Journal of Physical Chemistry, Vol. 64, Jan. 1960, pp. 19-29.

7.   Lengelle, G., "Thermal Degradation Kinetics and Surface Pyrolysis of Vinyl Polymers," AIAA Journal, Vol. 8, 1970, pp. 1989-1996.

8.   Krishnamurthy, L. and Williams, F.A., "On the Temperature of Regressing PMMA Surfaces," Combustion and Flame, Vol. 20, 1973, pp. 163-169.

9.   Madorsky, S. L., "Rates of Thermal Degradation of Poly-
      styrene and Polyethylene in a Vacuum, " Journal of Polymer
      Science, Vol. 9, No. 2, 1952, pp. 133-156.

10.   Grassie, N. and Kerr, W. W., "The Thermal Depolymeri-
      zation of Polystyrene, Part I, The Reaction Mechanism, "
      Transactions of the Faraday Society, Vol. 53, 1957, pp.
      234-239.

11.   Simha, R. and Wall, L. A., "Some Aspects of Depolymeri-
      zation Kinetics, " Journal of Polymer Science, Vol. 6, No. 1,
      1951, pp. 39-44.

12.   Frankel, J., Kinetic Theory of Liquids, Dover, New York,
      1955, p. 451.

13.   Rabinovitch, B., "Regression Rates and the Kinetics of
      Polymer Degradation, " 10th Symposium (International) on
      Combustion, The Combustion Institute, Pittsburgh, Pa.,
      1965, pp. 1395-1404.

14.   Grassie, N. and Kerr, W. W., "The Thermal Depolymeri-
      zation of Polystyrene, Part II, Formation of Weak Links, "
      Transactions of the Faraday Society, Vol. 55, Pt. I, 1959,
      pp. 1950-1955.

15.   Cameron, G. G. and Grassie, N., "Ther Thermal Depoly-
      merization of Polystyrene, Part IV, Depolymerization in
      Naphtalene and Tetralin Solutions, " Journal of Polymer
      Science, Vol. 2, 1961, pp. 367-373.

16.   Gordon, M., "Theory of Chain End Activated Degradation
      of Heterodisperse Polymers, " Transactions of the Faraday
      Society, Vol. 53, 1957, pp. 1662-1675.

17.   Gordon, M., "Recent Advances in the Theory of Thermal
      Degradation of Heterodisperse Polymers, " Monograph No.
      13, 1961, Society of Chemistry Ind. (London), pp. 163-183.

18.   Wall, L. A., "Condensed-Phase Combustion Chemistry,"
      Fire Research Abstracts and Reviews, Vol. 13, 1971,
      pp. 204-219.

19.   Browne, F. L., "Theories of Combustion of Wood and Its
      Control - A Survey of Literature", U.S. Forest Products
      Laboratory, Madison, Wis., Report No. 2136, 1958.

20.   Kilzer, F.J. and Broido, A., "Speculations on the Nature of
       Cellulose Pyrolysis", Pyrodynamics, Vol. 2, 1965, pp. 151-
       163.

21.   Halpern, Y., Riffer, R., and Broido, A., "Levoglucosenone
       (1,6-Anhydro-3,4-dideoxy-Δ^3-β-D-Pyranosen-2-one). A
       Major Product of the Acid-Catalyzed Pyrolysis of Cellulose
       and Related Carbohydrates," Journal of Organic Chemistry,
       Vol. 38, 1973, pp. 204-209.

22.   Kanury, A. Murty and Blackshear, P.L., "Some Considera-
       tions Pertaining to the Problem of Wood Burning", Combus-
       tion Science and Technology, Vol. 1, 1970, pp. 339-355.

23.   Browne, F.L. and Tang, W.K., "Thermogravimetric and
       Differential Thermal Analysis of Wood and of Wood Treated
       with Inorganic Salts During Pyrolysis," Fire Research
       Abstracts and Reviews, Vol. 4, 1962, pp. 76-91.

24.   Broido, A. and Martin, S.B., "Effects of Potassium
       Bicarbonate on the Ignition of Cellulose by Radiation,"
       Fire Research Abstracts and Reviews, Vol. 3, 1961, pp. 193-
       201.

25.   Broido, A., "Flammable-Whatever That Means," Chemical
       Technology, Vol. 3, 1973, pp. 14-17.

26.   Andersen, W.H. et al., "A Model Describing Combustion
       of Solid Composite Propellants Containing Ammonium
       Nitrate," Combustion and Flame, Vol. 3, 1959, pp. 301-317.

27.   Hansel, J.G. and McAlevy, R.F., "Energetics and Chemical
       Kinetics of Polystyrene Surface Degradation in Inert and
       Chemically Reactive Environments," AIAA Journal, Vol. 4,
       No. 5, May 1966, pp. 841-848.

28.   Chaiken, R.F. et al., "Kinetics of the Surface Degradation
       of Polymethylmethacrylate,   " Journal of Chemical Physics,
       Vol. 32, No. 1, Jan. 1960, pp. 141-146.

29.   McAlevy, R.F., Lee, S.Y., and Smith, W.H., "Linear
       Pyrolysis of Polymethylmethacrylate during Combustion,"
       AIAA Journal, Vol. 6, No. 6, June 1968, pp. 1137-1142.

30.   Fenimore, C.P. and Jones, G.W., Combustion and Flame,
       Vol. 10, 1966, p. 295.

31.   Kosdon, F.J., "Combustion of Cellulose", Ph.D. Thesis, University of California, San Diego, La Jolla, California, 1970.

32.   Blackshear, Jr., P.L. and Kanury, A. Murty, "Heat and Mass Transfer to, from, and within Cellulosic Solids Burning in Air," 10th Symposium (International) on Combustion, The Combustion Institute, Pittsburgh, Pa., 1964, pp. 911-923.

33.   Kosdon, F.J., Williams, F.A., and Buman, C., "Combustion of Vertical Cellulosic Cylinders in Air," 12th Symposium (International) on Combustion, The Combustion Institute, Pittsburgh, Pa., 1969, pp. 253-264.

34.   Kanury, A. Murty and Blackshear, Jr., P.L., "An X-Ray Photographic Study of the Reaction Kinetics of α-cellulose Decomposition," Western State Section, Combustion Institute, Paper No. 66-19, presented at Spring 1966 Meeting, Denver Research Institute, Colorado.

TABLE 4.1

Yield of monomer in the pyrolysis of
some organic polymers in a vacuum

Polymer	Temperature range, °C	Yield of monomer, % of volatiles
Polymethylene	335-450	0.03
Polyethylene	393-444	0.03
Polypropylene	328-410	0.17
Polymethylacrylate	292-399	0.7
Hydrogenated polystyrene	335-391	1
Poly(propylene oxide), atactic	270-550	2.8
Poly(propylene oxide), isotactic	295-355	3.55
Poly(ethylene oxide)	324-363	3.9
Polyisobutylene	288-425	18.1
Polychlorotrifluoroethylene	347-415	25.8
Poly-β-deuterostyrene	345-384	39.7
Polystyrene	366-375	40.6
Poly-m-methylstyrene	309-399	44.4
Poly-α-deuterostyrene	334-387	68.4
Poly-α, β, β-trifluorostyrene	333-382	72.0
Polymethylmethacrylate	246-354	91.4
Polytetrafluoroethylene	504-517	96.6
Poly-α-methylstyrene	259-349	100
Polyoxymethylene	Below 200	100

TABLE 4.2

Activation energies of thermal degradation of some organic polymers in a vacuum

Polymer	Molecular Weight	Temperature range, °C	Activation energy kcal/mole
Phenolic resin	---	331.5-355	18
Atactic poly (propylene oxide)	16,000	265-285	20
Polymethylmethacrylate	150,000	226-256	30
Polymethylacrylate	---	271-286	34
Poly(ethylene terephthalate)	---	336-356	38
Isotactic poly(propylene oxide)	215,000	285-300	45
Cellulose triacetate	---	283-306	45
Poly(ethylene oxide)	9,000 — 10,000	320-335	46
Polyisobutylene	1,500,000	306-326	49
Hydrogenated polystyrene	82,000	321-336	49
Cellulose	---	261-291	50
Polybenzyl	4,300	386-416	50
Polymethylmethacrylate	5,100,000	296-311	52
Polystyrene	230,000	318-348	55
Poly-α-methylstyrene	350,000	228.8-275.5	55
Poly-m-methylstyrene	450,000	318.5-338.5	56
Polyisoprene	---	291-306	58
Polychlorotrifluoroethylene	100,000	331.8-371	57
Polypropylene	---	336-366	58
Polyethylene	20,000	360-392	63
Poly-α-β-β-trifluorostyrene	300,000	333-382	64
Polymethylene	High	345-396	72
Poly-p-xylyene	---	401-411	73

TABLE 4.3

Kinetic Parameters for Pyrolysis of Cellulosics

Description	Contamination	Temp. Range °C	A g/cm³/ min First Stage	E kcal/ mole	Temp. Range °C	A min.⁻¹ Second Stage	E kcal/ mole	Author
Cellulose	None	270-340	---	17.4	340-370	---	36.5	Akita (1959)
Wood	None	280-325	1.93×10^7	23	325-350	3.92×10^{18}	54	Tang (1967)
α-Cellulose	None	240-308	3.85×10^{11}	35	308-360	2.37×10^{19}	56	Tang (1967)
Lignin	None	280-344	9.86×10^5	21	344-435	5.60×10^1	9	Tang (1967)
Beechwood	None	---			275	2.60×10^9	25	Roberts (1963)
Beechwood	None	---			435	9.10×10^4	15	Roberts (1963)
α-Cellulose	None	250-300	---	42		---		Lipska (1966)
α-Cellulose	None	276-360	---	42		---		Lipska (1968)
Douglas Fir Saw Dust	None		---		110-220	1.14×10^{11}	25	Stamm (1956)
α-Cellulose	None		---		110-220	2.88×10^{11}	26	Stamm (1956)
Hemi-cellulose	None		---		110-220	2.16×10^{12}	26.7	Stamm (1956)
Lignin	None		---		110-220	8.4×10^{11}	23	Stamm (1956)
Coniferous Wood	None		---		94-250	3.06×10^{13}	29.5	Stamm (1956)
Spruce	None		---		167-300	1.38×10^{13}	29.8	Stamm (1956)
Manila Paper	None		---		145-265	1.68×10^{14}	31.7	Kujirai (1925)

TABLE 4.3 (*Cont'd*)

Material	Treatment							Reference
Linen	None		---		165–265	2.7×10^{13}	31.0	Kujirai (1925)
Filter paper	None		---		165–265	4.5×10^{13}	31.0	Kujirai (1925)
Cotton	None		---		165–265	7.8×10^{13}	32.1	Kujirai (1925)
Cotton	None		---		275–305	---	50	Madorsky (1956)
Viscose Rayon	None		---		265–295	---	49	Madorsky (1956)
Cotton	6N HCl		---		285–305	---	47	Madorsky (1956)
Cellulose Triacetate	None		---		297–320	---	45	Madorsky (1956)
Wood	Sodium Tetraborate Decahydrate 2%	260–289	4.5×10^{7}	23	289–320	1.86×10^{19}	53	Tang (1967)
α-Cellulose	Sodium Tetraborate Decahydrate 2%	230–295	4.54×10^{10}	32	295–345	1.59×10^{19}	54	Tang (1967)
Lignin	Sodium Tetraborate Decahydrate 2%	280–356	1.58×10^{4}	16	356–440	5.79×10^{1}	9	Tang (1967)
Wood	Sodium Chloride 2%	250–276	6.35×10^{7}	23	276–315	8.19×10^{16}	46	Tang (1967)
α-Cellulose	Sodium Chloride 2%	260–286	1.74×10^{12}	35	286–325	2.76×10^{18}	51	Tang (1967)
Lignin	Sodium Chloride 2%	280–322	1.59×10^{6}	21	322–425	2.69×10^{1}	8	Tang (1967)
Wood	Potassium bicarbonate	245–271	1.32×10^{7}	21	271–305	3.62×10^{15}	42	Tang (1967)
α-Cellulose	Potassium bicarbonate	200–279	5.65×10^{6}	21	279–330	1.24×10^{15}	42	Tang (1967)
Lignin	Potassium bicarbonate	280–312	1.10×10^{4}	15	312–430	2.61×10^{1}	8	Tang (1967)
Wood	Aluminum Chloride Hexahydrate 2%	250–272	2.32×10^{7}	22	272–310	2.25×10^{15}	42	Tang (1967)
α-Cellulose	Aluminum Chloride Hexahydrate 2%	210–267	3.44×10^{11}	33	267–320	2.11×10^{20}	55	Tang (1967)
Lignin	Aluminum Chloride Hexahydrate 2%	280–349	1.87×10^{5}	19	349–430	2.44×10^{1}	8	Tang (1967)
Wood	Monobasic Ammonium Phosphate 2%	235–259	2.75×10^{6}	19	259–290	6.77×10^{20}	54	Tang (1967)

α-Cellulose	Monobasic Ammonium Phosphate 2%	190-260	2.24×10^8	19	260-300	1.18×10^{21}	55	Tang (1967)
Lignin	Monobasic Ammonium Phosphate 2%	290-336	3.50×10^4	17	336-435	4.63×10^1	9	Tang (1967)
Cotton	8% Sodium Chloride	---	---		255-285	---	35	Madorsky (1956)
Cotton	9% Sodium Carbonate	---	---		255-285	---	19	Madorsky (1956)
Viscose Rayon	9% Sodium Carbonate	---	---		245-275	---	30	Madorsky (1956)

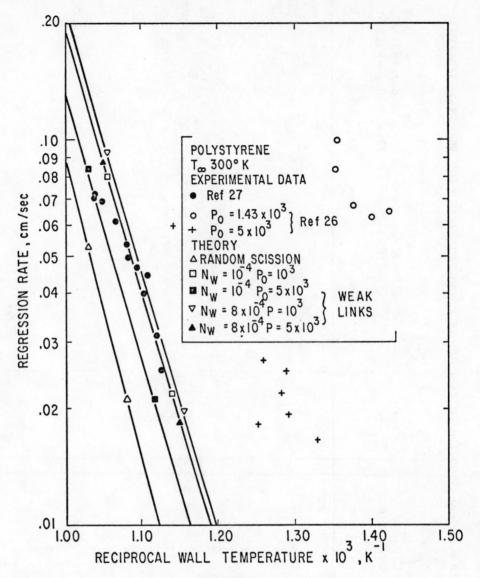

FIGURE 1

NOTE: Figure captions for this chapter are listed on page 237.

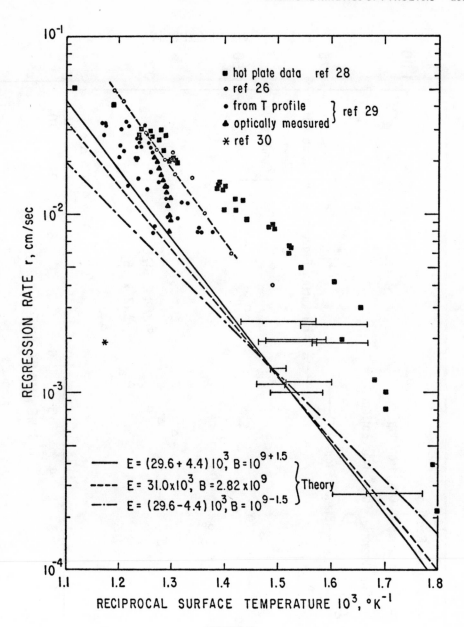

FIGURE 2

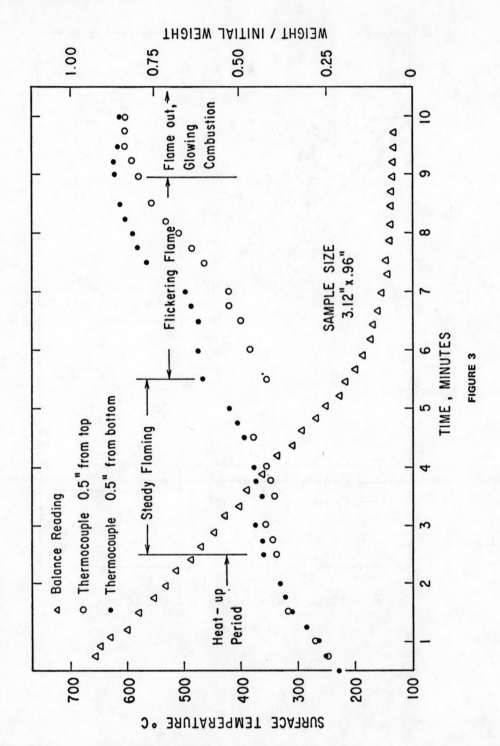

FIGURE 3

FIGURE CAPTIONS

Figure 1   Dependence of Regression Rate on Surface Temperature for Polystyrene.

Figure 2   Dependence of Regression Rate on Surface Temperature for Polymethylmethacrylate.

Figure 3 ·   Histories of Weight and of Surface Temperature for
Burning of a Vertically Oriented Cellulose Cylinder.

Chapter 5

VELOCITY DISTRIBUTIONS IN FIRES

R. C. CORLETT
University of Washington, USA

THE PRECEDING FOUR chapters have dealt primarily with condensed phase processes and their coupling with the gas-phase fire phenomena. Chapter 2 was devoted to the similarity between species concentration and sensible enthalpy (or temperature) fields in the gas phase. But this similarity can be demonstrated virtually without mention of the fluid dynamics of the fire. In Chapters 5 and 6 attention is directed to fire flow fields.

All fires, except perhaps the very smallest, are obviously mechani· cal phenomena inasmuch as motion of the flame gases is a manifest characteristic. There are a variety of motivations for some under standing of fire flow fields. To begin with, the patterns of heat and mass transport, and the heat release distribution in fires of any sig- nificant size, cannot be computed, or even realistically envisaged, without consideration of the flow field. Also, many fire characteris- tics of practical interest are direct properties of the flow field. One example is fire violence, which is a major contributor to the ha- zard and destructiveness of large fires. Another is the ability of a disposal fire to disperse combustion products in the atmosphere; this depends on the strength of the convection column induced by the fire.

Section 5.1 is a discussion of buoyant forces and their role in fire flow fields. The remaining two sections contain order-of-magni- tude treatments of low elevation flow phenomena for two types of area fires. One type is essentially unstructured and the other is struc- tured.

The distinction between structured and unstructured flow patterns is illustrated in Fig. 5.1 which contains sketches of pool fire flow fields for several values of pool diameter d . At d $\approx$ 1 cm or less,

NOTE: All figures quoted in text are at the end of the chapter.

product-rich combustion gases diffuse across the entire pool surface.
The visible flame is laminar as is the products-air column upwards for
many diameters. For such small diameter pools, the bulk motion of the
gases appears to play only a secondary role in determining the heat
release distribution and the pool burning rate. As d is increased
to ≃ 3 cm , a region of essentially pure fuel vapor appears over the
center of the pool. The visible flame is still laminar, as is the products-air column above for several diameters. The laminar diffusion
length D/v (D mean diffusion coefficient ~ 1 cm^2/sec[1] for flames in
standard air, v upward flow velocity at pool surface ~ 1 cm/sec , for
$d \simeq 3$ cm) is now less than the pool radius. As d increases to
$\simeq 10$ cm , the central region of cold fuel vapor grows somewhat,
especially for relatively rapidly burning fuels[2]; it is bounded by a
conical laminar flame boundary layer (see also Fig. 5.6). Transition
to turbulence occurs just above the flame boundary layer. But the entire flow field shows signs of instability. When viewed continually
through a shadowgraph apparatus, the general configuration of the flame
boundary layer is seen to fluctuate. Occasionally the central fuel
vapor region dissappears entirely. Waviness and general instability
are apparent on both the inner and outer surfaces of the flame boundary
layer. At still larger diameters, the laminar flow pattern breaks
down everywhere. The visible flame is thoroughly turbulent at all observable elevations for $d \simeq 100$ cm or larger. With further increase
of d , the height of the visible flames, expressed in numbers of diameters, tends to decrease. However, the general shape of the combined
visible flame and products-air column remains approximately constant.
The shape of the low elevation flow field, below the point where the
column begins to diverge with increasing elevation, is also roughly
the same for well-developed turbulent fires over a wide range of sizes
and fuelbed types and for laminar pool fires with $d \sim 2$ to 3 cm. This
leads to the surmise that the general magnitude of the turbulent diffusion coefficient in large fires and the average value of the laminar
diffusion coefficient D are the same, when expressed in the form of
a suitable dimensionless number. This equivalence is discussed in
Chapter 6.

In both the very small d (d ≃ cm or less) and very large d
(d ≃ 100 cm or more) regimes, there is diffusion throughout the low
elevation region of the fire. Under such circumstances it makes sense
to talk about such average quantities as temperature $\overline{T}$ or total products fraction Y_p . This is called an unstructured fire. Some of its
representative magnitudes are estimated in Section 5.2. On the other
hand, the flame boundary layer and the cold fuel central region, and

---

[1] As used in this chapter, the symbol "$\sim$" denotes order-of-magnitude
equality. Here the phrase "order-of-magnitude equality" should be
interpreted as substantially better equality than merely the nearest
power of 10. The nearest power of 2 is more typical of the intended
precision.

[2] In pool-like gas burner experiments [1] at burning rates somewhat higher than typical pool burning rates, the cold fuel region is a very pronounced feature of the flow field.

their configuration, are decisive features of area fires in the intermediate (d $\cong$ 10 cm) regime. Some of the main properties of the flame
boundary layer, together with a criterion for the existence of a cold
fuel region, are estimated in Section 5.3.

5.1 BUOYANCY IN FIRES

As discussed in Chapter 2, average temperatures of visible flames
are typically on the order of 1000°C. The average mass density of the
flame gases is much less than that of the surrounding atmosphere. This
results in buoyant forces which accelerate the flame gases upward, creating a pressure defect which tends to draw fresh air into the fire.

Precise quantitative analysis of such free convection driven flow
fields is generally not feasible. Except for a few special fire configurations amenable to boundary layer treatment, numerical flow field
analysis is required. Even the numerical approach is beset with serious
difficulties due to turbulence and the general geometric complexity of
real fires. Fortunately, available experimental experience and results
of numerical analyses confirm a widely accepted physical argument that
buoyant forces calculated on the basis of the hydrostatic equilibrium
principle can be used to estimate vertical accelerations in free convection flow fields of this type.

This principle states that an object with density ρ <u>statically</u>
immersed in a fluid of density ρ_∞ experiences a net upward force
sufficient for acceleration of magnitude g b where g is the local
gravitational acceleration and b , here called the buoyancy, is defined

$$b = (\rho_\infty - \rho)/\rho \qquad\qquad (5.1)$$

This force is called the equilibrium buoyant force. Strictly speaking,
the principle fails as soon as motion of the immersed object ensues.
The buoyant force is the vector integral of hydrostatic equilibrium
pressure forces over the surface of the object. With motion of the
surrounding fluid the pressure field is perturbed from this equilibrium
distribution. The pressure field perturbation is here called the dynamic pressure. Also, the object experiences viscous (or turbulent)
shear forces. Its actual acceleration therefore depends on a balance
between the hydrostatic equilibrium buoyant force, the dynamic pressure
force, and shear forces. The estimating principle to be exploited here
identifies parcels of hotter (or cooler) than ambient gas as the "immersed object". It is based on physical reasoning that for buoyancy
driven gas flows this resultant force is of the same general magnitude
as the hydrostatic equilibrium force. In flow fields where the boundary layer assumptions hold, it can be shown that the dynamic pressure
becomes negligibly small. For very small length scales, the viscous
shear force tends almost entirely to balance the equilibrium buoyant
force. What is meant by very small length scale is discussed in Chapter 8. Suffice it to say here that for flow dimensions in excess of a
few cm, and in general for turbulent buoyancy driven flows, the accelerating force and the viscous force are of the same general magnitude,
and the dynamic pressure force is generally no more than this magnitude.

Thus the vertical acceleration of the heated gases in normal fire flow fields can be estimated to order-of-magnitude accuracy $g\,\bar{b}$ where $\bar{b}$ is an average buoyancy. Letting ℓ denote a characteristic vertical acceleration length, the general magnitude u of the velocities thus attained is

$$\bar{u} \sim \sqrt{g\,\bar{b}\,\ell} \qquad (5.2)$$

A typical estimate of $\bar{b}$ for visible flame gases is 3. With $g \approx 10^3$ cm/^{2}sec and ℓ taken, for example, as 300 cm, there results $u \sim 10^3$ cm/sec .

There is a connection between $\bar{b}$ and the average total products fraction $\bar{Y}_p$. Recall that Y_p , as defined in Chapter 2, includes the fraction of nitrogen corresponding to stoichiometric oxygen. For combustion in ordinary air, nitrogen in fact dominates combustion products thus defined. In free burning fires, pressure perturbations are almost never on the same order-of-magnitude as the absolute pressure. Also, except in highly fuel rich regions, the mean molecular weight is almost identical with that of air; for H_2, CH_2 and C burning in standard air the mean molecular weight of the stoichiometric products is, respectively, 24.5, 28.7 and 31. Thus, in visible flames, and on the air side generally, the density - absolute temperature product $\rho\,T$ is essentially constant. If specific heat variations are ignored, Eq. (2.11) can be used to obtain

$$b = (\rho_\infty - \rho)/\rho \approx (T - T_\infty)/T_\infty$$
$$\approx (h_s - h_{s,\infty})/(\bar{c}_p\,T_\infty) = \theta\,b^* \qquad (5.3)$$

where

$$b^* - \Delta h^*/(\bar{c}_p\,T_\infty) . \qquad (5.4)$$

If a suitable mean value of the specific heat $\bar{c}_p$ is chosen and if the effective heat of reaction Δh is corrected for radiation loss, b^* for hydrocarbons burning in standard air has a value in the neighborhood of 4 to 7. Finally Eq. (2.19) then shows that

$$\bar{b} \approx b^*\,\bar{Y}_p \approx b^*\,\bar{\theta} \qquad (5.5)$$

In concluding this section, it should be pointed out that under some special circumstances, such as prevail for symmetrical and laminar fires of high molecular weight fuel vapor [1], fire flow fields do exhibit significant buoyant effects of other than thermal origin. These effects are the result of the existence of a large central region of cold and relatively dense fuel between the flames and the boundary from which issues the fuel vapor (e.g. Fig. 1.6). It is probable that cold fuel pockets occur in other fire situations. But it is doubtful these dominate the dynamics of any turbulent fire.

5.2 SOME CHARACTERISTIC MAGNITUDES FOR UNSTRUCTURED AREA FIRES

In this section attention is directed to a class of fuelbeds which provide gasified fuel from a well-defined and approximately symmetrical region on the ground plane, as sketched in Fig. 5.2.   A circular pool of flammable liquid is an example. Another would be an array of solid fuel elements stacked to a height small compared with the horizontal length scale d of the fuelbed.

It is also assumed that the fire is open, and not disturbed by such environmental factors as wind and rotation. By an open fire is meant one whose ground plane extends to a sufficient extent that there is negligible blockage or other disruption of the induced inflow of air.

Finally, it is assumed that the flow field is coherent. This means that the flames and the convection column generally are organized on the length scale of the complete fuelbed. For small fuelbeds burning sensibly (as opposed to smoldering) this is almost always the case. For larger fuelbeds, say with $d \simeq 10^4$ cm or more, there is a tendency for the fire to break up into a number of more or less independent columns.

At sufficiently large elevation z , the column becomes a plume. As the word "plume" is meant in this book, the mathematical approximations of a free boundary layer are valid in the plume region. These approximations, henceforth called plume approximations, are: (1) negligible pressure variation transverse to the direction (z) of plume flow, and (2) negligible inversible transport flux components parallel to this direction. Irreversible transport means transport of a quantity by turbulent or molecular diffusion $\propto$ to a gradient in some measure of the local density of this quantity. For example, in a laminar flow, the term $\rho D \, \partial Y_i / \partial y$ in Eq. (1.1) represents the y-component of irreversible flux of species i due to a gradient in its mass fraction Y_i . If the flow is turbulent, Y_i is interpreted as an average on a time or length scale large compared to that of the turbulence, and ρD is replaced by a coefficient which is determined primarily by the structure of the turbulence rather than the Y_i field. Other important irreversible fluxes are those of sensible energy and momentum; a flux of a momentum component transverse to the flux direction manifests itself as a shear stress. In the sense of the above stated plume approximations, physical quantities are said to be neglible if the terms representing these quantities are much smaller than the most important terms in the partial differential equations governing the fields of species concentrations, temperature and velocity.

Plume approximations must break down at sufficiently low elevation. There is a region just above the fuelbed where irreversible transport in the z direction is clearly important (otherwise there would be no heat and mass exchange with the fuelbed) and horizontal pressure variation is important (otherwise there would be no indraft). This region is here called the "core" of the convection column. The upper boundary of the core is the elevation at which the plume approximations become valid. This elevation cannot in general be precisely defined. But the plume approximations are probably at least qualitatively correct at the elevation where the column begins to diverge. On the basis of observed column geometry, the column begins to diverge at elevation z roughly on the same order-of-magnitude as the characteristic horizontal length d . Thus, for purposes of estimation within the core region, the buoyant acceleration length ℓ is here taken to be the same as d.

It is instructive to write an order-of-magnitude energy balance on the core of a turbulent fire. Let q_f denote the heating value of the fuel consumed expressed as a flux, i.e. power per unit fuelbed area. As noted above the shape of the core and in general of the complete column is substantially the same for a wide variety of turbulent burning conditions. Thus it is unnecessary to account for changes of geometry as q_f or ℓ is varied. Let $\bar{b}$ be the average buoyancy of the core. Then the average density of the core gases $\sim \rho_\infty/(\bar{b} + 1)$. The velocity of the gases leaving the core $\sim \sqrt{g\,b\,\ell} \sim \sqrt{g\,b\,d}$. The core cross-section area $\sim d^2$. Thus the mass throughput for the core, neglecting fuel mass, since the stoichiometric air requirement of typical fuels greatly exceeds unity, $\sim d^2 \rho_\infty \sqrt{g\,b\,d} /(\bar{b} + 1)$. The energy of combustion goes to raise the sensible energy of the gases passing through the core by an amount $\bar{c}_p \Delta T \sim \bar{c}_p \bar{b} T_\infty$. Thus the energy balance finally reads

$$Q \sim \bar{b}^{3/2}/(\bar{b} + 1) \qquad (5.6)$$

where

$$Q = q_f/(\bar{c}_p \rho_\infty T_\infty \sqrt{g\,d}) \qquad (5.7)$$

is a dimensionless fuelbed combustion heat flux. There are two limiting solutions to Eq. (5.6):

$$\bar{b} \sim Q^2 \quad [Q \gg 1] \qquad (5.8)$$

$$\bar{b} \sim Q^{2/3} \quad [Q \gg 1] \qquad (5.9)$$

For $Q \gg 1$, the flow at the core is said to be strongly buoyant and, for $Q \ll 1$, weakly buoyant. For weak buoyancy the average core buoyancy is much less than that of visible flames, meaning that a large amount of excess air is entrained into the core.

Before proceeding with interpretation of these results, the realistic range of magnitude of Q should be examined. The fuel consumption rate q_f expressed as sensible power per unit area is not particularly sensitive to the length scale d , at least for reasonably large fires. The heat release rate varies significantly with the type of fuelbed. Thus Q is roughly proportional to $d^{1/2}$ for a given fuelbed type, provided that essentially complete combustion does occur in the core region. If complete combustion does not occur in the core, then the value of q_f in Eq. (5.7) must reflect only the heat

release that does occur there. As sketched in Fig. 5.1, for sufficiently small d visible flames persist well up into the plume region.

_Assuming complete combustion in the core, the general range of
$q_f/(c_p \, \rho_\infty \, T_\infty)$ is 100 to 1000 cm/sec. Thus if $d \simeq 1000$ cm , the general range of Q would be 0.1 to 1.0. With $d \simeq 10^5$ cm , the range
of Q is 0.01 to 0.1. This means that the core of all fires with realistic burning rates and characteristic length greater than around
10^3 cm tend to be weakly buoyant. This point is elaborated later.

Now the foregoing estimates, in conjunction with Eq. (5.6) suggest
that $\bar{b}$ can be increased well above unity merely by lowering d another order of magnitude, say to 100 cm. But there is a limit to $\bar{b}$.
Even if the core were entirely filled with visible flames, $\bar{b}$ according to Eq. (5.5) has as an upper bound b^* , the characteristic buoyancy expressed in Eq. (5.4). A reasonable upper bound for the mean
total products fraction $\bar{Y}_p$ is 1/2 and for b^* is 7 (assuming standard air as an oxygen source). Thus the maximum possible value of
$\bar{b}$ is around 4. According to Eq. (5.6), this implies a maximum of Q
around unity. For q_f in the range 100 to 1000 cm/sec, $Q \sim 1$ corresponds to d in the range 10 to 1000 cm.

The preceding development is based entirely on order-of-magnitude
arithmetic. Nevertheless, the results correspond quite well with reality. For example, the $d \sim 10$ to 1000 cm range deduced in the preceding paragraph does bracket the range where visible flame heights are
approximately equal to the fuelbed horizontal length scale d . As
fire size is increased beyond this range, the fraction of the core
region occupied by visible flames tends to decrease, the remainder
being occupied by combustion products and air at relatively low temperatures.   (The region is of course visibly well-defined in practice
because of the smoke that normally accompanies fire combustion products.) For very large values of d it is a reasonable approximation
to neglect the visible flame height altogether in comparison to d .
In this limit, the gross structure of the fire flow field appears to
be well determined by a model which incorporates all sensible heat
release on the ground surface. For such very large length scale fires,
comprehensive temperature, species concentration and velocity field
data is lacking. But the limited results now available do suggest
that the general temperature range implied by Eq. (5.9) does prevail
above large area fires. For example, a test fire, [2] with $d \approx 4 \times 10^4$ cm
and $Q \sim 0.03$ to 0.1, produced mean temperatures $\sim 100°C$ above ambient
at elevation around 1000 cm^2.

The estimates of the mean buoyancy in the core of area fires makes
possible further estimates of the mean updraft velocity $\bar{u}$ via Eq.
(5.2). In the weakly buoyant region, since $\bar{b} \propto Q^{2/3}$, and $Q \propto d^{-1/2}$,
Eq. (5.2) indicates that $\bar{u} \propto d^{1/3}$ if the heat release rate q_f is
held constant. One of the striking features of such estimates of $\bar{u}$
is the modest magnitudes that result. For example, taking $\bar{b} = 0.2$
(which is probably unrealistically high for a very large fire) and tak-
ing ℓ (i.e. d) = 10^5 cm, gives $\bar{u} \sim 4 \times 10^3$ cm/sec. As discussed
in Section 5.1, the use of the hydrostatic equilibrium principle to
estimate accelerations incorporates errors, insofar as buoyancy alone
drives the flow. In general these errors consist of neglect of forces

(especially shear forces) which tend to retard the flow. Thus, Eq.
(5.2) is probably an upper bound for $\bar{u}$.

With an estimate of $\bar{u}$, it is possible also to construct an estimate of the mean low elevation indraft velocity $\bar{u}_{in}$ by balancing
the secondary flow into the core with the upflow out of it. If the core
is strongly buoyant, the density ratio of the entering and leaving gases
must be accounted for. This ratio $\sim \bar{b} + 1$. If the core is weakly
buoyant the ratio of $\bar{u}_{in}$ to $\bar{u}$ is approximately the ratio of the
upflow to inflow cross-sectional areas. This area ratio is on the order
of 1/5 to 1/3 depending upon the geometric model chosen. Thus, for the
example just cited, an estimate of $\bar{u}_{in}$ is 10^3 cm/sec. These velocity
estimates are also consistent with data from large scale field experiments [2], if the latter is interpreted in an average sense. However, very
much larger velocity magnitudes have been observed locally. This topic
is dealt with further in the discussion of fire violence in Section 6.1.

The above velocity estimates justify the neglect of kinetic energy
of the macroscopic flow and of viscous dissipation in Eq. (2.7), an
energy conservation statement. Viscous dissipation can be neglected in
comparison with the heat conduction term in the energy equation [6] if the
Eckert number $E = (u*)^2/1h* \ll 1$. Taking $u*$ as 4×10^3 cm/sec,
and Δh^* as 1000 cal/gm $\simeq 4 \times 10^{10}$ ergs/gm, yields $E < 10^{-4}$. Neglect
of kinetic energy convection is tantamount to the neglect of pressure
work. The criterion for neglect of convection of kinetic energy in
comparison with convection of sensible enthalpy h_s , also is $E \ll 1$.

In summary, the set of estimates developed in this section are applicable under circumstances where the core region of an area fire is
not too highly structured. No account is taken of variations in the
geometrical shapes of the flow column rising above the fire. That the
shape of the column is so constant for a wide variety of turbulent area
fires suggests that the structure of fire induced turbulence has some
self-adjusting features. As is discussed in more detail in Section
6.1, dealing with scale modeling, the effective turbulent diffusion
coefficient must vary in a special manner with fire length scale in
order to preserve this geometric similarity. If one searches for a
laminar fire which is also relatively unstructured and which has a
flow field shape similar to the shape of turbulent fire flow fields,
a $d \simeq 2$ cm pool fire is a good candidate. In this case, the core is
completely occupied by visible flames and the $Q \sim 1$ description of
the fire appears to hold fairly well.

A further observation is that very large area fire flow fields
tend to lose their coherence. Furthermore, area fires of any size tend
to lose their coherence as heat release rate q_f is independently reduced. For example, the convection column above a fire with $d \sim 10^3$
cm often breaks up into a number of smaller columns as the fire burns
out. In the latter case, q_f is independently decreased. In the former case d is independently increased. In either case, the flow field
tends to lose coherence as Q is decreased. Thus Q may have fundamental significance as a stability parameter.

This section has dealt with the core region of area fires. To
complete the fire flow field treatment would require consideration of

the other flow regimes indicated in Figure 5.2. There is a large and
well developed body of literature describing the mechanics of buoyant
plumes. Reference 3 is a comprehensive survey.[3] As far as concerns rigorous analysis of plume behavior from first principles, the big obstacle
is the treatment of turbulence. However, much is known experimentally
about the structure of plume turbulence and satisfactory assumptions
regarding air entrainment, and accurate profiles of field variables
such as velocity and temperature, have been developed. Reference 5 is a
useful example of the application of a simplified plume model to a
specific fire problem (flame height).

It is noteworthy that all plume solutions are in a sense "initial
value" problems. The initial values must be stated for the base of
the plume. Using experimental data, it has been possible to develop
plume initial conditions for a variety of specific fire configurations.
In principle, plume initial conditions could also be provided by a sufficiently detailed and reliable analysis of the core region flow, but
such analyses are presently not available.

The development of plume flow depends on the meteorological environment. A stable lapse rate (vertical gradient of ambient temperature)
tends to retard the plume. Of course, the plume also responds to ambient wind. Sometimes the flow induced by very large area fires seems
to alter the ambient meteorology. Meteorological interactions are beyond the scope of this book.

The secondary flow regime has hardly been investigated at all. It
may be possible to treat this flow in some manner as a hydrodynamic
sink flow. However, the extent to which viscous or turbulent shear
forces are significant in this flow regime has not been established.
If turbulent shear is important, then the turbulent structure of interest is that of the ambient atmosphere, thus introducing a new dimension of meteorological complexity.

5.3 FLAME BOUNDARY LAYER STRUCTURE IN POOL FIRES

In the preceding section, the mean core buoyancy $\bar{b}$, and
through $\bar{b}$, corresponding velocity components $\bar{u}$ (updraft) and
$\bar{u}_{in}$ (inflow) are estimated for the core region of unstructured fire.
The fire core is unstructured because of turbulent diffusion mechanisms
strong enough to smear energy and species throughout the region.

In general, the dimensionless heat flux Q tends to increase with
decreasing horizontal length scale d .   For sufficiently small d ,
complete combustion no longer occurs in the core. Diffusive
transport of oxygen into the core in inadequate. For still smaller
d , the low elevation flow is laminar. For laminar flow at the velocities of interest, diffusion lengths are only on the order of a few
mm. Pool fires and pool-like gas fires [1] are laminar at low elevations

---

[3] Reference 4 is a more recent survey, not available at the time of this
writing.

up to d $\stackrel{\sim}{=}$ 10 cm , as indicated in Fig. 5.1.   The result is a highly
structured core region in which a flame boundary layer surrounds a
body of cold fuel vapor.[4] Some structural characteristics of the flame
boundary layer are examined in this section. Again, the flow field is
so complicated that accurate analyses have not been carried out. What
follows is presented in the same order-of-magnitude spirit as the estimates in Section 5.2.

Figure 5.3 shows an idealized model of the flow configuration.
The flame boundary layer has characteristic tangential and normal velocity components u_t and u_n , respectively. From considerations of
continuity [7] (conservation of total mass), these quantities are related
to the boundary layer length $\sim d/\cos \phi$ and thickness δ ,

$$u_n/u_t \sim \delta/(d/\cos \phi) \qquad (5.10)$$

In this example, the flow acceleration vector in the flame boundary layer is not parallel to the gravity vector, indicating that some
of the forces ignored in Eq. (5.2) play a significant role in determining the gross flow pattern. However, the dominant accelerating force
within the flame boundary layer must be buoyancy, and a modification
of Eq. (5.2) can be applied.

Let $\bar{b}$ denote the mean buoyancy in the flame boundary layer. For
hydrocarbon fuels burning in air, $\bar{b}$ is probably around 4, based on
arguments essentially the same as those given in the preceding section
for a flame-filled fire core region. The buoyancy acceleration component parallel to the boundary layer is $g \bar{b} \sin \phi$. In analogy with
Eq. (5.2) there results

$$u_t \sim [(g \bar{b} \sin \phi) (d/\cos \phi)]^{1/2} \sim \sqrt{g \bar{b} d \tan \phi} . \quad (5.11)$$

The boundary layer thickness δ can be estimated from the laminar case
of Eq. (1.21), taking $d/\cos \phi$ again as the acceleration length, u_t
as the characteristic streamwise[5] velocity, and $\bar{D}$ as an average laminar diffusion coefficient for the flame. The result is

$$\delta \sim 5 [\bar{D}^2 d/(g b \cos \phi \sin \phi)]^{1/4} \qquad (5.12)$$

Finally, combining the last three equations yields for the normal
velocity u_n

$$u_n \sim 5 [\bar{D}^2 g \bar{b} \cos \phi \sin \phi/d]^{1/4}$$

---

[4] A cold fuel vapor region is not always observed. If the burning rate
is too low no such region appears. Under some conditions, the cold
fuel vapor region appears and disappears as the fire pulsates.

[5] Of course, u_t is not an external velocity as u^* is stated to be in
Chapter 1.   But what really belongs in Re in Eq. (1.21) is the
characteristic magnitude of the velocity component parallel to the
boundary layer.

With $\phi \simeq 45°$, $d \simeq 10$ cm , $g \simeq 1000$ cm/sec^2 , $\bar{b} \simeq 4$ and $\bar{D} \simeq 1$ cm^2/sec , typical numerical magnitudes are:

$$u_t \sim 100 \text{ cm/sec}$$

$$\delta \sim 1 \text{ cm}$$

$$u_n \sim 20 \text{ cm/sec}$$

These above results include a dependence on the flame lean angle ϕ despite the fact that the expressions have only order-of-magnitude validity, to indicate sensitivity. Provided ϕ is not too close to zero or 90°, the quantities δ and u_n appear relatively insensitive to variation of ϕ . Tangential velocity u_t is somewhat more sensitive. Determination of ϕ requires a balance of mass and momentum flow along the flame boundary layer, together with a radial momentum balance. The latter must include the momentum of the inflowing fresh air, and the pressure excess or defect in the cold fuel vapor region, depending on the fuel density. Because the flame boundary layer has a non-uniform thickness, and is subject to some pulsations and surface instability, a precise value of ϕ cannot be experimentally defined. Nevertheless, ϕ displays a clear experimental dependence on diameter d and on fuel vapor density, decreasing with d and increasing with vapor density. Experiments with pool-like gas burners [1] show that ϕ is nearly independent of burning rate, provided that the burning rate is greater than a critical minimum. This is essentially the same as the burning rate required to sustain the flame boundary layer.

The critical minimum fuel burning rate, expressed as a surface mass flux $(\rho v)_{crit}$, can be estimated on the basis of the following reasoning. On both the air and fuel sides there is mass flow into the flame boundary layer. The flame sheet model discussed in Chapter 2 is appropriate. Reaction zone thickening near the cold boundary (Fig. 2.1) is a local effect, here ignored. Consider now the disposition of fuel in the flame boundary layer. For purposes of discussion, the flame boundary layer is regarded as ending at the tip of the cold fuel region it surrounds. In general, some of the fuel vapor diffuses to the flame sheet and reacts, and some remains unreacted on the fuel side of the flame sheet until it reaches the end of the flame boundary layer. In principle, the same statement holds for air. But the stoichiometrically required air/fuel mass ratio r for hydrocarbon fuels is much greater than unity. In view of this fact, the flame sheet must be located close to the air surface of the flame boundary layer as indicated in Fig. 5.1. Thus the flux of air into the flame boundary layer and the flux of air into the flame sheet must be approximately the same. This flux $\sim \rho_\infty u_n/(\bar{b} + 1)$ where the factor $\bar{b} + 1$ again accounts for flame/ambient density ratio.

On the other hand, because $r \gg 1$ the mass balance, and hence the fluid mechanics, of the flame boundary layer are insensitive to the rate that fuel is taken up. Excess fuel has only a minor effect on the structure of the flame boundary layer. But a region of cold fuel cannot exist unless the rate of supply from the pool or burner exceeds the stoichiometric requirement of the air reaching the flame sheet. Thus, the critical fuel mass flux $(\rho \; v)_{crit}$ can be estimated

$$r \; (\rho \; v)_{crit} \sim u_n/\cos \phi \qquad (5.14)$$

The factor $\cos \phi$ is approximately the pool surface/flame boundary layer area ratio.

Using Eq. (5.13) for u the above result shows that $(\rho \; v)_{crit} \propto (\sin \phi/\cos^3 \phi)1/4^n$. This means that $(\rho \; v)_{crit}$ is not highly sensitive to the lean angle ϕ for the range of ϕ in which the concept of a cold fuel region is meaningful. Moreover the factor containing ϕ can be neglected when estimating $(\rho \; v)_{crit}$; i.e. in the range $\phi = 15°$ to $45°$, value of $(\sin \phi/\cos^3 \phi)1/4$ goes from 0.73 to 1.2

For $d \sim 10$ cm , with $\bar{D} \sim 1$ cm^2/sec , there results finally a criterion for the existence of a cold fuel region:

(molar stoichiometric air/fuel ratio) × (fuel vapor velocity at pool surface) $\gtrsim$ 20 cm/sec.

In experiments with pool-like gas burners, wherein the fuel vapor velocity is independently variable, this criterion is verified except that the numerical value turns out to be in the 5 to 10 cm/sec range. It also tends to increase slightly with increasing fuel molecular weight. Since the theory is based on order of magnitude arithmetic, this numerical discrepancy is not surprising. The burning rates of real pool fires in the diameter d range around 10 cm also are consistent with the criterion. With reasonable parameter selection, the experimental pool fire burning rates are in good agreement with the stable burning point indicated by pool-like gas burner data (e.g. Fig. 1.7) [8]. Such indicated stable burning points are always to the right of the convective heat transfer q_{CONV} versus burning rate $\rho \; v$ characteristic in Fig. 1.7. The peak of this characteristic correponds to the first appearance, usually intermittent, of a cold fuel region over the burner. In general, for higher burning rates, the cold fuel region grows. The heat transfer apparently decreases because the increasing flux of cold fuel vapor pushes the flames away from more and more of the burner surface. Conversely, when the burning rate falls below the point of maximum q_{CONV} , the mixture tends to become overlean near the parameter of the burner, so that the flame appears to shrink and q_{CONV} again decreases.

A further conclusion can be drawn from the estimates of u_t and u_n . Since the burning rate of liquid pools expressed as a fuel vapor velocity is only on the order of 1 cm/sec, the momentum of the fuel vapor leaving the pool is negligible as regards fire dynamics.

REFERENCES

1.   Ref. 9, Chap. 1.

2.   "Project Flambeau – An Investigation of Mass Fire (1964-1967)",
     Final Report in 3 Volumes, USDA Forest Service, Pacific Southwest
     Forest and Range Experiment Station, P.O. Box 245, Berkeley,
     California 94701, U.S.A.

3.   G. Briggs, "Plume Rise", U.S. A.E.C. Critical Review Series, M69,
     TIO-25075, U.S. Dept. of Commerce (1969).

4.   J.A. Fay, "Buoyant Plumes and Wakes", Annual Review of Fluid Me-
     chanics, 5 (1973).

5.   F.R. Steward, Combustion Science and Technology, 2 (1970) pp.
     203-212.

6.   Ref. 8, Chap. 1.

7.   Ref. 3, Chap. 1.

8.   R. C. Corlett, Combustion and Flame, 12 (1968) pp. 19-32.

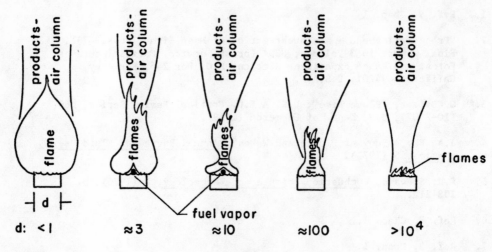

FIGURE 5.1

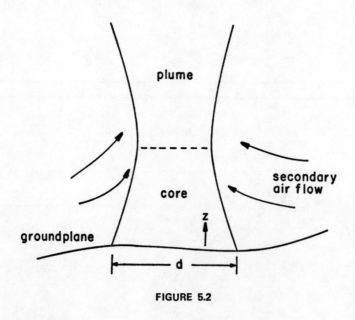

FIGURE 5.2

NOTE: Figure captions for this chapter are listed on page 254.

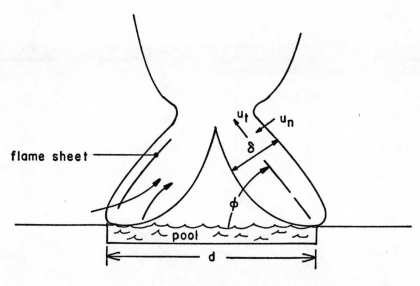

FIGURE 5.3

FIGURE CAPTIONS

Figure 5.1   Pool fire flow patterns at varying diameter d.

Figure 5.2   Flow regimes in unstructured area fire.

Figure 5.3   Flame boundary layer idealized geometry.

Chapter 6

FIRE VIOLENCE AND MODELING

R. C. CORLETT
University of Washington, USA

IN CHAPTER 5, SOME structural features of fires of simple configuration
are examined. Even for these relatively simple fires, there are no real
flow field theories except for the plume region, which by definition
is restricted to elevations high enough for the mathematical boundary
layer approximations to hold. It is possible to estimate some salient
properties of the remainder of the flow field, as is done in Chapter
7, but enormous mathematical complexity is encountered when further
progress is attempted.

But accidental fires are of predominant practical interest. Accidental fires have a high degree of randomness. The fuelbeds are likely
to be non-uniform and the geometry irregular. In view of the difficulties encountered in attempting to understand flow fields of simpler
fires, what can be done with such complex fire configurations?

First, it is necessary to ask whether or not fuelbed and geometric
complexity introduces significant new features to the phenomena or
merely adds detail. In other words, are the simplified models dealt
with in Chapter 5 capable of incorporating the essential features of
accidental fires? The answer is yes and no. Of course, some essential
features such as flame temperatures, and gross updraft and inflow velocities, are obtainable from such models. But the simplified models
introduced in Chapter 5 in no way incorporate realistic magnitudes of
fire violence. Entirely new physical mechanisms must be added. These
new mechanisms are the subject of Section 6.1.

A second issue that presents itself as a consequence of inability
to treat complex fire flow fields is the feasibility and potential
usefulness of experimental modeling techniques. Fully realistic fire

NOTE: All figures quoted in text are at the end of the chapter.

tests in principle can provide the data needed by the practitioner, even if a theoretical understanding of the phenomena is lacking. There are two problems posed by such a suggestion. One is the expense of fully realistic fire tests, particularly if the length scale exceeds 10^3 cm or so. Because of the random nature of accidental fires a large number of variables must be considered; the resulting requirement for a large number of tests greatly exacerbates the cost aspect. The other problem is a more subtle one. It is not always clear what is meant by "full realism". This is particularly true for open fires for which the meteorological environment is inherently poorly controlled, and for fires with elaborate and costly fuelbeds. Both of these problems demand consideration of the concept of experimental modeling.

Experimental modeling has the same philosophical meaning as theoretical or conceptual modeling mentioned in Chapter 1, namely the abandonment of some aspects of reality in the interest of simplicity or generality. The difference is in the set of feasible steps that might be taken, and in criteria for simplicity. The objective is to find models which incorporate the features of interest but which avoid other features entailing undesirable cost or complexity. Reduction of length scale is one of the most desirable experimental modeling steps. This is called scale modeling. The means and rationale of scale modeling are treated in Section 6.2. Experimental modeling steps other than scale modeling are not covered in this book, partly due to space limitations and partly due to the absence of an easily organizable body of information. But a common logic underlies all experimental modeling procedures. Hopefully, a brief consideration of scale modeling serves as a useful illustration of experimental modeling in general.

6.1 FIRE VIOLENCE

The estimates of flow velocity magnitudes developed in Chapter 5 on the basis of buoyant acceleration alone are not imposing. Maximum inflow velocities are on the order of 1000 cm/sec and core updraft velocities perhaps several times higher. Although the velocity magnitudes do tend to increase with horizontal length scale d , the dependence is a weak one. There is no hint, on the basis of simple buoyancy models, of "fire storms" occuring at some sufficiently large d . Indeed, for the anticipated range of heat release rates, the dimensionless parameter which appears to characterize the strength of buoyant flow above an area heat source [i.e. Q, Eq. (5.7)] actually decreases with increasing d . Depending on fuelbed non-uniformity and perhaps on the spectrum of ambient meteorological disturbances, fire flow fields tend with decreasing Q to lose coherence and thus to behave as a number of smaller and independent buoyant columns.

Yet there is a lot of evidence of violent fire behavior. The fire storms of World War II and so-called "blow-ups" of wildland fires suggest velocity magnitudes on the order of 10 times those cited in the preceding paragraph. A common phenomenon is aerial transport of firebrands (pieces of burning debris) with minimum length (e.g. diameter of a stick) up to the order of 2 or 3 cm. In some circumstances this is a controlling mechanism of fire propagation. High local velocities within fire flow fields are of practical interest because structural

damage and mechanical obstruction of fire-fighting activities are a
direct consequence, and because convective heating rates are enhanced.
The combination of mechanical damage and high heating rates may lead
to increased burning rates, leading in turn to more violent fire behavior. However, quantitative data describing such unstable coupling is
sparse.

Clearly, there exists an important class of violent fire phenomena whose explanation lies beyond buoyant acceleration alone. Broadly
speaking, two additional mechanisms have been put forward to explain
violent fire behavior. The first of these mechanisms is triggering of
atmospheric instability by the fire-induced flow. The second is the
interaction of rotational motion with buoyant accelerations.

Regarding atmospheric instability, it must be admitted that atmospheric processes not involving fire are capable of generating very violent flows. Tornados come immediately to mind. Weather violence is
frequently identified with instability, often connected with buoyant
flows resulting from solar heating of continental land masses.

The enhancement of flow velocities by rotation combined with buoyancy driven vertical motion is easy to envisage. If the flow external
to a fire is rotating, the air drawn into the convection column has
angular momentum. Conservation of angular momentum requires that tangential velocity increase as radius decreases. The result is a fire
whirl.

The remainder of the discussion in this section is devoted to fire
whirls. There are two broad problem areas associated with fire whirls.
One is the fluid mechanics of the whirls themselves, assuming a suitable
source of angular momentum in the boundary conditions. The other is
the source of the angular momentum. How much angular momentum will a
fire flow field develop, as a function of meteorological variables and
fuelbed descriptors?

Both of these areas receive some attention below. Violent weather
and weather instability in general are not discussed further except for
the following general observations. A sizable number of occurrences
of severely violent fire phenomena for which meteorological documentation is available are associated with atmospheric temperature profiles
near dry-adiabatic. In the absence of wind, a dry-adiabatic lapse rate
is neutrally stable. Although clear documentation seems to be lacking,
there probably exists some correlation between fire violence and unstable wind phenomena, particularly with intense atmospheric turbulence
with eddy scale on the same order as the fire length scale. On the
other hand, high wind velocities by themselves should not be expected
to enhance fire violence. High winds interfere with the operations of
firefighters and sometimes greatly accelerate fire spread rates, which
can be catastrophic in wildland fire situations. But they do not lead
to particularly high fire flow velocities. There is some evidence that
the most violent interaction between ambient wind and the convection
column of a fire is maximized for relatively light wind magnitudes.
Sufficiently strong winds essentially overwhelm the vertical convective
motion and may destroy the coherence of the flow.

Fire whirls have been obtained and described by many investigators. Rotating a cylindrical screen coaxial with a small fire or even stacking vertical panels so as to give a tangential component to the inflowing fresh air is sufficient to produce the phenomena. Reference 1 is a good example of relevant laboratory work. In that work an ii 225 cm diameter cylindrical screen was rotated about a 10 cm diameter acetone pool fire. Such a pool fire without rotation is described at various points in Chapters 1 and 5, particularly Section 5.3. The important rotational boundary value is the circulation Γ (circumference x tangential velcity) which ranged fromm 6×10^3 to 34×10^3 cm^2/sec. Increasing Γ from zero through this range has two salient effects. The burning rate increases to several times its $\Gamma = 0$ value and flame height markedly increases. Because of its increased height, the fire whirl flame is relatively well-described as a plume. In general, the fire whirl plume is narrower than that of a fire without rotation. Qualitatively, the increased flame height reflects the increased rate of fuel vaporization (the increased burning rate) together with a decreased rate of turbulent mixing. Since the gas-phase burning can only take place as fast as permitted by mixing, the flow length (flame height) required for complete combustion thus increases.

Laboratory investigations such as Ref. 1 have demonstrated that fire whirls are essentially inviscid hydrodynamic flow fields except for a thin cortex core[1] where shear stresses balance inertial forces. Indeed, for the boundary conditions imposed by a rotating screen, the flow outside the core closely approximates a hydrodynamic free vortex [2].

In the vortex core, turbulent momentum transport, and the diffusion flame mixing and heat transport processes take place. Except for the first pool (or, more generally, fuelbed) diameter or so of elevation, the vortex core satisfies the mathematical boundary layer conditions required for the development of a plume theory. (The pressure condition must be generalized from that stated in Section 5.2 to include tangential motion.) The primary difficulty in developing a plume theory of the vortex core arises in characterization of the turbulence. A frequent approach to doing this utilizes integral forms of the conservation equations, wherein turbulence appears only through an entrainment coefficient E, defined [3] as the ratio of the <u>rate of mass entrainment per unit plume length, at a given plume cross-section</u>, to a <u>suitable mean mass flux in the plume direction, times a local length scale (e.g. a plume width)</u>. The magnitude of E is on the order of 0.1. For non-rotating and weakly buoyant plumes, E is effectively a constant. In Ref. 1, it was found that E not only decreases with increasing rotation rate but tends to increase with elevation. It remains unclear how general is such variation of E with elevation. The turbulence levels in larger scale fire whirls have not been measured.

But some of the features of laboratory fire whirl phenomena, such as described in Ref. 1, almost certainly carry over to larger scale

---

[1] The vortex "core", as discussed in Section 6.1, has an entirely different meaning from the "core" of a fire column below the plume, as defined in Section 5.2.

fire whirls. The essential concept of a hydrodynamic free vortex surrounding a thin core supportive of turbulent stresses is widely accepted, as also is the suppression of turbulence by rotation. On the other hand the coupling of the rotational strength of a fire whirl and its thermal power (as determined by burning rate alone in Ref. 1 but conceivably also by the low elevation entrainment of hot combustion products from elsewhere in the fire, or of air warmed by contact with the sun-heated ground) is not known. The most that can be said is that all of the effects of rotation that come to mind appear to enhance the thermal power.

At this point in the discussion, one can conclude that the influence of rotation on a fire can result in the formation of whirls of much greater violence than can be explained on the basis of simple buoyancy effects. Although knowledge of the structure of fire whirls remains primitive, a more serious problem is posed by the rotational boundary conditions in a real fire situation, as opposed to laboratory experiments with clearly applied tangential flow components. Before discussing this problem it is worthwhile to rephrase it in terms of vorticity.

Vorticity is a vector quantity defined as the curl (or cross-derivative) of the velocity vector field. It is treated at length in fundamental fluid mechanics monographs (e.g. Ref. 4). Its vertical component ζ is expressed in terms of derivatives the horizontal velocity components v_1 and v_2,

$$\zeta = \partial v_2/\partial y_1 - \partial v_1/\partial y_2 \qquad (6.1)$$

when y_1 and y_2 are horizontal (Cartesian coordinates) in the directions of v_1 and v_2 (Fig. 6.1). Evidently, in a simple horizontal shear flow (e.g. $v_1 \equiv 0$, $v_2 \propto y_1$) the magnitude of ζ is the same as that of the velocity gradient. For a rigid body rotating about a vertical axis with angular velocity ω, $\zeta = 2\omega$. The strength of a vortex is proportional to the integral of $\rho \zeta$ (ρ mass density) over the cross section. For constant density flow, this integral is the same as $\rho \Gamma$ where Γ is the circulation (the line integral around the cross section, of the velocity component tangential to the path of integration).

Under some conditions, ζ satisfies a conservation equation similar to Eq. (2.16) which has the interpretation that ζ is preserved along streamlines except where diffusion is important. These conditions are not met for typical three-dimensional fire flow fields, especially if there are significant density variations. Nevertheless, it is conceptually useful to envisage fires as collecting vorticity more or less "intact" except in core regions where it is transported by diffusion, together with linear momentum, sensible energy, and species. The crucial question is, where does the vorticity come from?

There are two broad answers to this question. The vorticity may be present in the atmosphere independently of the fire, or the vorticity may be created by the fire. Vertical vorticity is always present in the atmosphere. Its time average at any point corresponds to the local horizontal earth rotation component. In addition there are vorticity fluctuations on all length scales up to continental dimensions. The view that such ambient vorticity contributes to fire violence is reviewed in Ref. 5.

Development of a vortex over a fire or other heat source requires time for the collection of vorticity. At the earth's rotation rate, an inordinately long time is required to develop vortex strength typical of violent fires. On the other hand there is good evidence that rotation on the order of one power of 10 greater is available on a length scale $\sim 10^5$ cm, with a resulting development time $\sim 10^4$ sec.

There is no reason to disbelieve any of the reasoning which supports ambient vorticity as a source of fire violence. In particular, if long burning times occur, this becomes an increasingly probable mechanism. But to obtain the required combination of energy release rate and duration requires a fuelbed typical of a densely built up city. Fortunately, there have been no fires of such magnitude since World War II, when intensive incendiary bombing brought on several severe urban fires. Under some circumstances, wildland fires approach the required conditions. But so far it has not been possible to clearly establish the source of vorticity in any large typical unwanted wildland fires.

On the other hand, there is strong evidence that ambient vorticity is not necessary for the generation of very intense fire whirls. Figure 6.2 is a schematic sketch of a test fire [6] carried out with piles of brush and small trees distributed over a rectangle with characteristic length $\sim 4 \times 10^4$ cm. The piles were ignited simultaneously, building up to peak average heat release rate (q_f, Chapter 5) 15 cal/cm^2/sec in time ~ 400 sec and then declining to ~ 3 cal/cm^2/sec over the next ~ 1000 sec. Ambient wind speed was gentle, on the order of 300 cm/sec. For the most part the fire burned as a large area fire as described in Chapter 5. Very quickly the fire established a coherent flowfield as sketched at the top of Fig. 6.2. Almost everywhere flame heights were on the order of 10^3 cm, velocities and temperatures consistent with the estimates developed in Section 5.2. But two very distinct fire whirls appeared at locations as sketched at the bottom of the figure. These visible flames in the whirls were several times as tall as flame heights elsewhere in the fire, and the smoke whirls above were visible to heights on the order of the fire horizontal length scale. Velocities within the whirls were estimated[2] to be on the order of 10^4 cm/sec. The whirls were not attached to specific piles of burning fuel, but wandered about the general locations indicated in Fig. 6.2. Late in the die-down phase of the fire, one was observed to wander outside the fuelbed for a time, thus becoming a dust whirl.

From the symmetry with respect to wind plane of the above-described whirl locations and directions, and the short time required for development, the conclusion is inescapable that the mechanism of generation was an interaction between the ambient wind and the rising gas

---

[2] From streaks, identified as moving fire brands, on photographs with known exposure time [7].

column induced by the fire. So far the detailed mechanism has not been
established. Probably the wind/ground boundary layer plays a crucial
role. But even if the boundary layer is disregarded, it is easy to
envisage velocity gradients appropriate for the generation of vertical
vorticity.

Since it is established that wind interaction is a viable (albeit
not unique) mechanism for vorticity generation leading to fire whirls,
the following conjecture respecting the effect of fuelbed irregularity
is speculatively put forward. Consider a highly irregular fuelbed, in
the absence of any ambient wind or vorticity. It is well known that
irregular fuelbeds can develop a coherent buoyant flow if total heat release rate is sufficiently high. The indrafts in such a coherent flow
can easily reach the 300 cm/sec level cited above. But because of the
irregularity of the fuelbed there will be local regions of intense
burning, with relatively strong buoyancy. The result would be an as-
semblage of small buoyant columns imbedded in the larger coherent field.
Such local columns should interact with the larger scale inflow to
generate vorticity in essentially the same manner as the relatively regular fuelbed depicted in Fig. 6.2 interacted with ambient wind.
Thus fuelbed irregularity may contribute directly to fire violence.
This conjecture is supported by the fact that experiments following
those described in Ref. 6 have been carried out with similar fuel type
and fuelbed length scale, but with non-uniform fuelbed arrangements for
such conditions, numerous small whirls were observed.[8]

6.2 SCALE MODELING

In this section, attention is directed to the problem of determining
some properties of interest of a postulated real fire, called the prototype or full-scale fire, on the basis of experimental measurements and
observations of another system on a different length scale, called the
model. The model need not necessarily be a fire. For example, if the
flow pattern and associated velocity magnitudes are of primary interest,
and if the prototype heat release is known to be concentrated near a
surface, electric resistance heating could be an acceptable substitute
for combustion. For fire problems of practical interest, the model
length scale is generally less than the prototype length scale.

It is naive to attempt to model a system as complex as a fire
merely by scaling down the complete prototype system. Not only are
serious fabrication problems often thus encountered but, more impor-
tantly, essential features of the phenomena may be badly distorted.
The result would then be model data which is difficult or impossible
to interpret in terms of the prototype phenomena.

A rational attempt at scale modeling requires an underlying theoretical model, if only implicitly. At the very least recognition of
which physical parameters are important is required. It is not necessary always to develop a complete mathematical formulation, and certainly not necessary to actually obtain solutions. Thus the decision
that the only parameter needed to determine irreversible momentum
fluxes (stresses) is the dynamic viscosity μ amounts to tacit accept-
ance of the appropriate version of the Navier-Stokes equation; the

fact that the Navier-Stokes equations are rarely solvable is not relevant. As another example, the decision that neither multi-step chemistry, finite rate kinetic parameters, nor chemical equilibrium parameters are important means that a mathematical formulation devoid of such information should imply correct answers. In this case the implicit chemical model would be a diffusion controlled flame sheet model. Alternatively, it could be postulated that such chemical parameters are important but in some sense are the same for model and prototype. A decision along such lines requires supportive theoretical analysis or strong empirical justification. For example, it might be argued that finite rate chemistry is important but that it can be characterized by an ignition temperature or by flammability limits, perhaps unknown but the same for model and prototype.

If a complete mathematical formulation has been developed, a procedure called similarity analysis [9,10] can be carried out. In essence, this amounts to restating the entire problem in dimensionless form, recognizing a set of dimensionless parameters, and a set of dimensionless dependent and independent variables. If all of the dimensionless parameters are preserved (i.e. the same between model and prototype), then the relationships between the variables must be the same for the model as for the prototype. Preservation of parameters leads to a set of rules for carrying out the experiments, and recognition of the dimensionless variables provides rules for data interpretation.

In general, the choice of parameters and variables is not unique. Also, geometric similarity is not generally required. In particular, some of the geometry may be included in the dimensionless parameters of the problem. In such cases some geometric distortion may be required.

If an explicit mathematical formulation is not available, parameters and variables are constructed by the methods of dimensional analysis [10], carried out in various degrees of formalism.

Neither similarity analysis or dimensional analysis provides useful results if the number of dimensionless parameters is too large. The scaling rules become impossible to follow. For fire phenomena treated in anything approaching full physico-chemical detail, the number of parameters is overwhelming [11]. Thus it is necessary to guess which are the essential dimensionless parameters on the basis of physical reasoning and order-of-magnitude estimates of the importance of various phenomena.

In what follows, the complete logical development outlined above is not recapitulated. Rather, the results are stated and interpreted, with emphasis on the convective phenomena in fires. For notation and terminology, the reader is referred to Chapters 2 and 5.

For simplicity consider first as a prototype an area fire which can be regarded as a distributed surface heat source of power density q_f . The model is another distributed surface heat source. The model and prototype are geometrically similar except for fine scale detail. This means that, with d the characteristic prototype horizontal length, prototype details of length scale $<< d$ are not incorporated in the model. The fluid heated by both model and prototype is standard air.

Unless the prototype flow is only weakly buoyant, kinematic similarity (accounting for fluid expansion) requires that the model and prototype fields of absolute temperature, and hence also buoyancy b [Eq. (5.1)], be similar functions of geometrically scaled position. This in turn requires that the dimensionless thermal loading Q [Eq. (5.7)] must be preserved. With gravitational acceleration g and mean specific heat c_p taken as constants, there results the scaling rule

$$
\begin{aligned}
q_f/q_f' &= (\rho_\infty'/\rho_\infty)\ (T_\infty'/T_\infty)\ \sqrt{d'/d} \\
&= (P_\infty'/P_\infty)\ \sqrt{d'/d}
\end{aligned}
\tag{6.2}
$$

where primed quantities refer to the model, and P_∞ , T_∞ and ρ_∞ , respectively, denote ambient absolute pressure, absolute temperature and mass density.

Neglecting for the moment the possibility of pressurizing the model, it is observed that the power density ratio q_f/q_f' is dictated by the scaling ratio d/d' . For very large outdoor fires, scaling ratios of well over 100 are of interest. It turns out the required power density reduction is not easily achievable with a combustion heated model. With free burning systems, it is difficult to achieve a factor of 10 or more reduction in power density. Even with a gas burning system, with controlled fuel supply rate, complete combustion is difficult to achieve at low power densities. The practical solution sometimes turns out to be installation of electrical resistance heaters.

The foregoing paragraph illustrates the futility of attempts to scale model fires in complete detail. Suppose the prototype consists of a large number of piles of wood with typical fuel element diameters on the order of cm. It is certainly possible to scale model such a fuelbed by using piles of shavings or splinters. But the use of such fine fuels in the model would result in an energy release rate q_f' much higher than indicated by Eq. (6.2). Indeed, $q_f' > q_f$ would be plausible. Thus Q' would greatly exceed Q   Not only would the prototype be inaccurately scaled due to mismatch of the scaling parameter Q , but the original assumption of heat release effectively restricted to the ground surface would probably be violated. Thus, although geometric similitude is necessary on length scales ∿ d and d' , rigid similitude at finer length scales would not only be unnecessary, but very likely deleterious.

But preservation of Q is not a sufficient condition for rigorously correct scale modeling. An additional dimensionless parameter which comes out of similarity analysis is the Grashof number Gr ,

$$
Gr \simeq g\ b^*\ d^3/D^*
\tag{6.3}
$$

where b^* is a characteristic buoyancy and D^* a characteristic diffusion coefficient. Strictly speaking, the denominator of Eq. (6.3) should contain instead of D^* the kinematic viscosity. However, the latter is nothing more than a momentum diffusivity and for gases is numerically never far from D^* . In view of the assumptions underlying the adoption of a universal diffusion coefficient (Chapter 1), it is appropriate to ignore the difference. (Formally, this means that the Prandtl number is taken as unity.)

Preserving Gr introduces enormous difficulties, particularly for large scaling ratios. For strongly buoyant flow fields, b^* must also be preserved. Also, it is experimentally impractical to vary ambient absolute temperature T_∞ to any significant extent. Since $D^* \propto P_\infty^{-1}$, the scaling rule becomes thus reduces to

$$(P_\infty'/P_\infty)^2 \; (g'/g) = (d/d')^3 \qquad (6.4)$$

The gravity ratio g'/g can be increased by mounting the experiment on a large centrifuge [11]. Further benefit results from pressurizing the model. But even if the pressure and gravity ratios were each 100 (which would require a facility of considerable sophistication and operating expense), the achievable scaling ratio would be just 100. If gravity scaling is abandoned, there is a possibility of meaningful pressure scaling for modest scaling ratios, say 10 or less. Such scaling would suffice for practical studies of many enclosed fires, and is in fact being developed for that application [12], but is inadequate for large open fires.

For weakly buoyant flow fields, the characteristic buoyancy b^* need not be preserved. Unfortunately, this does not add any significant additional freedom for preserving Gr , since to do any real good the model buoyancy (b^*) would have to be increased well past the weakly buoyant range.

On the basis of the preceding discussion, it can be concluded that Grashof number preservation is utterly hopeless for scaling ratios in excess of 100 and, in any case, introduces considerable experimental expense. Thus there is a strong incentive to abandon the requirement for preserving Gr . The justifying arguments require some consideration of the physical meaning of the Grashof number.

The Grashof number is a measure of the relative importance of buoyancy forces to viscous forces. This can be demonstrated as follows. Suppose the flow in a small region of characteristic dimension d has a density ρ and buoyancy b . In the absence of viscous retardation, velocities reached by buoyant acceleration $\sim \sqrt{g\,b\,d}$ [Eq. (5.2)]. The corresponding velocity gradient $\sim \sqrt{g\,b\,d}/d$. So the shear stress $\sim \rho D \sqrt{g\,b\,d}/d$ (for a gas the viscosity $\mu \sim \rho D$). The shear force on the region $\sim [\rho D \sqrt{g\,b\,d}/d]\, d^2$. This is on the same order of magnitude as the buoyant force $\rho g b d^3$, when $G_r = g b d^3/D^2 \sim 1$. Alternatively Gr can be regarded as the square of a Reynolds number with the characteristic velocity taken as $\sqrt{g\,b\,d}$, corresponding to buoyant acceleration over length d . More careful analysis, and empirical data, show that viscous forces severely retard the flow when $Gr < 1000$ or so. This is the basis of the statement in Chapter 7 that in the small diameter d limit, fire flow fields tend to be controlled by diffusion instead of buoyancy.

If the flow is turbulent, the shear stresses are of course much larger than indicated in the estimates above. Characterizing the turbulence by an effective diffusion coefficient D_T (for order-of-magnitude reasoning only!!) allows the estimate of the effective turbulent Grashof number Gr_T

$$Gr_T = g \, \bar{b} \, d^3/D_T \qquad\qquad (6.5)$$

for large area fires. This is done by observing that laminar area fires with $d \approx 2$ cm have roughly the same flow geometry as turbulent area fires. Thus

$$Gr_T \sim (Gr)_{(d \approx 2 \text{ cm})}$$

Taking $g \approx 10^3$ cm^2/sec , $\bar{b} \approx 3$, $d \approx 2$ cm and $\bar{D} \approx 1$ cm^2/sec yields $Gr_T \sim 2 \times 10^4$. This is consistent with computer calculations of the laminar flow induced by a heated horizontal flat plate, with the diffusion coefficient adjusted to as closely as possible match experimental measurements of the velocity and temperature fields in the turbulent flow above a 60 cm diameter heated plate [13]. The latter approach yields $10^4 < Gr_T < 10^5$. The turbulent diffusion processes must somehow adjust themselves to maintain this magnitude of Gr_T for a wide range of coherent turbulent flows. Otherwise the geometry of convection columns above large fires could not be as consistent as is observed.

From the standpoint of scale modeling, the significance of this reasoning is that turbulent transport phenomena tend to self-adjust as length scale varies. To an unknown degree of accuracy, the resulting fluxes are independent of the value of the laminar diffusion coefficient D . This is the rationale for neglect of Gr as a scaling parameter.

Surely this argument fails at some level of detail. The situation is analogous to the neglect of Reynolds number Re in forced convection modeling. Many forced convection turbulent flows self-adjust to some extent as Re increases. But there are always flow features which do not. For large scale buoyant flows induced by fires, the prototype data base is too poor to permit an evaluation of errors introduced by failing to scale Gr. Results of experiments with a 60 × 60 cm heated flat plate model [14] of the large scale field burn [6] discussed in Section 6.1 revealed no significant errors. However, the only prototype data are low elevation (~ 0.1 d) velocities and temperatures, and this data suffers from considerable scatter. Except for general location and the above-cited velocity estimate, no prototype data is available for the imbedded fire whirls; no attempt has been made to model the wind interaction. It is likely that Gr becomes increasingly important in determining such relatively fine scale features as local fire whirls.

Neglect of Gr implies that the phenomena of interest are independent of the magnitudes of laminar transport properties, here characterized by the diffusion coefficient D . This is certainly not true as regards convective heating of condensed-phase surfaces. As D is varied independently at constant pressure, the convective unit conductance (i.e. α in Chapter 1) for turbulent flow usually varies as 1/D

to the 1/5 to 1/3 power. For heat transfer determination using model
experiments without preservation of Gr , a practical procedure is the
following. Use a model experiment to determine the external flow velocity and temperature experienced by the prototype surface to which
heat is transferred. Then use this information as input for calculation of α from engineering heat transfer correlations. If model heat
transfer is measured, the data can be used to adjust the correlation
to the specific circumstance.

Before turning to other aspects of fire flow field scaling, one
further consequence of the self-adjusting feature of turbulent convection should be noted. To the extent that Gr_T is well-defined and
constant, the characteristic turbulent diffusion coefficient $D_T \propto d^{3/2}$
It grows without limit as size is increased. This leads to the conclusion that for large enough fires, convective transport always dominates radiative transport. Large fires (or large smoke columns) must
be optically thick. In the optically thick limit radiative transport
is described by a radiative diffusion coefficient (the Rosseland [15] diffusivity) which depends only on the local mean thermodynamic state. It
is estimated that for $d \sim 10^4$ cm , turbulent convective transport within a fire and its smoke plume exceeds radiative energy transport by a
factor of 10 or more. However, a point that must be carefully considered in scale modeling is that, if the model is a fire, radiative
transport may play an important but essentially spurious role in the
model phenomena.

To recapitulate, in most scaling procedures of practical interest,
the Grashof number Gr is not preserved. In essence this is a matter
of experimental necessity although there is some theoretical justification. In general, the justification fails unless both model and prototype flows are turbulent, as is assumed henceforth.

To the extent that Gr need not be preserved, the problem of
scale modeling a fire flow field reduces to the problem of proper modeling of the heat release distribution. For large fires which can be
approximated as distributed surface heat sources, this problem is
straightforward and has already been discussed. Indeed, since large
area fires tend to be weakly buoyant except in the relatively thin combustion region (e.g. Fig. 6.2, top) which is not modeled anyway,
scaling rules stated in Eq. (6.2) need not be followed. Instead the
only requirement is that the model remain weakly buoyant. Of course,
if the prototype heat source is not uniform, the model heat release
distribution must be similar.

In general, flow velocity magnitudes at analogous points scale as
$(\bar{b} \, d)^{1/2}$. Scaling rules for all other quantities, with units only
of length and time, are easily derived. Thus time scales $(d/\bar{b})^{1/2}$ and
circulation (Γ in Section 6.1) as $(\bar{b} \, d^3)^{1/2}$. Application of these
scaling rules requires consideration of $\bar{b}$. If the flow is weakly
buoyant [Q << 1, Eq. (5.7)], $\bar{b}$ scales as $Q^{2/3}$. Otherwise $\bar{b}$ must
be preserved, and thus drops out of the rules for scaling velocity,
time, etc.

For fires of reasonable length scale (i.e. $d \sim 10^3$ cm or less),
the flow is often strongly buoyant in important regions; visible

flames occupy a significant fraction of the flow field volume. On the
other hand, only moderate scaling ratios are required since little test
economy, if any, is realized by working with model length scales less
than $\sim$ 100 cm . For scaling ratios $\sim$ 10 or less, a considerable body
of knowledge now exists for devising fuelbeds, particularly cribs [16],
whose burning rates satisfy the heat release scaling requirement.
(This is not accomplished by full geometric scaling of the cribs!)

In scaling enclosed fires, numerous practical problems arise. A
type of information of interest is the thermal response of walls, doors,
and various room contents. It has already been pointed out that the
convective unit heat conductance α does not follow a simply derivable
scaling rule. But α does not vary rapidly with length scale d , and
the length scale dependence is fairly well understood empirically (i.e.
$\propto$ d^{-n} for 1/5 < n < 1/3, which corresponds to a maximum uncertainty
of about 30 percent for a scaling ratio of 10). On the other hand, the
thermal time constant of various heated objects generally does not scale
with the fire time constant. For example, the heat conduction time
constant of a wall of thickness s and thermal diffusivity $(\lambda/\rho\ c_p)_{wall}$
varies as $s^2/(\lambda/\rho\ c_p)_{wall}$. Thus special adjustments of wall thickness and thermophysical properties would be required to achieve temperature preservation even if α itself were preserved exactly. In
the absence of such special adjustments, prediction of prototype thermal response from model thermal response usually is a tractable problem
of engineering analysis.

The degree to which enclosure geometry must be preserved is also
an important practical problem. Strictly applied, the scale modeling
principle requires that enclosures be geometrically similar. Satis-
faction of this requirement is certainly feasible in principle. In
practice, experiments are carried out in a test enclosure equipped to
safely control the fire and to carry out the necessary observations
and measurements. Overly-detailed modification of the test enclosure
could result in considerable unnecessary expense and inconvenience.
One of the most important characteristics of test enclosure geometry
is the provision for ventilation. If the nature of the enclosure is
such that the fresh air flows in through one channel and combustion
products (including excess air) flow out through a different one, then
throughflow which satisfies scaling rules can be obtained with fans
or flow obstacles such as baffles, based on suitable engineering analysis. The problem is more difficult if the same opening (e.g. a window) serves as a fresh air inlet and combustion products outlet. Here
the required window area is a function of the window width/height ratio
as well as its elevation above the floor. Some empirical relations
have been developed, notably the recognition of A $H^{1/2}$ (where A is
window area and H height) as an important parameter [17,18], and considerable progress is being made toward the understanding of the general
relationships of enclosed fire dynamics [19]. But reliable and generally
applicable rules covering distortion of ventilation geometry are not
yet available.

REFERENCES

1.   H.W. Emmons and S.-J. Ying, Eleventh Symposium (International) on Combustion, The Combustion Institute, Pittsburgh (1967) pp. 475-488.

2.   R.C. Binder, "Fluid Mechanics" 2nd Ed., Prentice-Hall, New York (1949) pp. 157-161.

3.   B.R. Morton, Tenth Symposium (International) on Combustion, The Combustion Institute, Pittsburgh (1965) pp. 973-982.

4.   L.M. Milne-Thomson, "Theoretical Hydrodynamics", Macmillan, New York (1955) Chapters 3, 4, 13, 18, 19.

5.   R.R. Long, Fire Research Abstracts and Reviews, 9 (1967) pp. 53-68.

6.   Ref. 2, Chap. 7.

7.   C.P. Butler, "Operation Flambeau", Report NRDL-TR-68-143, U.S. Naval Radiological Defenses Laboratory, San Francisco (19 June 1968).

8.   T.Y. Palmer, "Convergence and Vorticity in a Mass Fire Experiment", Proc. Mass Fire Symposium, Canberra, Australia, 10-12 Feb. 1969, Vol. II, Defence Standards Lab, Maribyrnong, Victoria, Australia.

9.   L.I. Sedov, "Similarity and Dimensional Methods in Mechanics", 4th ed., Trans. by M. Friedman and M. Holt, Academic Press, New York (1959).

10. S.J. Kline, "Similitude and Approximation Theory", McGraw-Hill, New York (1965).

11. F.A. Williams, Fire Research Abstracts and Reviews, 11 (1969) pp. 1-23.

12. J. de Ris, A.M. Kanuri and M.C. Yuen, "Pressure Modeling of Fires", 14th Symposium (International) on Combustion, The Combustion Institute, Pittsburgh (1973), in press.

13. T.M. Fu, "An Exploratory Study of the Turbulent Free Convection Flow Above a Horizontal Heated Circular Flat Plate", PhD dissertation, University of Washington, Seattle, Washington, USA (March 1970); see also R.C. Corlett and T.M. Fu, "Turbulent Convection Above a Heated Flat Plate", 1969 Technical Session, Central States Section/The Combustion Institute, University of Minnesota, Minneapolis, Minnesota (March 18-19, 1969).

14. B.T. Lee, Combustion Science and Technology, 4 (1972) pp. 233-239.

15. J.W. Bond, Jr., J.M. Watson and J.A. Welch, Jr., "Atom Theory of Gas Dynamics", Addison-Wessley, Reading, Mass. (1965) art. 10-5.

16. G. Heskestad, "Modeling of Enclosure Fires" 14th Symposium (International) on Combustion, The Combustion Institute, Pittsburgh (1973), in press.

17.   D. Gross and A.F. Robertson, <u>10th Symposium (International) on Combustion</u>, The Combustion Institute, Pittsburgh (1965) pp. 931-942.

18.   A. Tewarson, <u>Combustion and Flame</u>, <u>19</u> (1972) pp. 101-112 and 363-371.

19.   P.H. Thomas, "Behavior of Fires in Enclosures - Some Recent Progress", <u>14th Symposium (International) on Combustion</u>, The Combustion Institute, Pittsburgh (1973), in press.

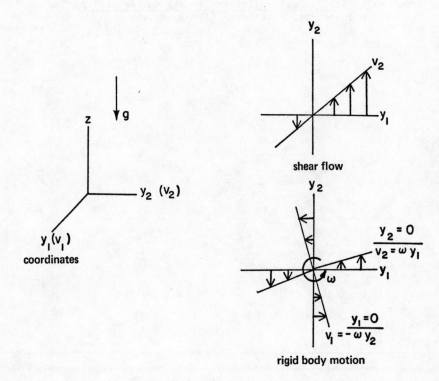

coordinates

shear flow

rigid body motion

FIGURE 1

NOTE: Figure captions for this chapter are listed on page 272.

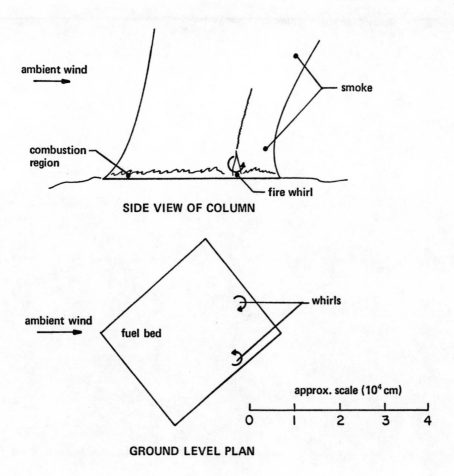

SIDE VIEW OF COLUMN

GROUND LEVEL PLAN

FIGURE 2

FIGURE CAPTIONS

Figure 6.1   Illustration of vorticity.

Figure 6.2   Observed fire whirl formation due to wind/column interaction.

PART IV

RADIATIVE HEAT TRANSFER ASSOCIATED WITH FIRE PROBLEMS

INTRODUCTION

The ignition, development and suppression of fires is obviously a complicated combination of heat transfer, mass transfer, chemical kinetics and fluid dynamics. Fire phenomena contain all the usual complications of industrial combustion problems with the important additional one that essentially no control can be exerted over the rate at which fuel and oxygen are fed to the system. This additional complication adds to our basic ignorance with respect to fire problems. The fire itself determines the rate of fuel consumption by generating its own heat transfer.

This feed back mechanism makes it difficult to describe even such a simple case as a small flame propagating down a single match splint. What is the relative importance of convective heat transfer through the gas phase and conductive heat transfer through the match splint itself? What are the chemical reactions involved in the destructive distillation of the match splint in the presence of oxygen? By what mechanism does the volatile material generated within the match splint reach the surface? What are the mechanism by which the volatile combustible material and oxygen mix in order to undergo chemical reaction? At the present time it is only possible to make semiquantitative assessments of each of these questions with more or less confidence depending on the particular situation and the nature of the material. In certain problems one or more of these mechanism can be neglected in comparison with those which are predominant. On the other hand in some cases the heat transfer aspect of the problem must be considered in conjunction with the mass transfer or fluid dynamics since the interaction is of primary importance. However, it is essential to understand the various processes individually before one can attempt to combine them intelligently for a particular problem of interest.

This manuscript is primarily concerned with the radiative transfer associated with fire problems. The first chapter introduces the basic concepts of radiation and the laws by which these are governed. In this section each concept is introduced in as general terms as possible to give a feel for the complexity of radiative transfer. Simplifying assumptions are introduced in a rapid manner so that methods of practical use can be derived without considering refinements not presently justified for the complicated geometries normally encountered in fire problems. Several examples are given for illustrative purposes.

The two remaining chapters consider two specific fire problems in some detail: fire spread through a fuel bed and ignition of cellulosic materials by irradiation. In each of these examples radiative transfer is an important mechanism of heat transfer. However, other heat transfer mechanisms as well as other physical and chemical processes which are believed to be important are also considered.

Chapter 1

BASIC PRINCIPLES OF RADIATIVE TRANSFER

F. R. STEWARD
Fire Science Centre, University of New Brunswick
Fredericton, N. B., Canada

1.1 THE NATURE OF RADIATIVE TRANSFER

One of the most important concepts of radiative transfer is that the heat is normally transferred over distances
which are large relative to the heat transferred by either
conduction and convection.

Radiation consists of electromagnetic energy in the
form of waves which are produced when a particle undergoes
a transformation from a high energy state to a lower one.
Conversely the wave is assimulated when a particle is
transformed from a lower energy state to a higher one.
When the transition of the particle to the high energy
state from which the wave originates is produced by the
molecular motion associated with temperature the radiation
is termed thermal. This is the type of radiation which
transfers heat between materials at different temperatures.

Electromagnetic waves pass unimpeded through a vacuum
and thus a transfer of heat by radiation does not require
the presence of mass as do conduction and convection
which depend on molecular or fluid motion. Since conduction and convection depend on the mass of the material itself to carry out the transfer of heat these phenomena are
local in nature. Conversely radiation only depends on the
local material to capture or absorb it. The source of the
radiation could be and normally is located at a distance
very much greater than either the molecular distances associated with conduction or the size of the turbulent fluctuation associated with convection. Only in the limiting
case where the material involved is highly absorbing can

NOTE: All figures quoted in text are at the end of the chapter.

radiative transfer be treated as a local phenomenon. In
most practical problems associated with fire the radiation
within the system must be treated as a whole. All parts
of the system are exchanging radiation with all other parts
of the system. In mathematical terms this yields integral
equations rather than the differential equations given by
conduction and convection. If radiation is considered together with conduction and/or convection the result is
integrodifferential equations. The solutions to these
type of equations present particular difficulties both
analytically and numerically.

1.2 BASIC LAWS GOVERNING RADIATIVE TRANSFER

The historical development of the basic laws of radiative transfer is a fascinating subject in itself. A number of eminent physists in the latter part of the nine-
teenth century spent a considerable amount of effort on
the theoretical aspects of electromagnetic waves. Since
during this period it was possible to investigate thermal
radiation experimentally in some detail many of the
theoretical developments in electromagnetic wave theory
were substantiated or rejected on the basis of thermal
radiation measurements. The development of these fundamental laws can be followed in a translation of Planck's
original treatise (13).

The basic relation describing radiative heat transfer
is the familiar Stefan-Boltzmann equation which states that
the maximum amount of radiation which can be emitted from
a surface per unit area is directly proportional to the
fourth power of its absolute temperature.

$$q = \sigma T^4 \tag{1-1}$$

where σ is a physical constant commonly known as the Stefan-
Boltzmann constant. The value of this constant in some
common units is

$$0.1713 \times 10^{-8} \; Btu/ft^2 hr.\,^\circ R^4$$
$$1.356 \times 10^{-12} \; cal/cm^2 \; sec.\,^\circ K^4$$
$$4.88 \times 10^{-8} \; kcal/m^2 hr.\,^\circ K^4$$
$$5.67 \times 10^{-12} \; watts/cm^2 \,^\circ K^4$$

A surface which has this maximum emissive power for a
given temperature is termed a black surface. No actual
material can achieve this maximum emission. However, a
cavity at thermal equilibrium contains black radiation at
the temperature of the cavity walls and will emit black
radiation through an infinitely small hole in its surface
to the surroundings.

Since a black surface emits the maximum amount of radiant energy at a given temperature it must be emitting
the maximum amount in any given direction. Therefore, the

emission per unit area of black surface "viewed" is constant and per unit area of normal black surface deminished by the cosine of the angle between the normal and the direction of emission. This concept can be visualized as a surface of certain brightness which maintains the same brightness when viewed from all angles, but the area viewed is reduced by the cosine of the angle between the normal and that of the point from which the surface is viewed.

This fourth power relation was first postulated by Stefan (1879) on the basis of experimental evidence. It was later derived by Boltzmann (1884) on the basis of thermodynamics using the radiation pressure deduced from Maxwell's electromagnetic wave equations.

The relation which gives the distribution of energy in the spectrum of a black body has been named Planck's Law (1900). This relation can be written in terms of wavelength or frequency as

$$W_{B\lambda} = \frac{2\pi hc^2 \lambda^{-5}}{e^{hc/k\lambda T} - 1} \tag{1-2}$$

$$W_{B\nu} = \frac{2\pi h\gamma^3 c^{-2}}{e^{h\gamma/kT} - 1} \tag{1-3}$$

where h is Planck's constant, 6.6256×10^{-27} erg-sec.
c is the velocity of light in vacuum, 2.9979×10^{10} cm/sec
k is Boltzmann's constant, 1.3805×10^{-16} ergs/°K
λ is the wavelength of the radiation, cm
γ is the frequency of the radiation, $\sec^{-1}$

The quantities of the left hand side of the above relations are the monochromatic emissive power or a black body at the wavelength λ or the frequency γ.

In order to derive this relationship it was necessary for Planck to postulate a model which assumed that radiant energy was emitted in discrete quantities, quanta, by a system of linear oscillators which undergo step changes between energy levels. The experimental verification of this spectral distribution was the first success of the quantum theory.

Considering the spectral distribution over wavelength division of both sides of equation (1-2) by the absolute temperature to the fifth power yields

$$\frac{W_{B\lambda}}{T^5} = \frac{2\pi hc^2 (\lambda T)^{-5}}{e^{hc/k\lambda T} - 1} = \frac{C_1 (\lambda T)^{-5}}{e^{C_2/\lambda T} - 1} \tag{1-4}$$

where $C_1 = 3.740 \times 10^{-5}$ erg cm^2/sec
$C_2 = 1.4388$ cm °K

This relation inidcates a unique relation between $W_{B\lambda}/T^5$ and λT and is presented in Fig. 1-1 The spectral distribution is zero only at wavelengths of zero and infinity and passes through a maximum at $\lambda T = 0.2898$ cm°K.

An alternative diagram of the spectral distribution is presented in Figure 1-2 where $W_{B\lambda}$ is plotted vs the wave length λ for various absolute temperatures. This diagram clearly illustrates two important aspects of the spectral distribution. The monochromatic intensity of black radiation increases at all wavelengths with an increase in temperature and the maximum in the spectral distribution shifts to shorter wavelengths as the temperature increases. The shift in the maximum of the distribution explains the change in color of a body from red to yellow to white as it increases in temperature.

In so far as fire problems are concerned the spectral distribution of black radiation is important due to the following considerations. The visible region of the spectrum is between 0.4 and 0.7 microns. A black flame with a temperature of $2000^\circ K$ would emit visible radiation in a region between λT 800 to 1400 microns - $^\circ K$ on the abscissa of Fig. 1-1. The black body emissive power in this region varies between approximately 1% to 20% of the spectral maximum and accounts for only about 0.5% of the total emission. The radiation from a fire which you see is insignificant compared to the major emission in the infrared region. This also means that the color of a surface which is a rough indication of its reflectivity in the visible region of the spectrum is not necessarily a good indication of its reflectivity of radiation emitted from a fire.

Equation (1-4) can be integrated over all wavelengths to give the total black body emissive power.

$$W_B = \int_0^\infty W_{B\lambda} d\lambda = T^4 \int_0^\infty \frac{C_1 (\lambda T)^{-5}}{e^{C_2/\lambda T} - 1} d(\lambda T) \qquad (1-5)$$

or

$$W_B = \sigma T^4 \qquad (1-6)$$

where

$$\sigma = \int_0^\infty \frac{C_1 x^{-5}}{(e^{C_2/x} - 1)} dx \qquad (1-7)$$

which is of course the Stefan-Boltzmann Law, Eq. (1-1).

Additional information on the basic laws governing radiation can be found in recent books published on radiative transfer (8, 16, 18, 25).

1.3 RADIATIVE PROPERTIES OF SURFACES

The Stefan-Boltzmann Law, Eq. (1-1), gives the maximum amount of radiation which a surface can emit. No real surface can achieve this limiting emission. The emissivity of a surface, ε, is defined as the ratio of the rate of emission of that surface to the rate of emission of a black body at the same temperature. In general the

emissivity of a surface will be a function of the direction of emission, wavelength and the temperature of the
surface. Thus

$$\varepsilon = \oint(\theta,\phi,\lambda,T) \qquad (1\text{-}8)$$

where ϕ and θ are the azimuth and planar angles required
to define the direction of the emission.

In many practical radiative calculations the total
hemispherical emissivity is used. This quantity is the
ratio of the rate of emission of a surface to that of a
black surface in all directions and all wavelengths. The
total radiant energy emitted by a surface at a temperature
T per unit area with a total emissivity ε is given by

$$q = \varepsilon \sigma T^4 \qquad (1\text{-}9)$$

It is more convenient to measure a quantity called
the total normal emissivity of a surface which is the
ratio of the rate of emission of a surface to that of a
black surface at the same temperature in a direction normal to the surface.

The total normal emissivities of a number of surfaces which are encountered in fire problems have been
tabulated in Table 1-1.

Radiation impinging on a surface may be absorbed,
reflected or transmitted. The absorptivity, reflectivity,
and transmissivity of a surface are defined as the fraction of radiation incident on the surface which are absorbed
reflected or transmitted respectively. The absorptivity
of a surface is in general a function of the direction
and wavelength of the incident radiation and the temperature of the surface. The reflectivity and transmissivity
of a material are in general a function of the same variables as the absorptivity with an additional variable being the direction at which the quantity is measured.

For practical radiative calculations it is often
convenient to use the total quantities, that is the fraction of the total radiation incident on the surface which
is absorbed, reflected or transmitted. By definition the
total quantities must satisfy the relation

$$\alpha + \rho + \tau = 1.0 \qquad (1\text{-}10)$$

According to Kirchhoff's Law (1859) the total absorptivity of a surface must equal the total emissivity of
the same surface if the surface is in thermal equilibrium
with its surroundings

$$\alpha = \varepsilon \; at \; thermal \; equilibrium \qquad (1\text{-}11)$$

The thermal equilibrium restriction is based on an argue-
ment which uses the second law of thermodynamics. However,
if the spectral distribution of the radiation incident on
the surface is similar to that leaving the surface the
above relation can be used as an approximation for cases

other than thermal equilibrium. Another way of stating
the same restriction is that the temperature of the source
of the radiation incident on the surface must not greatly
differ from the temperature of the receiving surface.

If this restriction is satisfied the values of the
emissivities presented in Table 1 may also be used as the
absorptivities of the same surfaces.

An important case where this restriction is not ful-
filled is solar radiation impinging on a surface at
terrestial temperatures. The absorptivity of a surface
for solar radiation may be quite different from the
emissivity of the surface at a relatively low temperature.
Solar radiation entering the atmosphere is well approximated by that of a black body at 6000°K.

TABLE 1- 1
Total Normal Emissivity of Various Surfaces

Asbestos	$T°F$	ε_N
board	100	0.96
cement	100	0.96
cloth	100	0.90
paper	100	0.93-0.95
slate	100	0.96

Bricks	$T°F$	ε_N
red, rough but no gross irregularities	70°	0.93
grog glazed	2012°	0.75
building	1832°	0.45
fireclay	1832°	0.75
red refractory	100°	0.88
silica	1800°	0.80-0.85
ordinary refractory	2000°	0.59
white refractory	2000°	0.29

Carbon	$T°F$	ε_N
candle soot	206-520	0.952
graphite, pressed and filed	480-950	0.98
lampblack, rough deposit	212-932	0.84-0.78
rough plate	212-608	0.77

Cloth (Black velvet)	100	0.97

Concrete (Rough)	100	0.94

Forest	70	0.90

Glass	$T°F$	ε_N

TABLE 1-1 (*Cont'd*)

polished	100°	0.95
quartz (2mm)	500°	0.92
pyrex	500°	0.94
smooth	100°	0.94
Grass (Dry)	70	0.90
Gypsum		
0.02 mm thick on smooth or blackened plate	70	0.903
Ice	$T°F$	ε_N
rough crystals	32	0.985
smooth	32	0.966
hoarfrost	0	0.99
METALS		
Aluminum		
polished	212	0.095
rough polish	212	0.18
commercial sheet	212	0.09
heavily oxidized	200-940	0.20-0.31
Copper		
polished	212-240	0.018-0.023
commercial scaped shiny	72	0.072
Gold		
pure, highly polished	440-1160	0.018-0.035
Iron and Steel		
steel, polished	212	0.066
iron, polished	800-1880	0.14-0.38
iron, freshly emeried	68	0.24
cast iron, newly turned	72	0.44
steel, groundsheet	1720-2010	0.55-0.61
iron, smooth, sheet	1650-1900	0.55-0.60
ironplate, completely rusted	67	0.69
iron oxidized	212	0.74
steel oxidized at 1100°F	390-1100	0.79
steel plate, rough oxidized	100-700	0.94-0.97
iron wrought dull oxidized	70-680	0.94
Lead		
pure (99.96%) unoxidized	260-440	0.057-0.075
oxidized at 300°F	390	0.63

TABLE 1-1 (*Cont'd*)

Nickel

electroplated, polished	74	0.045
plate, oxidized by heating at $1110^\circ F$	390–1110	0.37–0.48

Platinum

pure, polished plate	440–1160	0.054–0.104
strip	1700–2960	0.12–0.17
silver polished	100–1160	0.020–0.032

Stainless Steel

polished	212	0.074
type 316, A	450–1600	0.57–0.66
type 301, A	450–1740	0.57–0.55
type 347, A	450–1650	0.52–0.65

Tin

Commercial tin plated sheet iron	212	0.07–0.08

Zinc

commercial (99.1% pure) polished	440–620	0.045–0.053
galvanized sheet	212	0.21

Paints

	$T^\circ F$	ε_N
oil, all colors	212	0.92–0.96
lacquer, flatblack	100–200	0.96–0.98
snow-white enamel varnish on rough iron plate	73	0.906
black shiny lacquer sprayed on iron	76	0.875
black shiny shellac on tinned iron sheet	70	0.821
black matte shellac	70–295	0.91
aluminum (varying age and Al content)	212	0.27–0.67
10% Al, 22% lacquer body on rough or smooth surface	212	0.52
Al lacquer, varnish binder on rough plate	70	0.39
pigments all colors	125	0.74–0.99
national lead high heat black paint	300	0.925
white paint	190	0.880
black paint on aluminum substrate	77	0.91

TABLE 1-1 (*Cont'd*)

Paper

white	100	0.95-0.96
writing paper	100	0.98
any color	100	0.92-0.94
roofing	100	0.91
Plaster	100	0.91-0.92
Porcelain (Glazed)	100	0.92-0.93

Rubber

hard	68	0.92-0.94
soft, rough gray	100	0.86

Sand

dry	70	0.90
wet	70	0.95
Slate	100	0.67-0.80
Snow (New)	32	0.82
Water (Deep)	32-212	0.96

Woods

oak (planed)	70	0.90
beech	158	0.94
spruce (sanded)	100	0.82
walnut (sanded)	100	0.83
sawdust	100	0.75

The data presented in this table were selected from the following previous tabulations: (6, 8, 15, 16, 17, 18, 21, 22, 23).

According to Fig. 1.1 the maximum in the spectral distribution for solar radiation occurs at 0.48 micron. Since the maximum for 300°K radiation would occur at 9.65 micron a surface whose emissivity varied significantly with wavelength could have a very different absorptivity for solar radiation than its low temperature emissivity.

The reflectivity of a surface which does not transmit is according to Eq. (1-10) one minus the absorptivity. The reflectivity of a surface can be strongly directional dependent. The two limiting cases are a surface which reflects all radiation with the angle of reflectance equal to the angle of incidence, a spectral or mirror reflector, and a surface which reflects all radiation equally in all directions, a diffuse of lambert reflector. Real surfaces do not follow either of these special cases although some surfaces closely approach these limiting cases. Highly polished metal surfaces reflect primarily as a specular surface while rough surfaces of all types are nearly diffuse reflectors.

Additional information can be found on the radiative
properties of real surfaces in recent engineering texts
on radiative transfer (8, 16, 18, 25).

1.4 DIRECT RADIANT EXCHANGE BETWEEN SURFACES

The direct radiant exchange between surfaces is essentially a geometry problem. In general it is a function of
the radiative properties of the two surfaces exchanging
radiation discussed in the previous section. Since the
emissivity or absorptivity of a surface is a function of the
wavelength, the two angles required to define direction
of emission or incident impingement and the temperature of
the surface, the direct radiant exchange will also be a
function of these quantities. In addition to the surface
property variables the direct radiant exchange will also
be a function of the geometry of the two surfaces in relation to each other. In order to restrict the analysis
to geometry alone this section considers the direct radiant
exchange between <u>black</u> surfaces. The following section is
devoted to the <u>total</u> radiant exchange among surfaces and
indicates methods which can account for surface property
variations.

The <u>total</u> radiant energy leaving a black surface is
given by the Stefan-Boltzmann Law equation (1-1). This
radiation leaving any elemental area, dA_1, on a black surface in the direction of an elemental surface, dA_2, is
proportional to the cosine of the angle ϕ_1 between the
vector r connecting the two elemental areas and the vector
normal to dA_1, $\cos \phi_1$, and the solid angle subtended by
dA_2, $d\omega_2$ (see Fig. 1-3).

$$dQ_{1 \to 2} = I_1 dA_1 \cos\phi_1 d\omega_2 \qquad (1-12)$$

I_1 is a proportionally constant defined by equation (1-12)
usually called the intensity of radiation.

The solid angle subtended by dA_2 is given by

$$d\omega_2 = \frac{dA_2 \cos \phi_2}{r^2} \qquad (1-13)$$

where ϕ_2 is the angle between the vector connecting the
two surfaces r and the vector normal to dA_2. Thus

$$dQ_{1 \to 2} = \frac{I_1 dA_1 \cos\phi_1 dA_2 \cos \phi_2}{r^2} \qquad (1-14)$$

This quantity of radiation will impinge on the element
dA_2 and be absorbed if the space between the two elements
is vacuo and surface 2 is black. It is apparent that the
above consideration will yield the identical result if the
emitted radiation were followed from the element dA_2 instead of dA_1 except the proportionally constant would be
the intensity of radiation of surface 2, I_2. Therefore,

the net exchange of radiation between the two elements is given by

$$dQ_{1 \neq 2} = \frac{dA_1 \cos\phi_1 \, dA_2 \cos\phi_2}{r^2} (I_1 - I_2) \qquad (1-15)$$

The term outside the brackets in the above equation is only a function of the geometry of the surface elements. This must be the case since if the temperature of the two surfaces are identical and thus the emissive powers, the net rate of radiant energy exchange between the two elements must be zero in order to satisfy the second law of thermodynamics.

Since all the radiation leaving the element dA_1 must pass through a hemisphere of radius r with its centre located at dA_1 the integration of the element dA_2 over this hemisphere is equal to the black body emissive power of the element dA_1

$$W_{B1} dA_1 = I_1 \int_{A_2} \frac{\cos\phi_1 dA_1 dA_2}{r^2} \qquad (1-16)$$

From the geometry shown in Fig. 1-4.

$$dA_2 = r \sin\phi_1 d\theta \, r \, d\phi_1 \qquad (1-17)$$

Substitution and integration over the proper limits yield a relation between the intensity of radiation and the emissive power of a black surface.

$$W_{B1} = \pi I_1 \qquad (1-19)$$

The net exchange of radiant energy between two black surface elements separated by vacuo can be written in terms of the black body emissive powers of the elements using Eq. (1-15).

$$dQ_{1 \neq 2} = \frac{dA_1 \cos\phi_1 dA_2 \cos\phi_2}{\pi r^2} (W_{B1} - W_{B2}) \qquad (1-20)$$

The direct exchange of radiation between two finite black surfaces can be found by integrating the above expression over the limits of the two surfaces.

$$Q_{1 \neq 2} = \int_{A_1} \int_{A_2} \frac{dA_1 \cos\phi_1 dA_2 \cos\phi_2}{\pi r^2} (W_{B1} - W_{B2}) \qquad (1-21)$$

The quantity under the double integral in the above equation is hereafter termed the "direct-interchange area", $A_1 F_{12}$.

$$A_1 F_{12} = \int_{A_1} \int_{A_2} \frac{dA_1 \cos\phi_1 dA_2 \cos\phi_2}{\pi r^2} \qquad (1-22)$$

It is apparent that for this expression reciprocity holds.

$$A_1 F_{12} = A_2 F_{21} \qquad (1-23)$$

In most instances it is more convenient to tabulate the quantity, F_{12}, hereafter called the "view factor" for the numerous geometrical configurations of interest. This quantity is the fraction of radiation emitted by a black surface 1 which is intercepted and absorbed by a black surface 2.

The solution of Eq. (1-22) normally involves the evaluation of either a double or quadruple integral depending on whether the direct interchange area desired consists of one infinitesimal surface and one finite surface or two finite surfaces. The determination of the multiple integrals arising from this equation can be evaluated by all the normal integration techniques, analytical, numerical, and graphical. A good summary of the various methods can be found in reference (8).

Since fire problems present a special set of geometric configuration for which radiative heat transfer calculations are required a tabulation of View Factors Associated with Fire Problems is presented in Appendix I.

One set of view factors associated with fire problems which deserve special attention is that which arise when one dimension in the system goes to infinity. A particularly simple expression can be derived from first principles for this case and it is not necessary to solve the cumbersome integral equation (1-22).

Consider the two diagrams in Fig. (1-5) which are infinite in the dimension normal to the drawing. Let a string be stretched across each surface in Fig. 1.5a to represent an effective surface from which a straight line cannot be drawn which will intersect the same surface at another point. This yields three effective surfaces which cannot exchange radiation directly with themselves. Since all the radiation leaving any one of the surfaces must be intercepted by the other two.

$$A_1 = A_1 F_{12} + A_1 F_{13}$$
$$A_2 = A_2 F_{21} + A_2 F_{23} \qquad (1-24)$$
$$A_3 = A_3 F_{31} + A_3 F_{32}$$

The direct interchange area is identical between any two surfaces, of Eq. (1-23), , $A_1 F_{12} = A_2 F_{21}$, $A_1 F_{13} = A_3 F_{31}$, and $A_2 F_{23} = A_3 F_{32}$. Solution of the three equations for the three unknowns yields

$$A_1 F_{12} = \frac{A_1 + A_2 - A_3}{2} \qquad (1-25)$$

Now consider the general two dimensional enclosure given in Fig. 1-5b. The radiation emitted from the surface AB and intercepted by the surface DC is the total amount emitted from surface AB minus that intercepted by the surfaces BD and AC. The latter two interchange factors can be evaluated from the respective three sided enclosures ABD and ADC according to Eq. (1-25).

$$AB \rightarrow BC = \frac{AB + BC - AD}{2}$$

$$AB \rightarrow AC = \frac{AB + AC - BC}{2} \qquad (1-26)$$

Therefore

$$AB \rightarrow CD = AB - (AB \rightarrow BC) - (AB \rightarrow AC) \qquad (1-27)$$

$$AB \rightarrow CD = \frac{BC + AD - BD - AC}{2} \qquad (1-28)$$

The direct interchange area between any two surfaces in a two dimensional system can be found by the sum of the distances traced by strings stretching from the opposite ends of the surface minus the sum of the distances traced by strings stretching from adjacent ends of the surface divided by two. This important simplified method has often been termed the "crossed strings" method.

Many fire problems are well approximated by assuming one dimension is much larger than the other two. A system in which one dimension is ten times larger than either of the other dimensions falls into this catagory. An example which receives attention later in this text is a line fire. Direct interchange areas which are associated with line fires can use this simple method directly.

Another method which yields direct interchange areas in a relatively simple fashion is the use of algebraic combinations of those already tabulated to determine new ones which may not be available. Consider the surface arrangement given in Fig. (1-6). It is desired to determine the direct interchange area between the squares labelled 1 and 8. The direct interchange area for opposing parallel rectangles is given among those tabulated in Appendix I.

The radiation emitted by surface I and intercepted and absorbed by surface II can be broken down into parts. For instance, this quantity will be equal to four times the radiation leaving the surface 1 and intercepted and absorbed by surfaces 5, 6, 7 and 8. Therefore

$$A_1 F_{18} = \frac{A_I F_{I\ II}}{4} - (A_1 F_{15} + A_1 F_{16} + A_1 F_{17}) \qquad (1-29)$$

By similar reasoning

$$A_1 F_{16} = \frac{A_{1+2}\ F_{1+2,5+6}}{2} - A_1 F_{15} \qquad (1-30)$$

Since $A_1 F_{17}$ is identical to $A_1 F_{16}$ the direct interchange area desired is given in terms of only the direct interchange area between opposing parallel rectangles.

$$A_1F_{18} = \frac{A_1F_{1\ 11}}{4} - A_{1+2}F_{1+2,5+6} + A_1F_{15} \qquad (1\text{-}31)$$

Some consideration of Fig. (1-6) would allow this expression to be written directly.

The above procedure in combination with the view factors tabulated in Appendix I as well as other sources (9, 16, 19) will yield most of the view factors one can conceivably wish to use in fire problems.

1.5 TOTAL RADIANT INTERCHANGE AMONG SURFACES

The direct radiative interchange between two surfaces is independent of the nature and geometry of the other surfaces in the system. However, the total radiative interchange between two surfaces depends on the radiative characteristics and geometry of all the surfaces included within the defining system. Radiation emitted by one surface which is absorbed by a second surface will pass through numerous paths which may include one or more reflections. It is necessary to account for all possible paths of transfer to determine the total radiant interchange between two surfaces within an enclosure.

It is apparent that if it is possible to determine the radiant energy which leaves one surface and is absorbed by a second surface the solution is general for any two surfaces in the enclosure. Therefore, assign surface i a unit black body emissive power, $W_{Bi} = 1.0$, and all other surfaces which absorb radiation an emissive power of zero, $W_{Bk} = 0$, k = 2 to n. Any adiabatic surfaces in the system will attain their equilibrium emissive power, W_{Bn}, such that the surface reemits all the radiation which it absorbs.

Let the flux density incident on an element of surface j streaming from all surfaces in the system be $_iH_j$ where the first subscript indicates that the radiation was originally emitted by surface i. The flux density leaving an element on surface j is defined as $_iR_j$ which can be related to the incident radiation, $_iH_j$, through the surface reflectivity.

$$_iR_j = \rho_j \, _iH_j \qquad (1\text{-}32)$$

Thus the total amount of radiant energy emitted by surface i which is absorbed by surface j defined as $A_i\mathcal{F}_{ij}$ and often termed the total interchange area will be

$$A_i\mathcal{F}_{ij} = \alpha_j A_j \, _iH_j \qquad (1\text{-}33)$$

where α_j is the absorptivity of the surface j.

In general each of the above quantities will be a function of wavelength since the original emission and subsequent reflections are dependent on the properties of the various surfaces in the system. Since the original emissionn is from surface i the following equation can be written for each wavelength.

$$A_i{}_iH_j = \frac{A_i{}_iR_j}{\rho_j} = \int_{A_j}\left[\int_{A_i}\varepsilon_i\frac{d^2(A_iF_{ij})dA_i}{dA_idA_j}\right.$$

$$\left.+\sum_{k=1}^{n}\int_{A_k}{}_iR_k\frac{d^2(A_kF_{kj})dA_k}{dA_kdA_j}\right]dA_j$$

$$(1\text{-}34)$$

where the quantities $d^2(A_iF_{ij})$ and $d^2(A_kF_{kj})$ are the
interchange areas between the differential elements of
area designated by the subscripts.

This equation applies to each of the source or sink
surfaces in the system.

An exception must be made here for the adiabatic
surfaces for which

$$_iH_j = {}_iR_j \quad \textit{adiabatic surfaces only} \qquad (1\text{-}35)$$

All the radiation received by an adiabatic surface
must be reflected or reemitted. In the latter case a
redistribution of the radiation over the spectrum depending on the emissive properties of the surface would also
occur. Thus Eq. (1-34) applies for adiabatic surfaces with the restriction, Eq. (1-35).

The first term in the brackets of Eq. (1-34)
is the original emission from surface i and the second
term accounts for the radiation reflected or reemitted
by all the surfaces in the system including both i and j.
The differential view factors in general would not be the
view factors between black surfaces since the emission
and reflection of a general surface would not follow
Lambert's law. Also the quantities $_iR_k$, would be a function of position of the surface k and the angle between
the vector connecting the elemental areas and the normal
to the surface.

In order to solve the above integral equations the
following three simplifying assumptions are often made.

1. The radiative properties of the surfaces
    are independent of wavelength.

2. All surfaces in the system reflect according
    to Lambert's Law i.e. diffusely.

3. The radiation emitted by surface i is diffuse.

These three assumptions introduce the concept of
what is normally termed a gray surface. A gray surface
has the emissive characteristics of a black surface except that the emission is reduced by a constant fraction
at all wavelengths and angles. This reduces the complexity of the relation considerably and a number of
solutions exist for systems with relatively simple geometry.
Typical examples of these types of solutions can be found
in references (16) and (18).

For many practical engineering or fire problems involving radiative transfer an analytical or even numerical solution of the set of Eqs.(1-34) is not practical because of the complex geometry of the system. An alternative procedure of analysis has been developed by Hottel (6) and (8), often called the "zone" method.

The zone method is identical in principle to the rigorous derivation which yields the set of Eqs. (1-34). However, in addition to the three simplifying assumptions listed above the zone method adds a fourth.

4.   The radiation leaving each of the surfaces in the system $_iR_k$ is uniform over the entire surface k.

This simplification allows the use of the direct interchange areas discussed in the previous section in place of the area integrals in the set of Eqs. (1-34).

$$A_j \ _iH_j = A_j \ \frac{_iR_j}{\rho_j} = \epsilon_i A_i F_{ij} + \sum_{k=1}^{n} A_k F_{kj} \ _iR_k \qquad (1\text{-}36)$$

This relation is a set of simultaneous equations in which the unknowns, $_iR_j$, are equal to the number of zones in the system. Since an equation can be written on each zone the number of equations equals the number of unknowns.

The solution to this set of equations is standard and can be obtained by Crammers Rule (5).

$$A_j \ _iR_j = \frac{-\epsilon_i A_i A_j \ (-1)^{i+j} D_{ij}}{\rho_i \qquad D} \qquad (1\text{-}37)$$

where D is the determinant

$$
\begin{vmatrix}
A_1 F_{11} - \dfrac{A_1}{\rho_1} & A_2 F_{21} & A_3 F_{31} & \cdots A_R F_{R1} & A_S F_{S1} & \cdots \\[2ex]
A_1 F_{12} & A_2 F_{22} - \dfrac{A_2}{\rho_2} & A_3 F_{32} & \cdots A_R F_{R2} & A_S F_{S2} & \cdots \\[2ex]
A_1 F_{13} & A_2 F_{23} & A_3 F_{33} - \dfrac{A_3}{\rho_3} \cdots A_R F_{R3} & A_S F_{S3} \\[2ex]
\vdots & \vdots & \vdots \qquad \vdots & \vdots \\[1ex]
A_1 F_{1R} & A_2 F_{2R} & A_3 F_{3R} & A_R F_{RR} - A_R & A_S F_{SR} & \cdots \\[1ex]
A_1 F_{1S} & A_2 F_{2S} & A_3 F_{3S} & A_R F_{RS} & A_S F_{SS} - A_S & \cdots \\
\vdots & \vdots & \vdots & \vdots & \vdots
\end{vmatrix}
$$

and D_{ij} is the minor determinant of D found by crossing out the ith row and the jth column or the jth row and the ith column, D being symmetrical. The numbered surfaces in the determinant refer to those which are absorbers or emitters of radiation while the lettered surfaces refer to those which are adiabatic and thus follow Eq. (1-35).

The total interchange area between any two surfaces defined by Eq. (-33) will be given by

$$A_{i\,ij} = \frac{-\varepsilon_i A_i \varepsilon_j A_j (-1)^{i+j} D_{ij}}{\rho_i \rho_j \quad D} \qquad (1-39)$$

The total interchange factor of a surface for itself is given by

$$A_{i\,ii} = \frac{-\varepsilon_i A_i}{\rho_i} \left| \varepsilon_i \quad + \frac{\varepsilon_i A_i D_{ii}}{\rho_i D} \right| \qquad (1-40)$$

This total interchange area can be used in combination with the Stefan-Boltzmann equation to yield the net radiative heat flux between two surfaces in an enclosure.

$$Q_{i\neq j} = A_i \overline{\mathcal{F}}_{ij} \; \sigma (T_i^4 - T_j^4) \qquad (1-41)$$

Further details of this procedure can be found in the original development (8) as well as subsequent presentations (8, 16, 18, 25).

An example as to how this method can be used on a simple fire problem is given in Appendix III.

1.6 RADIATIVE PROPERTIES OF GASES AND SOLID SUSPENSIONS

Radiative heat transfer associated with fire problems obviously involves radiation in an absorbing and emitting medium. The combustion of most fuels consumed in fires will produce carbon dioxide, water vapor, carbon monoxide, and particulate matter all of which emit radiation at the temperatures generated in ordinary flames.

The essential difference in radiative transfer among surfaces and among absorbing and emitting gas systems in so far as the mathematical analysis is concerned is the emission in depth and absorption by attenuation. The decrease in intensity of a beam of collimated monochromatic radiation on passing through a differential layer of attenuating material is proportional to the intensity of the beam.

$$-dI = kIdx \qquad (1-42)$$

The constant k is the total extinction coefficient and includes attenuation both by absorption and scattering. If the attenuation occurs over a finite length x and the intensity of the incident radiation is I_0 the solution to the above differential equation yields the exponential law associated with numerous types of processes including radiative transfer

$$I = I_0 e^{-\int_0^L kd\ell} \qquad (1-43)$$

If k can be assumed independent of path length

$$I = I_0 e^{-kL} \qquad (1-44)$$

A particularly important example associated with fire problems is the absorption of a collimated beam of radiation passing through a volume of uniformly suspended black particles which have dimensions large compared to the wavelength of the radiation. The absorption coefficient in this instance is directly proportional to the number of particles per unit volume and the mean projected area of a single particle. It will also be independent of the wavelength of the incident radiation.

$$k = a_p c \qquad (1-45)$$

The mean projected area of a particle which has only convex external surface and is randomly oriented can be obtained from the following consideration.

The area projected toward a beam of incident radiation which makes an angle ϕ with an element of area, dA_1 on the surface of the particle is given by $dA_1 \cos \phi$. The mean projected area is obtained by weighting the projected area of the element $dA_1 \cos \phi$ by the solid angle about the normal $\sin \phi d\phi d\theta$ where ϕ and θ are the normal polar and azimuth angle in spherical coordinates. Integration of the weighted quantity should be carried out over the hemisphere where incidence occurs while the total orientation involves the entire sphere since the incident beam can strike only one half the particle at any time.

$$dA_{mean} = \frac{\int_0^{\pi/2} \int_0^{2\pi} dA_1 \cos \phi \, \sin \phi d\phi d\theta}{\int_0^{\pi} \int_0^{2\pi} dA_1 \sin \phi \, d\phi d\theta} = \frac{dA_1}{4} \qquad (1-46)$$

The mean projected area is thus given by one fourth of the surface area of the particles.

The attenuation law for a suspension of black particles with dimensions large compared to the wavelength of the radiation is therefore given by

$$I = I_0 e^{-\frac{a' c L}{4}} \qquad (1-47)$$

This expression is also valid for the attenuation of radiation passing through a bed of uniform randomly oriented fuel particles such as one might encounter in a bed of pine needles, an area of brush or foliage or other similar situations.

For gas radiation it is often desirable to relate the total extinction coefficient directly to the partial pressure or concentration of the absorbing gas

$$k = k_p p = k_c c \qquad (1-48)$$

which yields an attenuation law

$$I = I_0 e^{-k_p p L} = I_0 e^{-k_c c L} \qquad (1-49)$$

This makes the absorption coefficient nearly independent of the concentration of the absorbing material for some commonly encountered gases.

The emission and absorption of radiant energy by matter in the gas phase occurs by a process of energy transitions between the various energy levels of the gas molecules. According to quantum theory the change in energy level of a molecule is related to the frequency of the emission or absorption by the relation postulated by Planck.

$$\Delta E = h\gamma \qquad\qquad (1-50)$$

where ΔE is the change in energy level
  h is Planck's constant
  γ is frequency of radiation

Molecules undergo three types of transitions which result in emission or absorption of radiation. Electronic transitions which yield frequencies in the ultraviolet and visible, simultaneous vibrational and rotational transitions which yield frequencies in the near infrared and rotational transitions only which yield frequencies in the far infrared. The emission and absorption of thermal radiation is primarily concerned with simultaneous vibrational and rotational transitions.

Each particular transition between two energy levels produces an absorption line at a characteristic frequency. The absorption coefficient associated with the line varies over a narrow range of frequency with a maximum at the characteristic frequency. The variation of the absorption coefficient around the line is called broadening and is produced by the natural uncertainty in the energy levels, the interference of molecular collisions with emission and absorption and the Doppler effect. Therefore the emission spectra of a typical radiating gas is a large number of broadened lines grouped together in narrow wavelength regions called bands. In so far as the process of thermal radiation is concerned these bands are primarily in the infrared. Additional information on the theoretical details of the emissions and absorption of gases can be found in references (4) and (11).

For practical fire problems the total emissivity and absorptivity of the radiating gas volume is the quantity required for heat transfer calculations. Measurements have been made on a number of common gases which emit and absorb radiation at typical flame temperatures. The three gases of primary importance in fire problems are carbon dioxide, water vapor, and carbon monoxide. Diagrams for these gases and the procedures for using them are presented in Appendix IV.

In general the emissivity of a volume of a particular radiating gas depends on the temperature of the gas, the optical path and the total pressure. The optical path in engineering calculations is usually based on the partial pressure of the radiating gas, while meteorologists and physicists normally use concentrations. In any case it is important to remember that unlike the radiative properties of a surface the radiative properties of a gas depend on its optical thickness and therefore on the geometry of the gas volume and the concentration or partial pressure of the gas. The effect of total pressure is normally of secondary importance and is usually accounted for by a correction factor to the value at one atmosphere.

Particle radiation may also contribute substantially to the radiant emission from a fire. The particles may be carbon or very heavy hydrocarbons produced by burning with incomplete mixing between the fuel and air or much longer sparks and embers which are physically swept up into the flame. The rigorous analysis of the interaction of a particle with incident radiation involves the solution of Maxwell's wave equations. The parameters associated with these equations are the complex refractive index of the material, the ratio of a characteristic particle dimension to the wavelength of the radiation and the particle shape. The solution to these equations have been obtained for a few simple shapes, spheres, infinite cylinders, spherical shells and infinite cylindrical shells. However, the solutions are complex and difficult to evaluate. An additional problem is the availability of the complex refractive index for the particles formed in the flame. Although this quantity has been determined for a number of pure materials (1, 14, 24) its value for particles which are generated or swept into the fires can only be estimated. Additional information can be found on this complex problem for the case where general solutions have been obtained (Mie equations) in reference (24).

The type of particles which are likely to be generated in fires, particularly those swept into the flames physically, are large compared to the wave length of the radiation and relatively black. Assuming this is the case the particles can be treated as blackbody emitters or absorbers in the same manner as black surfaces. The absorption coefficient for this type of particles is equal to the mean projected area of all the particles per unit volume. The transmittance of such a volume of suspended particles is given by Eq. (1-47). Therefore the emissivity of a volume of uniformly suspended black particles is given by

$$\varepsilon_\rho = 1 - e^{-ca'L/4} \qquad (1-51)$$

where c is the concentration of particles per unit volume
a' is the surface area of the particle
L path length

Other semiempirical relations for luminous industrial furnace flames which may apply to the fire situation can be found in reference (8).

1.7 DIRECT RADIANT EXCHANGE IN SYSTEMS WITHIN AN ABSORBING AND EMITTING MEDIUM

The direct exchange of radiation in a system which contains an absorbing and emitting medium will involve three types of direct interchange areas; surface to surface, surface to volume and volume to volume. The geometry of elements for these three situations can be seen in Fig. 1-7.

First consider the direct exchange of monochromatic radiant energy between an element of volume dV_1, in which the absorption coefficient of the gas is k and an element of black surface dA_2, Fig. (1-7a). The emission of the elemental gas volume in all directions will be proportional to its volume, $C'dV_1$, with the fraction headed toward dA_2 proportional to the solid angle subtended by the surface element, $d\omega_2$. The fraction which reaches the surface to be absorbed is proportional to the transmissivity of the path of travel $\tau(r)$.

$$\overline{dg_1 s_2} = C'dV_1 d\omega_2 \, \tau(r) \qquad (1-52)$$

The solid angle subtended by the surface element dA_2 is given by $\dfrac{dA_2 \cos\phi_2}{4\pi r^2}$

$$\overline{dg_1 s_2} = \frac{C'dV_1 \cos\phi_2 dA_2 \tau(r)}{4\pi r^2} \qquad (1-53)$$

The monochomatic transmissivity of an absorbing gas is given by Eq. (1-43).

$$\tau(r) = e^{-\int_0^r k dr} \qquad (1-54)$$

If the absorption coefficient is uniform over the path traversed

$$\tau(r) = e^{-kr} \qquad (1-55)$$

The proportionality constant C' can be obtained from the following consideration. The surface element dA_1 is located at the center of a hemisphere of radius r of a gas emitting monochromatic radiation uniformly according to an absorption coefficient k, Fig. (1-8). Let the monochromatic emission of the gas volume to the surface element be ε_g. An increase in the emission of the hemisphere to the surface element by adding a differential amount to the radius of the hemisphere would be given by $(d\varepsilon_g/dr) dr$ which is also the emission of a hemispherical shell of gas of thickness dr at a uniform radius r to the same surface element through the absorbing gas. The direct exchange of a volume element dA_2 dr within this hemispherical shell and the area element dA_1 through the absorbing gas is given by

$$\overline{dg_1 s_2} = \frac{\cos\phi \, dA_1 d\varepsilon_g/dr \; dr dA_2}{\pi r^2} \qquad (1-56)$$

This quantity is identical to that given by Eq. (1-53). The absorbing gas along the path traversed by the radiation has a constant absorption coefficient, k, so the equality yields

$$de_g/dr = C'/4 \; e^{-kr} \qquad (1-57)$$

The emission of a hemisphere of radius r of a uniformly emitting gas with a monochomatic absorption coefficient k to the center of its base must be

$$\varepsilon_g = 1 - e^{-kr} \qquad (1-58)$$

and

$$de_g/dr = -ke^{-kr} \qquad (1-59)$$

The proportionallity constant C' is given by

$$C' = 4k \qquad (1-60)$$

The direct exchange of monochomatic radiant energy between a volume element of emitting gas, dV_1 and a surface element dA_2 is given as

$$\overline{dg_1s_2} = \frac{kdV_1 \; cos\phi_2 dA_2 \tau(r)}{\pi r^2} \qquad (1-61)$$

Second consider the direct exchange of monochromatic radiant energy between two elemental volumes of emitting and absorbing gas (Fig. 1-7b). As shown above the emission of the elemental volume, dV_1, in all directions is equal to $4kdV_1$ with the fraction headed toward dV_2 proportional to the solid angle $d\omega_2$. The transmissivity along the path r is given as $\tau(r)$ with the absorption by the element itself proportional to the coefficient, $k_2 dr$.

$$\overline{dg_1g_2} = 4k_1 dV_1 d\omega_2 \tau(r)k_2 dr \qquad (1-62)$$

The solid angle subtended by the elemental area of the second volume facing the first volume is $\dfrac{dA_2}{4\pi r^2}$.

$$\overline{dg_1g_2} = \frac{4k_1 dV_1 k_2 dA_2 dr\tau(r)}{4\pi r^2} \qquad (1-63)$$

Since $dV_2 = drdA_2$ the direct exchange of monochromatic radiation between two elemental volumes of absorbing and emitting gas volumes is given by

$$\overline{dg_1g_2} = \frac{k^2 \; dV_1 \; dV_2 \; \tau(r)}{\pi r^2} \qquad (1-64)$$

The direct exchange of monochromatic radiation between two surface elements through an absorbing medium is given by the same geometry as that for black surface elements equation (1-20), multiplied by the appropriate transmissivity $\tau(r)$

$$\overline{d\delta_1\delta_2} = \frac{dA_1\cos\phi_1 \; dA_2\cos\phi_2\tau(r)}{\pi r^2} \qquad (1-65)$$

The three Eqs. (1-61), (1-64) and (1-65) represent the defining equations for evaluating the direct interchange areas of monochromatic radiation in an absorbing and emitting gaseous medium.

As indicated in the previous section the absorption coefficient of a radiating gas is a strong function of wavelength. However, it is often useful to neglect this wide variation and use a mean value. A gas with an absorption coefficient independent of wavelength has been often termed a "gray" gas. For such a material the three direct exchange equations derived above represent the direct exchange for all radiant energy.

It has been demonstrated (6, 8, 11) that the radiative properties of carbon dioxide and water vapor can be well represented by a summation of gray gases with absorption coefficients and spectral weighting factors determined from either spectral line and band theory (11) or empirical curve fitting (8). The latter is preferable for most practical radiative calculations. Therefore, an analysis of radiant exchange based on the gray gas assumption can be used as a basis for more complex "real" gas analysis.

The direct interchange areas between finite surfaces and volumes are obtained by integrating Eqs (1-61), (1-64) and (1-65) over the appropriate limits.

$$\overline{g_1\delta_2} = \int_{V_1}\int_{A_2} \frac{k \; \cos\phi_2 (r) dA_2 dV_1}{\pi r^2} \qquad (1-66)$$

$$\overline{g_1{}_2} = \int_{V_1}\int_{V_2} \frac{k^2\tau(r) \; dV_2 \; dV_1}{\pi r^2} \qquad (1-67)$$

$$\overline{\delta_1\delta_2} = \int_{A_1}\int_{A_2} \frac{\cos\phi_1\cos\phi_2\tau(r)dA_2dA_1}{\pi r^2} \qquad (1-68)$$

where $\tau(r)$ for a uniform concentration of gray gas given

$$\tau = e^{-kr} \qquad (1-69)$$

For the case of a gray gas in which the absorption coefficient of the radiating gas is independent of wave length the net direct interchange of radiant energy between two finite surfaces and/or gas volumes is given by the above three relations when multiplied by the difference in the black body emissive power of the two regions.

$$Q_{1\rightleftharpoons 2} = \overline{g_1\delta_2} \; (W_{B1} - W_{B2})$$

$$\frac{or}{\overline{g_1 g_2}} \qquad\qquad (1\text{-}70)$$

$$\frac{or}{\overline{s_1 s_2}}$$

The relations (1-66), (1-67) and (1-68) involve fourth, fifth and sixth order integrals with a combination of algebraic and exponential terms. These types of integrals cannot be solved analytically and numerical methods must be used. In most instances it is useful to make one or more changes of variables in the process of evaluation so that the multiple integrals can be reduced to a series of double or single integrals as recommended in references (3) and (10).

A number of direct interchange areas for gaseous systems obtained for geometric configurations useful in fire problems have been tabulated in Appendix II. There are obviously a significant number of direct interchange areas for other useful geometries not presently available.

In order to reduce the amount of information in figure or tabular form which it is necessary to have available for the direct interchange area calculations a quantity known as the "mean beam length" has been introduced (6, 16, 18, 20). The mean beam length is defined as the length required to yield the emissivity of a gray gas volume, which when multiplied by the direct interchange area of a black surface covering the gas volume and the other surface or volume exchanging radiation, yields the desired direct interchange area.

$$\overline{g_1 s_2} = (1 - e^{-k L_m}) \; \overline{s_1 s_2} \qquad\qquad (1\text{-}71)$$

where L_m is the mean beam length and $s_1 s_2$ is the black body direct interchange area between a surface covering the gray gas volume and the surface in question.

In general the mean beam length is a function of the geometry of the gas and surface exchanging radiation and the optical thickness of the gas based on a characteristic dimension of the system. However, in many geometric configurations it is only a weak function of the optical thickness so that a single value is often sufficient. If the surface exchanging radiation is adjacent to the gas volume the black body interchange factor becomes unity and the mean beam length can be visualized as a hemisphere of radius L_m which exchanges the identical amount of radiation with the black surface as the more complex geometry. A number of mean beam lengths are given in Appendix II along with the direct interchange areas.

1.8 TOTAL RADIANT EXCHANGE IN SYSTEMS WITH AN ABSORBING AND EMITTING MEDIUM

The <u>direct</u> radiative interchange between two finite surfaces, volumes, or a surface and a volume is independent of the geometry of the other areas and volumes in the system except for those which lie on a direct path between the exchanging bodies. However, the <u>total</u> radiative interchange between any two surfaces, volumes, or a surface and a volume depends on the radiative characteristics and geometry of the entire enclosure defining the system. Radiant energy emitted by a surface or volume which is absorbed by a second surface or volume will pass through numerous paths which may include one or more reflections. If scattering of radiation is included, the paths of possible transfer would include the additional paths made possible by this phenomenon. This complexity is neglected in the following discussion although it may be important in some fire problems.

As indicated in section 1.5 when the total radiant interchange among surfaces was under consideration it was necessary to determine only the radiant energy which leaves one surface and is absorbed by a second surface since the solution is general for any two surfaces. An identical argument holds for the total radiative exchange involving volume and surface elements.

Assign volume i a unit black body emissive power $W_{Bi}=1.0$ and all other volumes or surfaces in the system which absorb radiation an emissive power of zero, $W_{Bk}=0$, k = 2 to n. As before any adiabatic surface in the system will attain its equilibrium emissive power, W_{Bn}, such that the surface reemits all radiation which it absorbs.

The incident flux density on an element of surface j streaming from all volumes and surfaces in the system will again be $_iH_j$, where the first subscript indicates that the radiation was originally emitted by to volume i. The flux density leaving an element on surface j is defined as through the surface reflectivity, ρ_j

$$_i R_j = \rho_j \, _i H_j \qquad (1-72)$$

Each of the above quantities is in general a function of wavelength since the original emission and subsequent reflections are dependent on the emitting properties of the gas volume and the properties of the various surfaces in the system. Therefore, the amount of radiation emitted by volume i and absorbed by volume j at each wavelength is given by

$$\overline{G_i G_j} = \int_{V_j} \left[\int_{V_i} \frac{d^2 (\overline{g_i g_j})}{dV_i dV_j} \right] dV_i$$

$$+ \sum_{k=1}^{n} \int_{A_k} {}_{i}R_k \; \frac{d^2(\overline{s_k g_j})}{dA_k dV_j} \; dA_k \Bigg] \; dV_j \qquad (1\text{-}73)$$

The quantity $d^2(\overline{g_i g_j})$ is the interchange area between the differential elements of volumes designated by the subscripts and the quantity $d^2(\overline{s_k g_j})$ is the interchange area between a differential surface element or surface k and a volume element within volume j. The first term represents the original emission from the volume i while the second term accounts for radiation reflected or reemitted by all the surfaces in the system.

The distribution of radiation leaving the various surface elements in the system can be determined from the relation

$$\frac{A_j {}_{i}R_j}{\rho_j} = \int_{A_j} \Bigg[\int_{V_i} \frac{d^2 \overline{g_i s_j}}{dV_i dA_j} dV_i + \sum_{k=1}^{n} \int_{A_k} {}_{i}R_j \frac{d^2 \overline{s_k s_j}}{dA_k dA_j} \; dA_k \Bigg] \; dA_j$$
$$(1\text{-}74)$$

Once again an exception must be made for the adiabatic surfaces for which

$$_iH_j = \, _iR_j \quad \text{adiabatic surfaces} \qquad (1\text{-}75)$$

It should also be remembered that the emission of the radiation from the adiabatic surfaces would be redistributed over the spectrum according to the emissive properties of the particular surface.

Equation (1-74) with the modification for adiabatic surfaces (1-75) applies to each of the n surfaces in the system. The differential view factors $d^2 \overline{s_k s_j}$ depends on the reflective properties of the various surfaces and the absorption coefficient of the gas over the path traversed by the radiation. The radiant energy leaving the surface k, $_iR_k$ is a function of the position on the surface k and the angle between the vector connecting the elemental surface and the elemental volume and the normal to the surface.

For many problems of engineering interest and undoubtedly also for a number of fire problems the following simplifying assumptions are made.

1. The radiative properties of the surfaces and the emitting and absorbing gas is independent of wavelength.

2. All surfaces reflect according to Lambert's Law i.e. diffusely.

3. The radiation leaving each of the n surfaces in the system, $_iR_n$, is uniform over the entire surface.

Assumption (2) and (3) are for the purpose of allowing a zone method of analysis and have been discussed previously, Section 1.5, and will not be further elaborated upon here. Assumption (1) reduces the system to a gray gas and gray surfaces. As indicated in section 1.6 emitting gases do not normally have absorption coefficients independent of wavelength. However, as indicated in the same section it is possible to treat real gases as a summation of several gray gases for many engineering calculations. In any case, the assumption of wavelength invarient properties is extremely attractive for analytical purposes.

The introduction of the above three assumptions reduces the system of integral equations to a set of simultaneous algebraic relations.

$$\overline{G_i G_j} = \overline{g_i g_j} + \sum_{k=1}^{n} \overline{s_k g_j} \, {}_i R_k \qquad (1\text{-}76)$$

$$\frac{A_j \, {}_i R_j}{\rho_j} = \overline{g_i s_j} + \sum_{k=1}^{n} \overline{s_k s_j} \, {}_i R_k \qquad (1\text{-}77)$$

The quantity $\overline{G_i G_j}$ is the total amount of radiation which is emitted by the volume zone i and is absorbed by the volume zone j. The quantities $g_i g_j$ $g_i s_k$, and $s_k s_j$ are the direct interchange areas defined by Eqs. (1-66), (1-67) and (1-68). The total amount of radiation emitted by the volume zone i and absorbed by the surface zone j, defined as $\overline{G_i S_j}$, is given by

$$\overline{G_i S_j} = A_j \, {}_i R_j \qquad (1\text{-}78)$$

The solution to the set of simultaneous algebraic relations is standard and can be obtained by Crammer's Rule (9).

$$_i R_j = \frac{D_j}{D} \qquad (1\text{-}79)$$

where D is the determinate

$$
\begin{vmatrix}
\overline{s_1 s_1} - \dfrac{A_1}{\rho_1} & \overline{s_2 s_1} & \overline{s_3 s_1} & \cdots \overline{s_R s_1} & \overline{s_S s_1} & \cdots \\[2ex]
\overline{s_1 s_2} & \overline{s_2 s_2} - \dfrac{A_2}{\rho_2} & \overline{s_3 s_2} & \cdots \overline{s_R s_2} & \overline{s_S s_2} & \cdots \\[2ex]
\overline{s_1 s_3} & \overline{s_2 s_3} & \overline{s_3 s_3} - \dfrac{A_3}{\rho_3} \cdots \overline{s_R s_3} & \overline{s_S s_3} & \cdots \\[2ex]
\vdots & \vdots & \vdots & \vdots & \vdots & \\[1ex]
\overline{s_1 s_R} & \overline{s_2 s_R} & \overline{s_3 s_R} & \overline{s_R s_R} - A_R & \overline{s_S s_R} & \cdots \\[1ex]
\overline{s_1 s_S} & \overline{s_2 s_S} & \overline{s_3 s_S} & \overline{s_R s_S} & \overline{s_S s_S} - A_S & \cdots
\end{vmatrix}
$$

$$(1\text{-}80)$$

and D_j is the determinate form by replacing the jth column with the column

$$\begin{array}{c} -\overline{g_1 \delta_1} \\ -\overline{g_1 \delta_2} \\ -\overline{g_1 \delta_3} \\ \vdots \\ -\overline{g_1 \delta_R} \\ -\overline{g_1 \delta_S} \\ \vdots \end{array} \qquad (1-81)$$

The surfaces with numbered subscripts refer to surfaces which absorb radiation while letter subscripts refer to adiabatic surfaces.

The total gas volume zone to gas volume zone interchange area is given by

$$\overline{G_i G_j} = \overline{g_i g_j} + \sum_{k=1}^{n} \overline{\delta_k g_j} \; \frac{D_k}{D} \qquad (1-82)$$

The total surface to surface interchange area in the presence of absorbing gas is given by

$$\overline{S_i S_j} = -\frac{A_i \varepsilon_i}{\rho_i} \; \frac{A_j \varepsilon_j}{\rho_j} \; (-1)^{i+j} \frac{D'_{ij}}{D} \qquad (1-83)$$

This expression is identical to that for total surface to surface interchange, Eq. (1-39) except the direct interchange areas included in the determinate (1-80) include absorption along the path of transmission according to Eq. (1-68),

The total interchange areas can be used in combination with the Stefan-Boltzmann equation to yield the net radiative heat flow between any two surfaces, two volumes, or a surface and a volume in the enclosure.

$$Q_{i \rightleftarrows j} = \overline{G_i G_j} \qquad \sigma (T_i^4 - T_j^4) \qquad (1-84)$$
$$\text{or}$$
$$\overline{G_i S_j}$$
$$\text{or}$$
$$\overline{S_i S_j}$$

Further details of this method can be found in the original development (6) as well as later presentations (7) and (8).

An example as to how this method can be used on a simple fire problem is given in Appendix V.

REFERENCES

1. Beer, J.M. and Howarth, C.R., "Radiation from Flames
    in Furnaces", Twelfth Symposium (International)
    on Combustion, p. 1205, The Combustion Institute,
    (1968).

2. Bramson, M.A., "Infrared Radiation", Translation by
    Rodman, R.B., Plenum Press, New York (1968).

3. Fang, J.B. and Steward, F.R., "A Radiative Inter-
    change Factor for Fire Propagation in a Porous
    Medium", Combustion Science and Technology, 4,
    187 (1971).

4. Goody, R.M., "Atmospheric Radiation", Oxford
    University Press, London (1964).

5. Hildebrand, F.B., "Methods of Applied Mathematics",
    Prentice-Hall Inc., Englewood Cliffs, N.J.
    (1952).

6. Hottel, H.C., "Radiant Heat Transmission", Chapter 4
    of "Heat Transmission", Third Edition, by W.H.
    McAdams, McGraw-Hill Book Company New York (1954).

7. Hottel, H.C. and Cohen, E.S., "Radiant Heat Exchange
    in a Gas Filled Enclosure", A.I.Ch.E. Journal,
    4, 3 (1958).

8. Hottel, H.C. and Sarofim, A.F., "Radiative Transfer",
    McGraw-Hill Book Company, New York (1967).

9. Moon, P., "Scientific Basis of Illuminating
    Engineering", Dover Publication Inc., New York
    (1961).

10. Oppenheim, A.K. and Bevans, J.T., "Geometric Factors
    for Radiative Heat Transfer through an Absorbing
    Medium in Cartesian Co-ordinates", Trans.
    A.S.M.E., Series C, 82, 360 (1960).

11. Penner, S.S., "Quantitative Molecular Spectroscopy
    and Gas Emissivities", Addison-Wesley, Reading,
    Mass. (1959).

12. Penner, S.S. and Olfe, D.B., "Radiation and Re-
    entry", Academic Press, New York (1968).

13. Planck, M., "The Theory of Heat Radiation", Second
    Edition, translation by M. Masius (1914), Dover
    Publications, Inc., New York (1959).

14. Plass, G.N., "Mie Scattering and Absorption Cross
     Sections for Aluminum Oxide and Magnesium
     Oxide", Applied Optics, $\underline{3}$, 7 (1964).

15. Reifsnyder, W.E. and Lull, H.W., "Radiant Energy
     in Relation to Forests", USDA, Forest Service
     Technical Bulletin No. 1344 (1965).

16. Siegel, R. and Howell, J.R., "Thermal Radiation
     Heat Transfer", McGraw-Hill Book Company, New
     York (1972).

17. Singham, J.R., "Tables of Emissivity of Surfaces",
     Int. J. Heat Mass Transfer, $\underline{5}$, 67 (1962).

18. Sparrow, E.M. and Cess, R.D., "Radiation Heat
     Transfer", Brooks/Cole Publishing Company,
     Belmont, California (1966).

19. Stevenson, J.A. and Grafton, J.C., "Radiation Heat
     Transfer Analysis for Space Vehicles", Rep.
     SID61-91, North American Aviation (AFASDTR 61-
     119 pt 1) (1961).

20. Steward, F.R., "Flame Spread Through Solid Fuel",
     D.Sc. Thesis, Chem. Eng. Dept., M.I.T.,
     Cambridge, Mass. (1962).

21. Touloukian, Y.S. and DeWitte, D.P., "Thermophysical
     Properties of Matter", Volume 7, 1FI/Plenum Data
     Corporation, New York (1970).

22. Touloukian, Y.S. and DeWitt, D.P., "Thermophysical
     Properties of Matter", Volume 8, 1FI/Plenum
     Data Corporation, New York (1972).

23. Touloukian, Y.S., DeWitt, D.P. and Hernicz, R.S.,
     "Thermophysical Properties of Matter", Volume 9,
     1FI/Plenum Data Corporation, New York (1972).

24. Van der Hulst, H.C., "Light Scattering by Small
     Particles", Wiley, New York (1957).

25. Wiebelt, J.A., "Engineering Radiation Heat Transfer",
     Holt, Rinehart and Winston, New York (1966).

PERCENT OF TOTAL ENERGY FOUND BELOW λ,
AS A FUNCTION OF λT

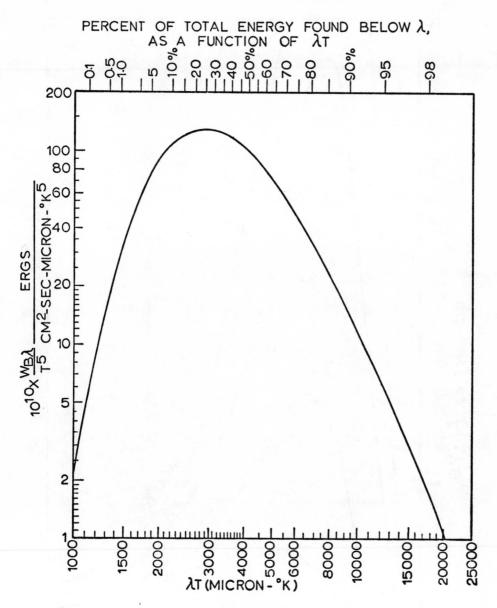

FIGURE 1.1

NOTE: Figure captions for this chapter are listed on page 314.

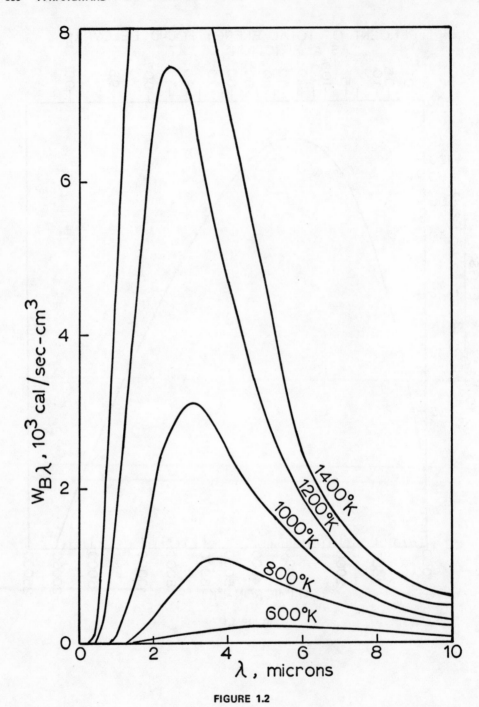

FIGURE 1.2

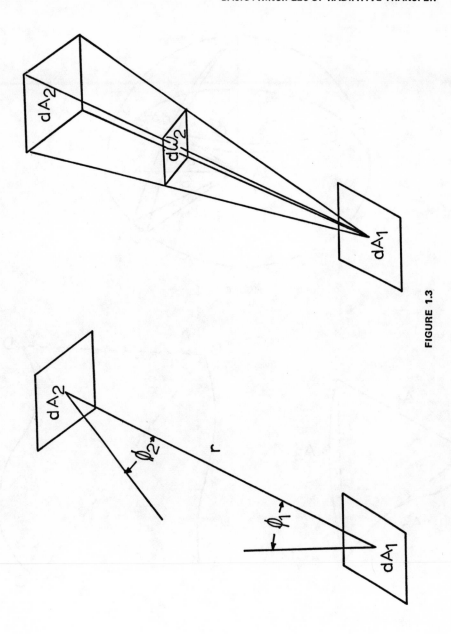

FIGURE 1.3

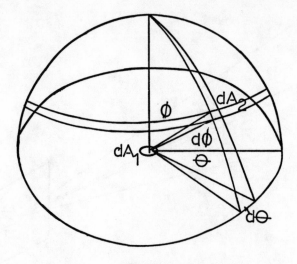

FIGURE 1.4

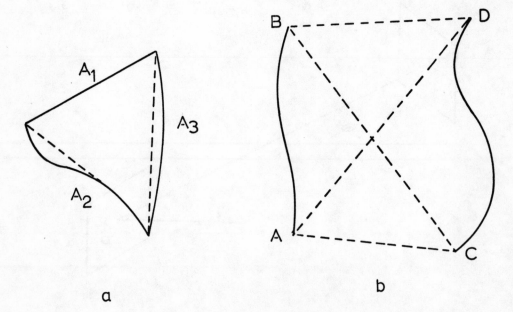

a

b

FIGURE 1.5

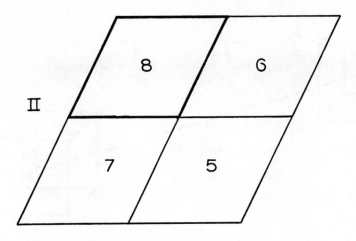

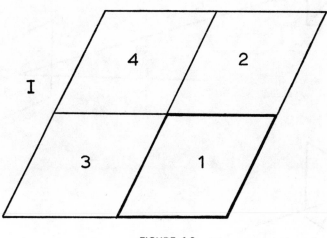

FIGURE 1.6

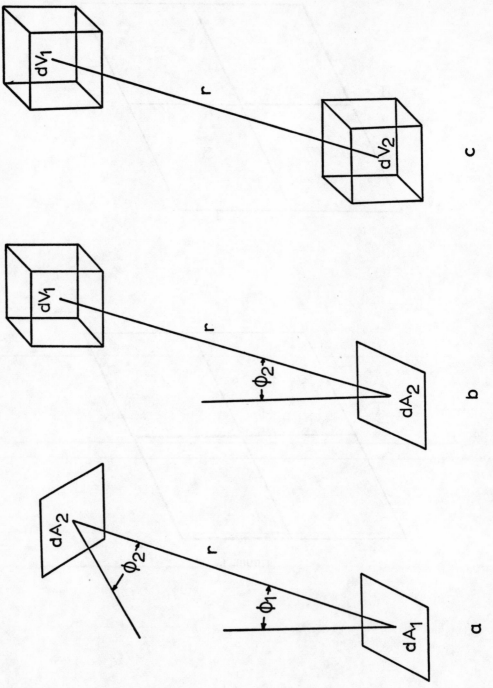

FIGURE 1.7

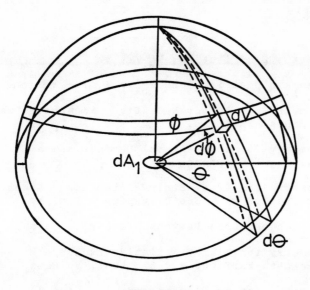

FIGURE 1.8

FIGURE CAPTIONS

1-1   Plank's Distribution, $W_{B\lambda}/T^5$ vs λT.

1-2   Planck's Distribution, $W_{b\lambda}$ vs λ, For Various Temperatures

1-3   Geometry for Direct Radiative Exchange Between
       Elemental Surfaces.

1-4   Geometry for an Elemental Surface and a
       Covering Hemisphere

1-5   Geometry for Cross String Interchange Areas
       for Systems with One Dimension Infinite.

1-6   Interchange Area Between Off Centre Squares.

1-7   Geometry for Direct Radiative Exchange Between
       Elemental Surfaces and Elemental Volumes.

1-8   Geometry for an Elemental Surface and a
       Covering Hemisphere of Absorbing Material.

Chapter 2

FIRE SPREAD THROUGH A FUEL BED

F. R. STEWARD
Fire Science Centre, University of New Brunswick
Fredericton, N. B., Canada

2.1 INTRODUCTION

A fire problem of obvious interest which has been receiving increasing attention in recent years is fire spread through a fuelbed. Almost any forest, brush or grass fire falls into this category. An urban fire which is large enough to be consuming numerous dwellings simultaneous can also be considered of this type. It is possible to investigate certain aspects of such a fire in the laboratory by burning small fuelbeds such as pine needles, wood shavings (excelsior), shredded newspaper, match splints, wood splints, wood cribs, etc. Although the scale factor between laboratory fires and field fires may be great, in some cases the basic equations governing the spread are identical. However, the scale of the fire undoubtedly markedly effects the relative importance of the mechanisms of heat transfer involved.

This chapter first discusses in some detail a line fire which is a fire with one dimension very much larger than the other two. A fire spreading through a fuel complex can in many instances be well approximated in this fashion. It has the distinct advantage of reducing the analysis to a two dimensional problem with all the subsequent simplifications and allows the definition of a number of quantities which can be used to characterize a fire quantitatively.

The next section gives a brief discussion of the types of heat transfer to and from the burning and potential fuel and the mechanisms by which these transfers

NOTE: All figures quoted in text are at the end of the chapter.

occur in a spreading fire. Some indication of the relative importance of the various mechanisms are given.

The following section lists the independent variables which are required to define a fuelbed, the individual fuel particles and the external conditions. Each of these variables is discussed in some detail as to its effect on the rate of fire spread and fire intensity. Data is presented when available.

The last section presents a discussion of the mathematical modeling of fire spread through a uniform fuel bed. The various heat transfer mechanisms are derived in mathematical terms and used to formulate equations to predict the rate of spread and the fire intensity.

2.2 CHARACTERISTICS OF A LINE FIRE

A fire in a naturally occurring fuelbed would normally start from a single source or several independent sources. It would therefore, at least initially spread radially. However, after the fire reaches a reasonable size or becomes driven by the wind its radial nature disappears and the fire develops a front which propagates through the fuel bed.

In a brush or forest fire this front can be and normally is irregular, burning more rapidly in one place than another and with greater intensity in one place than another. This is primarily due to two causes variations in the fuelbed and variations in the wind pattern. This phenomena can be readily produced with small fires in the laboratory by making nonuniform beds and variable wind patterns. However, even with an irregular fire front there will be a mean linear rate of spread and a mean burning rate and thus a mean fire intensity.

In the laboratory it is possible to produce a reasonably uniform fire front, if the fuelbed is constructed so that: 1. the individual particles are small compared to the size of fire, 2. the individual particles are arranged in some type of uniform fashion either randomly or in a well defined pattern, 3. noncombustible sides are placed next to the bed to prevent lateral air intake, 4. the bed is ignited evenly at one end and 5. if wind is desired it is supplied at a uniform velocity normal to the direction of spread. A fire in such a fuelbed will rapidly develop to a steady state situation with a uniformly spreading front. If the fuelbed is sufficiently wide the rate of spread becomes independent of bed width and the fire front becomes a truly two dimensional fire or line fire. Thus a line fire can be described in two dimensions, the third dimension being reproduced to infinity. A picture of such a fire is drawn in Fig. 2-1.

The advantage of producing a line fire in the laboratory is that reproducible data is more readily obtained

and the measured quantities more easily interpreted. It
is possible to measure a number of well defined quantities
which are discussed below. The advantage of considering
a fire in the field in terms of a line fire is that these
same well defined quantities are applicable even though
it is possible to consider only mean values. Therefore,
the discussion which follows throughout the remainder of
this chapter will be focused on a line fire except where
indicated.

It is useful to define certain quantities which
characterize and are associated with such a fire.

The rate of fire spread is obviously the distance
the fire front spreads in a fixed amount of time. Since
the fire is moving uniformly at all points there is no
ambiguity. In the case of an irregular fire front it is
possible to define the mean rate of fire spread over a
fixed section of the front.

The fire intensity (15) is defined as the rate of
energy liberation by combustion per unit length of fire
front. This quantity is a certain measure of the size of
a fire and ranges from as little as 5 Btu/sec.ft, for a
small laboratory fire to over 20,000 Btu/sec.ft.$^{(15)}$ for
a forest wild fire. This quantity refers to all the
energy liberated by the combustion of fuel from the particle of fuel just igniting at the front of the fire to
the last glowing piece of fuel at its rear.

$$I' = \Delta H_c\, mRf \qquad\qquad (2\text{-}1)$$

where ΔH_c is the heat of combustion of the fuel, h/m
      m is the fuel loading density, m/l^2
      f is the fraction of fuel consumed by the fire
      R is the rate of fire spread, l/t

The intensity of the fire defined above gives the
total rate of energy liberated by the fuel per unit length,
but does not describe its distribution within the fire.
The rate of consumption of fuel varies across the depth
of the fire normal to its direction of spread. This
quantity is termed the intensity distribution and is given
in terms of the energy release rate per unit area by combustion through the fire front. Several such distributions
measured by Frandsen and Rothermel $^{(29)}$ are presented in
Fig. 2-2. The are under the intensity distribution
curve is equal to the fire intensity. It is apparent that
two fires with similar intensities could have significantly
different intensity distribution curves and thus be considerably different in nature.

The various types of heat transfer to the burning or
potential fuel associated with a line fire can be given in
several ways. The total heat transfer yielded by any
particular mechanism is the rate of transfer per unit
length of fire front, because of its two dimensional nature.
It is also often useful to consider the distribution in

the rate of heat transfer per unit ground area. This
quantity thus varies with the distance normal to the direction of fire spread both within the potential and burning
fuel.

In a detailed analysis of fire spread it is often useful to consider the heat transfer to a single particle of
fuel. Relations of this type are normally given in terms
of the surface area of the fuel particle. These quantities
would in general be a function of the position of the particle in the fuelbed and possibly its orientation. This
type of detailed heat transfer analysis must be related
to the above defined distribution to be useful in discussing overall heat transfer rates.

2.3 HEAT TRANSFER FROM THE FIRE TO THE FUEL

For a fire spreading g through a fuelbed the heat
transfer to the burning and potential fuel are the two
important quantities to be determined. The former supplies the energy required to destructively distill the
fuel and thus generate the volatile matter which is consumed in the flame. The latter supplies the energy required to bring the fuel to its ignition point and thus
bring new fuel into the conflagration.

This heat transfer process in a spreading fire involves conduction, convection, radiation, and the movement of masses of hot or burning material. The relative
importance of these mechanisms from one type of fuelbed
to another can be immense so that it is necessary to
consider different mechanisms for different types of beds.
For instance radiative transfer from a flame is a strong
function of its size since the optical length is a
critical measure of its emissivity. Therefore, radiation
undoubtedly plays a much larger role in large fires than
the small ones normally produced in laboratories.

It is also often necessary to consider several situations for the same general type of heat transfer as it
can be occurring by entirely different mechanisms in different sections of the fuelbed. For instance the burning fuel could be receiving heat by convection due to
local flame impingement on its surface while fuel heated
by radiation ahead of the fire front may be cooled convectively by air drawn into the flame. The various means
by which each of the basic types of heat transfer can
occur in fire spread will now be outlined.

Conductive heat transfer in a spreading fire is
primarily concerned with transfer in the solid fuel. The
various means by which it can occur is shown in Figure
2-3. Heat is transferred to the surface of the fuel by
a number of processes but the important mechanism of
transfer from the surface into the interior of the fuel
is primarily by conduction. The largerr the dimension of
the fuel the more influence this mechanism exerts until

a certain size is reached at which point the fuel particle
becomes essentially infinite. The interior of a burning or
heating fuel particle acts as a heat sink drawing heat from
the surface and thus reducing the destructive distillation
of the material which supplies volatiles to support combustion or provide ignition.

It is also apparent that heat transfer by conduction
can occur along the burning fuel. This process is probably more important in the case of small fuel particles.
If a flame is attached to a fuel particle at one end the
high temperature of the particle below the flame will
produce conduction along the particle toward the opposite
end. This will help propagate the fire down the particle
and is a transfer mechanism to unburned fuel.

Depending on how one draws the line between conduction and convection heat transfer to both burning and unburned fuel heat transfer can occur over short distances
by conduction i.e. transfer of heat by molecular motion
in the gas phase. A flame stablilized on a fuel particle
below it and to the material just ahead of the flame front
as it advances along the particle. This type of transfer
within the gas phase is sometimes referred to as convection in the presence of laminar flow.

Convection heat transfer is associated with the bulk
motion of fluids and transfer of heat to or from a solid
surface. A number of convective mechanisms are shown
schematically in Fig. 2-4 a and b in the absence and
presence of wind. Convection heat transfer to the fuel
aside from the mechanism discussed above as conduction in
the gas phase can occur by direct flame impingement. The
flame or flames directly engulf the burning or unburned
fuel. This mechanism seems to be present in most fires
except possibly extremely small ones (a flame burning
down a single match splint).

Heat transfer by convection to the unburned fuel in
fuelbeds continuing numerous individual particles seems
to be from a number of small flames attached to individual
particles at the fire front (23, 62). These flames are
distinctly separate from the large overhead flames and
flicker and dance around among the particles providing
direct flame impingement.

Photographic evidence of this mechanism is presented
in Figures 2-5 a, b, c and d. These photographs were
taken with a motion picture camera through a telephoto
lens and represent an area about 3 inches square. The
fuel is red pine needles. Figure 2-5a was taken at the
surface of the fuel bed at the base of the flame in the
absence of wind. Figure 2-5b was taken through a glass
into the bed as the fire front was passing. It is apparent that both on and below the surface of the bed small
flames are attached to one or more individual fuel particles. The temperature gradients both within the gas

and solid particles in the vicinity of these small flames
must be large, thereby contributing a substantial amount
of heat transfer to the fuel as it approaches ignition.
Figure 2-5c and d is a repeat of Fig. 2-5a in the presence of light wind (1.5 ft/sec.) c and in the presence of
a stronger wind (3.0 ft/sec.) d. It is apparent that the
nature of the small flames is substantially affected by
the wind. The light wind makes the flames considerably
more active (63).

At the higher wind velocity the overhead flame is
swept out over the fuel bed ahead of the burning material
(the orange in Fig. 2-5d) allowing direct impingement
by a substantial part of the flame on the potential fuel.
The turbulence generated by wind also increases the activity of the burning gases at the base of the flame and
probably increase heat transfer in this region as well.

In light wind or its total absence the fire draws
ambient air across and through the fuelbed cooling the
fuel particles which have been receiving heat by radiation
from the flame and burning particles. In the case of a
back fire, a fire moving into the wind, the wind supplies
ambient air which will produce the same convective cooling
more effectively.

A natural or free convective mechanism will also produce a cooling effect on fuel particles heated by radiation. The warm particles heat air in their vicinity
generating upward currents due to the buoyancy of the
warmer air.

A final mechanism mentioned here can occur within the
fuel particle itself even though it may not be strictly
convective heat transfer in the usual sense. The volatile
material formed below the surface of the fuel as it is
destructively distilled passes to the surface or deeper
into the fuel modifying the conductive heat transfer in
the solid. In a live tree the movement of water along the
cambium layer by transpiration can be significant indeed
(73).

Radiative transfer will occur in various places at
and around the fire front as shown in Fig. 2-6. The hot
burning, glowing or smoldering embers will interchange
radiant energy with each other thus reinforcing the destructive distillation of the fuel. The overhead flame
radiates down into the burning fuel bed to supply additional energy to generate volatiles for combustion.

Both the hot embers and the overhead flame will radiate to the fuel ahead of the fire front. This type of
heat transfer can occur over much larger distances than
the conductive or convective mechanisms discussed previously. A fuel bed of small particles absorbs radiation
in depth and can be treated mathematically as an absorbing
gas.

As the fuel particles are heated by radiation they will immediately start to redistribute heat among themselves or the surroundings by their own emission and reflection thus providing additional transfer within the potential fuel.

The large scale movement of masses of hot or burning material occur primarily in large fires. Loose material within the fuelbed is swept into the flame by the incoming air and enters the convection column where it is transferred by the wind into the fuelbed ahead of the fire. This phenomena is often known as spotting in forest fires, but anyone who has observed urban fires knows that sparks and burning debris can be hazardous to other dwellings in the area of a large building fire. This is primarily an aerodynamics problem and has received treatment as such (65).

The movement of hot masses also takes place at the base of the fire front on a smaller scale. In some types of fuel beds as the particles begin to be consumed by the advancing fire curling and twisting motion occurs sometimes advancing the burning particles into the unburned section of the fuel bed.

The above described mechanisms are known to occur in fires of various types of fuel beds. However, the relative importance of these phenomena depend greatly on the type of fuelbed, the nature of the individual fuel particles and external conditions imposed by the surroundings.

2.4 VARIABLES WHICH AFFECT THE RATE OF FIRE SPREAD AND FIRE INTENSITY

The variables which affect the rate of fire spread and fire intensity are numerous and complicated. As will be demonstrated by the following discussion it is likely that some variables have not even been identified. For the purpose of organization these variables have been divided into three groups, those describing the fuelbed, those describing the individual fuel particles and those describing the surroundings.

Variables which describe the fuelbed.

1. Fuel loading density
2. Fuelbed voidage
3. Fuel arrangement within the bed

Variables which describe the individual fuel particles.

1. Type of fuel material
         or
2. Physical properties of the material
    a) fuel density
    b) fuel thermal conductivity
    c) fuel specific heat
3. Chemical composition of the material

4. Moisture content and related properties
5. Fuel particle dimension
    a) cross sectional area
    b) cross section shape
    c) particle length

Variables which describe the surroundings

1. Relative humidity and rainfall
2. Air temperature
3. Wind velocity and direction with respect to fire
    spread
4. Slope of the fuelbed with respect to the gravitation-
    al field.

Each of these factors can affect the various dependent variables associated with a spreading fire. The
problem is essentially the same as other physical problems in that it is necessary to define the system under
investigation. The above variables are required to define
the fuelbed and the external conditions which are imposed
on the system.

Variables which Describe the Fuel Bed

In order to define a fire spreading through a fuel
bed it is obviously necessary to define the fuelbed.
Given a particular type of fuel particle or distribution
of fuel particles it remains to define the fuelbed
structure.

An overall quantity which can readily be determined
is the fuel loading density, the weight of fuel per unit
area. In a forest this quantity is normally given in tons
per acre where 20 tons per acre is a high loading density
and 1 ton per acre is a low one.

In the laboratory the fuel loading density is measured in pounds per square foot and would normally be in
the vicinity of 0.05 to .5 lbs/ft.^{2} which is approximately
in the same range as the field values.

The data in Fig. 2-7 for two types of fuelbeds
poplar wood shavings (excelsior)[23] and shredded newspaper[60] indicates that the rate of fire spread in still
air is independent of fuel loading density over a wide
range when other variables are held constant. Once an
area is sufficiently covered with fuel to allow a fire to
spread the thickness of the bed does not effect the rate
or spread since the fire front is not influenced by the
depth of the fire. This situation could break down when
a strong wind is present since the depth of the fire
would effect the size of the overhead flame which is
forced over the fuelbed and contributes substantial convective and radiative transfer to the potential fuel.

It is apparent that if the rate of spread in still
air is independent of fuel loading density the intensity
of the fire will be directly proportional to the fuel

loading density over the same range if the fraction of
fuel consumed is not influenced by the loading density.
This was found to be the case in the two studies presented
in Figure 2-7 where a high percentage of the fuel was
consumed.

The fuelbed voidage is an important variable in
determining the rate of fire spread through a fuelbed in
either the presence or absence of wind. Figure 2-8 presents some data on the rate of fire spread vs. voidage
for wood sticks, pine needles and wood shavings. Over
the range of voidage investigated, .65 and .97, the rate
of spread increases continuously as the voidage increases
in all cases.

Figure 2-9 shows the same plot for a fuelbed of uniformly spaced poplar match splints over a wider range of
voidage (64). This diagram indicates that there is an
upper and lower limit of voidage for which fire spread
can occur. All fuelbeds must have similar behavior
although the limits would change from one type of fuel
bed to another.

In the case of the uniformly spaced match splints
the lower limit was reached when the flame could not
propagate from the upper surface of the splint s down
into the bed. This is probably due to the inability of
oxygen to penetrate into the bed in sufficient quantities
to support combustion.

Once the voidage is high enough so that the fire
will propagate as the voidage increases the rate of
spread increases. The data indicates that the rate of
spread from one row of match splints to the next row
remains essentially constant, but since the rows are
further apart the rate of spread increases. This increase in the rate of spread continues until the fire
has difficulty in advancing to the next row. The next
row is at too great a distance for the heat transfer from
the burning row or rows to ignite it. The voidage has
become too high for a fire to spread.

Since the rate of fire spread increases with voidage
over a substantial portion of the range of interest the
intensity of the fire will also increase by a corresponding amount for the same fuel loading density if the fraction of the fuel consumed remains unchanged. However,
data on wood shavings (21) indicate that the per cent of
fuel consumed decreases with decreasing voidage. Therefore, the fire intensity would increase with an increase
in voidage over and above that predicted only on the
increase in rate of spread. The fire intensity vs fuel
bed voidage for poplar wood shavings shown in Fig. 2-10
demonstrates the above. It is also very noticeable in
the laboratory that an increase in voidage produces a
larger overhead flame.

The arrangement of the fuel within the bed can also produce a substantial change in the rate of fire spread. Table 2-1 gives the rate of fire spread for two types of fuel particles poplar match splints and beach dowls for random and uniform arrangement with a voidage and fuel length of approximately the same values. In this instance the random beds were prepared by spreading the fuel particles evenly over a flat surface while the uniform beds were produced by placing the individual particles in a matrix aligned straight up. The data indicates that the rate of spread is substantially higher for randomly oriented particles in all cases.

TABLE 2.1

Rate of Fire Spread in Uniform and Random Fuel Beds

Type of Fuel	Arrangement	Voidage	Length in	Rate of Spread ft/hr
Poplar match splints	random	0.79	1.693	13.8
	uniform	0.79	1.600	10.0
Beach dowls	random	0.935	5.750	21.2
	uniform	0.951	4.13	13.3
Beach dowls	random	0.86	2.875	12.6
	uniform	0.804	2.00	7.5

Many types of fuel arrangements can be produced from ordered cribs and matrices of various types to systematic randomness. Each has its unique characteristics which will determine the rate of fire spread. More attention should be given to this important parameter.

Variables which Describe the Individual Fuel Particles

The variables which describe the individual fuel particle cannot be completely separated from those which describe the fuelbed. For instance the range over which it is possible to vary the voidage of the fuelbed is limited in many cases by the nature of the individual fuel particles. However, it is useful to make this distinction since there is a definite division in the type of variables.

It is possible to describe the fuel particle by two different methods. One can define the fuel particle as a fuel type or by describing its physical properties. For instance the fuel particles can be defined as red pine needles, or poplar match splints of a given size and shape. Alternatively the fuel particles can be defined in terms of its physical properties such as density, heat of combustion, cross sectional area and shape, thermal conductivity etc. In most instances some combination of the two is used as appropriate to the situation. This section will concern itself primarily with discussing the effect of physical properties since the classification of fuels by types although useful for descriptive purposes is rather limited in so far as analysis is concerned.

The physical properties of individual fuel particles are numerous and it is difficult to study the effects of each property independently.

Fons, et al. al (25, 27) found that the rate of fire spread through wood cribs of identical geometry and moisture content decreased as the density of the wood increased. However, it is conceivable that the data could have been correlated with another physical property such as the thermal conductivity, heat of combustion, or the kinetic characteristics for decomposition.

The thermal conductivity of the fuel is undoubtedly an important variable in determining the rate of fire spread as well as the rate of burning after ignition if the individual fuel particle is large enough so that a significant amount of heat can be transferred into its interior. As will be shown in the next chapter the ability to ignite a particular wood by irradiation is influenced by its thermal conductivity. Since the thermal conductivity of woods are directionally dependent the situation is complex. However, a recent compilation of thermal conductivities of various woods can be found in reference (45) along with kinetic parameters for pyrolysis.

Since the heating of fuel to ignition is an unsteady state process the thermal capacity of the individual fuel particle is an important variable. The heat capacity or specific heat of the fuel is the physical property which determines thermal capacity. Dunlap (17) has indicated that the specific heat of dry wood on a mass basis is similar for all species and given by

$$C_p = 0.266 + 1.16 \times 10^{-3} T \qquad (2-2)$$

where T is in °C and C_p is in cals/gram-°C.

As with other unsteady state heat transfer problems it is probably more useful to consider the thermal diffusivity, $k/\rho C_p$, as the fundamental variable rather than the heat capacity as the temperature distribution in a solid which is being heated is a function of this quantity rather than the heat capacity alone.

The heat of combustion of the fuel would undoubtedly exert some influence on the rate of fire spread and fire

intensity, the latter being directly proportional to it.
However, the heat of combustion of most woods are similar
on a mass basis (15).

The chemical composition of the various species of
wood is basically similar consisting primarily of cellulose, hemicellulose and lignin (11), although other compounds are known to be present in small amounts (5). It
is doubtful that the three major constitutents substantially change the nature of burning from one species to
another as demonstrated by the heat of combustion on a
mass basis being similar from one species to another.

However, Rothermel, et al.   (56) found by thermograve-
metric analysis that the rate of the loss of mass on
heating plant material was reduced by an increase in its
silica free mineral content. This presumably means that
its rate of burning would also be reduced by an increase
in the silica free mineral content of a fuel. King and
Vines (35) found that the "flammability" of leaves is
strongly influenced by the total mineral content. It is,
of course, well known that certain inorganic salts when
impregnated in wood act as fire retardants.

In the same paper King and Vines also point out that
the burning rate of a bed of leaves is strongly influenced
by a small increase in oil content. When over dried leaves
were allowed to take up eucalypt vapor their rate of burning was substantially increased. More information on the
effect of trace elements on the rate of burning and fire
spread is evidently needed.

Certainly one of the most important quantities related to fire spread and fire intensity associated with the
fuel particle is its moisture content. Whereas papers
and cloths have a well defined absorption isotherm so that
the moisture content of the material can be directly related to the relative humidity and temperature at equilibrium conditions, natural wood products do not. The equilibrium moisture content for a wood specie for an increasing relative humidity is different from that of a decreasing
one. If the additional factor of a living specie as
opposed to dead material is introduced the moisture content
is also a function of the time of year.

Although it is possible to bring fuels to an equilibrium moisture content in a laboratory with humdity and
temperature control the fuels in the open are always in a
dynamic situation as the relative humidity and temperature
are constantly changing. Add to this the irregular rainfall and the moisture content of fuels in the field is
indeed complicated. Due to this complex relationship the
normal procedure in a fire test is to measure the moisture
content of the fuel and report it.

Figure 2-11 shows the rate of spread vs fuel moisture
content for several types of fuel. The trend is as one

would expect from experience. The rate of fire spread is reduced as the moisture content is increased. Since the ignition of the fuel occurs above 100°C moisture in the fuel must be evaporated before ignition and any addition of water requires additional energy to bring the fuel to ignition. The data in Fig. 2-11 indicate that the decrease in rate with moisture content is somewhat more than that associated with merely evaporating the moisture. Taking a mean heat capacity for dry wood of 0.34 Btu/lb°F (16) and an ignition temperature of 500°F (21,55) the enthalpy required to heat dry fuel to ignition would be 146 Btu/lb dry fuel. The enthalpy required to evaporate the moisture is given as 28.8 + 1116 M (25) Btu/lb moisture where M is the moisture content of the fuel and 28.8 is the heat of wetting for wood. If the effect of moisture on the rate of fire spread is merely the additional enthalpy required for ignition per unit mass, Rh_{ig}, should be a constant. Considering the curve for Ponderosa Pine needles $R(C_p(T_{ig}-T) + 28.8 + 1116 M)$ for 5% moisture equals 11,300 Btu.ft/hr while 13% moisture equals 9,250 Btu.ft/hr. indicating that the moisture reduces the rate of fire spread by a greater extent than merely the evaporative effect.

Other possible moisture effects are the vapor leaving the fuel as it evaporates can absorb radiation which would normally have been absorbed by the fuel, the movement of water within the fuel as it is heated and the increase in vapor being swept into the flame which can influence the combustion process (33). One possibility with respect to the latter is the reduction in carbon particle formation in the flame due to the carbon steam reaction. This would reduce the emissivity of the flame and thus the radiative transfer to the burning and potential fuel.

The influence of moisture content on fire intensity can be seen in Figs. 2-10 and 2-12. The fire intensity is reduced as the moisture content increases. For the cases shown this is primarily due to the change in rate or spread, the fraction of fuel consumed remaining essentially unchanged.

The rate of fire spread and fire intensity is also a function of the geometry of the individual fuel particles. Most fuel types are much longer in one dimension than the other two so that it is appropriate to define the fuel in terms of cross section area, cross sectional shape and length.

The effect of the cross sectional area of the rate of fire spread can be seen in Fig. 2-8 from the different curves for the Ponderosa Pine sticks. As the cross sectional area increases for the same voidage the rate of fire spread decreases. This is undoubtedly due to the increase in heat transfer to the interior of the fuel per unit area of surface as the cross section increases thereby requiring more energy to bring the fuel to ignition.

The influence of the length of the individual fuel particle on the rate of fire spread is shown in Fig. 2-13. Over the small range investigated the data indicate a substantial effect in the case of poplar wood shavings. A smaller but definite effect (approximately 5%) is also demonstrated by poplar match splints even though the cross sectional area of the splints were larger for the longer splints. The increase in cross sectional area would tend to decrease the rate of spread according to the data in Fig. 2-8.

Figures 2-5(a-c) indicate that in beds of fine fuel there are a number of small individual fires at the leading edge of the fire front stabilized on one or more fuel particles. These small fires appear to be the major source of convective transfer to the fuel immediately ahead of the fire in the absence of wind. The heat transfer by convection or conduction along a fuel particle is interrupted by gaps between particles. Longer particles therefore seem to enhance this heat transfer mechanism.

The cross sectional shape of the particle will undoubtedly influence the conductive heat transfer into its interior. For instance a particle with a square or round cross sectional area would probably require a larger amount of heating to reach ignition than a particle whose cross section was rectangular or elliptical. The surface area of the particle for the same cross sectional area being greater for the two latter shapes than the two former. There is no conclusive evidence to the author's knowledge of this effect although the principle on which it is based is well known in conductive heat transfer theory.

Variables which Describe the Surroundings

The variables which describe the surroundings are those which are imposed on the fuelbed by the elements which vary from day to day and the terrain which essentially remains unchanged.

The effect of relative humidity and/or rainfall on the rate of fire spread have been discussed above in terms of the moisture content of the fuel. The equilibrium moisture content of dead fuels is controlled by the relative humidity and temperature of the atmosphere. The moisture content in living vegetation is also influenced by the relative humidity, but is primarily a function of the time of year which controls the growing cycle. As demonstrated in Figs 2-11 and 2-12 the moisture content of the fuel exerts a strong influence on the rate of fire spread and fire intensity. The mechanisms by which this occurs is discussed in the previous section.

The temperature of the surroundings obviously controls the temperature of the fuel particles. As the temperature of the fuel increases less energy is required

to heat to ignition and consequently the rate of spread
would increase by a corresponding amount. Another external quantity which influences the equilibrium temperature
of the fuel is the solar radiation impinging on the fuel
particles. A fuelbed in strong sunlight on a calm day
can reach temperatures significantly above the surrounding
air temperature.

One of the most important environmental variables is
obviously the wind. In the field the wind is difficult
to describe as it is continuously varying in velocity and
direction. In the laboratory however, it is possible by
use of a wind tunnel to impose a uniform velocity wind
on a propagating fire. Figure 2-14 gives data on the rate
of fire spread vs wind velocity for several types of fuel
beds obtained in this manner. It is apparent that when
the fire is spreading in the direction of the wind the
rate of spread is continuously increased as the wind
velocity increases over the range studied. When a fire
is "driven" by a wind the overhead flame is forced forward
and increases the heat transfer by both radiation and convection to the potential fuel. It is also apparent that
fires spreading into the wind are not greatly influenced
by the wind velocity over the same range. Although the
various types of fuelbeds presented in Fig. 2-14 show
the rate of spread is substantially different from one
fuelbed to another the variation of the rate of spread
with wind velocity is strikingly similar. Increasing the
wind velocity from zero to five miles per hour increases
the rate of spread from four to ten times for the fuel
beds shown.

Figure 2-15 gives some data obtained on the rate of
fire spread vs wind velocity based on field experiments.
It should be remembered that it is difficult to control
conditions in the field but nevertheless the curves show
behavior similar to that obtained in the laboratory. The
curves shown for Douglas Fir Slash and Lodge Pole Pine
Slash are based on numerous data points which show considerable scatter.

Data on the fire intensity as a function of wind
velocity is presented in Figs. 2-16 and 2-17. In general
the fire intensity increases substantially with wind
velocity in the forward direction but is little influenced
by a wind against the direction of spread. The data in
these two diagrams where obtained from weighing the fuel
beds as burning occured in order to determine the rate of
weight loss. The fire intensities developed in these
laboratory fires are approximately those encountered in
the field with prescribed or controlled burning.

The slope of the terrain can also influence the rate
of fire spread as shown by the data in Fig. 2-18. When
the fire spreads up a slope the flame will be leaning
toward the potential fuel as in the case of a head wind
thus increasing the heat transfer and thereby the rate of

spread as the angle increases. The rate of fire spread
down a slope does not seem to be affected by the angle of
the fuelbed.

The data presented in this section is believed to be
a reasonable summary of what is presently available. It
is apparent that most of the variables affecting the rate
of fire spread and fire intensity have not yet been fully
investigated. Other types of fuelbeds with various fuel
particle arrangements should be studied in order to obtain a more comprehensive picture.

2.5 MATHEMATICAL ANALYSIS OF FIRE SPREAD

The mathematical analysis of any physical problem is
an attempt to define the physical processes in such a
manner that mathematical equations can be written. It is
the intention of such an analysis to derive relations
which are of quantitative value. It is important to
remember that a mathematical analysis can never yield results which are superior to the physical understanding or
physical data upon which it is based.

The mathematical analysis of fire spread through fuelbeds has been attempted in a number of ways; empirical,
statistical, theoretical or some combination of these.

A purely empirical approach takes a set of data giving the rate of fire spread as a function of some independent variable such as fuel moisture content or wind
velocity and fits a predetermined function to the data.

A purely statistical approach takes a set of data
giving the rate of fire spread for a whole set of circumstances and determines the correlation coefficients for
each of the recorded independent variables.

A purely theoretical approach attempts to determine
the rate of fire spread on the basis of a physical understanding of the process. This involves a detailed understanding of the heat and mass transfer, the fluid dynamics
and the chemical reaction kinetics at the fire front and
its vicinity.

Each of the above methods has severe limitations.
Relations derived empirically cannot be extrapolated with
confidence. Statistical methods can only indicate that
a variable is of a certain level of importance. It can
never give an indication as to the reason for its importance
and thus no physical understanding. A purely theoretical
analysis of such a complex problem on the other hand is a
tall order indeed. It should also be pointed out here
that it is rather easy to write complex equations based
on dubious assumptions which never attempt to explain
physical data.

An intelligent mathematical analysis of fire spread
depends on the use that one intends to make of the analysis. In general, a combination of empiricism, theory,

and statistics is probably better than using any one independently. However, the relative amounts of each depends
on the type of relation desired and its intended use.

The remainder of this chapter attempts to give some
of the fundamental relations which one encounters in the
fire spread problem concentrating primarily on the heat
transfer aspects.

The Basic Equation of Fire Spread

As indicated in Section 2.3 fire spread through a
fuelbed is primarily a heat transfer controlled phenomenon. In order for the fire to spread into new fuel
areas fuel must recieve enough heat to generate sufficient
volatile combustibles to ignite. This heat is supplied
by the flame and burning fuelbed through various transfer mechanisms.

The volatile matter undergoing combustion in the
flame is supplied by the decomposition of the fuel within
the burning zone. The overall reaction involved in the
pyrolysis of the solid material is endothermic and therefore, requires the addition of heat. This heat is also
supplied by the overhead flame and the transfer of heat
within the burning section of the bed itself.

The overhead flame is a result of the combustion of
the volatile material generated within the burning section
of the fuelbed. The fluid flow in the vicinity of the
flame is thus induced by this combustion process.

The physical problem therefore involves a closed
cycle with the flame and burning zone interacting with the
potential fuel and the burning material which supplies the
volatile combustibles. Presumably a unique solution to
this problem exists for each defined fuelbed and set of
external conditions although Emmons and Shen (20) have
shown that the solution may depend on the initial conditions i.e. the manner in which the fire was initiated.

The following relations are developed in order to
describe the rate of spread and the intensity of a line
fire in accordance with the above discussion. The line
fire was described in Section 2.2 and is shown schematically in Fig. 2-1.

The rate of fire spread is governed by an energy
balance on the potential fuel ahead of the fire front.

$$Rmh_{ig}1 = \sum_j Q/L)_j \qquad (2-3)$$

R = rate of fire spread, $1/t$
m = mass of fuel per unit volume of bed, $m/1^3$
h_{ig} = enthalpy required to bring the fuel from ambient
conditions to the point where it enters the conflagration, h/m
1 = thickness of the bed, 1

$\Sigma\ Q/L)_j$ = the sum of the heat transfer rates to and from
   the unburned fuel as the fire front approaches per
   unit length of fire front, h/lt

This equation is a statement of the simple fact that
the sum of the heat transferred by all mechanisms to and
from the potential fuel as the fire front approaches will
be equal to the rate at which the fuel is heated to the
point where it passes into the fire front.

This equation has been written as a total energy balance on the potential fuel. The heat transfer rates by
the various mechanisms would in general be a function of
the distance from the fire front and the distance below
the surface of the fuelbed. The quantities in the
equation represent the integrated values of the local heat
transfer rates. The enthalpy required to heat the fuel
to the point where it enters the conflagration is a function of heating rate and thus in general varies with the
distance below the surface of the fuelbed.

This relation can be looked upon as either an energy
balance integrated over the thickness of the bed and thus
a total energy balance on the potential fuel or an energy
balance at a particular level in the bed. The two concepts are possible since the fire must spread at the same
rate at all levels in the fuelbed although not necessarily
arriving at the same horizontal position simultaneously.

The rate of evaporation of volatile material which
supplies the fire with combustible material is determined
by an energy balance on the burning zone within the fuel
bed.

$$\frac{M_v}{L}\ h_v = \sum_k Q/L)_k \qquad (2\text{-}4)$$

where $\dfrac{M_v}{L}$ = mass rate of liberation of volatile material
   within the burning section of the fuelbed per
   unit length of fire front, m/lt

h_v = enthalpy required to liberate a unit mass of
   volatile material from the solid fuel, h/m

$\Sigma\ Q/L)_k$ = the sum of the heat transfer rates to and from
   the burning fuel per unit length of fire front
   h/lt.

Although there is no doubt that an energy balance of
this form exists within the burning zone the terms in the
equation are somewhat more ambiguous than the previous
formulation giving the rate of fire spread. The enthalpy
required to liberate a unit mass of volatile material from
the solid fuel is an extremely difficult quantity to
evaluate as the chemical reactions of the decomposition of
wood are not well understood, see Chapter 3, and the rate
of evolution of volatile products from pyrolysis vary with
temperature, rate of heating, and the geometry of the fuel
particle and fuelbed (8, 9, 10, 31, 45, 47, 48, 52, 70).
The geometric influence would include the ability of oxygen to penetrate the burning zone.

Once again a total energy balance has been written for simplification of presentation. The heat transfer rates by the various mechanisms and the enthalpy required to liberate volatile matter would be a function of horizontal and vertical position within the burning zone. The heat transfer rates are the overall integrated values while h_v represents a mean value in the equation as written. However, this balance as written must yield the rate of mass liberation of volatiles per unit length of fire front, M_v/L, or the rate of consumption of solid fuel per unit length of fire front which is essentially the same quantity.

The quantity M_v is related to the fire intensity discussed in Section 2-2 according to the equation

$$I = M_v/L \ AH_c \qquad\qquad (2-5)$$

where AH_c is the heat of combustion of the volatile material per unit mass, h/m.

Another relation is required to define the problem mathematically. The heat transfer rates by the various mechanisms on the right hand side of Eqs. 2-3 and 2-4 are determined by the temperature distribution and flow patterns in the burning zone and the overhead flame. These quantities are in term a function of the mass liberation of volatiles in the burning zone and the depth over which this liberation occurs, D_B. The flame will also be influenced by external wind while the burning zone characteristics would be influenced by both geometry and wind.

Flame temperature distribution and flow pattern

$$\delta(M_v/L, D_B, \ wind \ velocity) \qquad\qquad (2-6)$$

Burning zone temperature distribution and flow pattern

$$\delta(M_v/L, D_B, \ wind \ velocity \ and \ fuel \ bed \ geometry) \quad (2-7)$$

These equations have been written in functional form only since the relations would change immensely from a small laminar fire to a large turbulent one including fires occurring over large areas which do not give a single coherent flame.

Equations (2-3), (2-4) (2-6) and (2-7) mathematically define the problem. The solution of this set of equations for even the simplestt type of fuelbed structures has not been achieved in the sense that one can make useful predictions by such a solution. The mathematical solutions which have been developed use various parts of the overall problem and make simplifying assumptions to circumvent the difficult aspects.

The Energy Balance on the Potential Fuel

The energy balance on the potential fuel into which the fire is spreading has received most of the attention in the past for the very good reason that this predicts the rate of fire spread. This relationship has been written by numerous workers in various forms for fuelbeds ranging from small laboratory fires of shredded newspaper, pine needles, excelsior, etc. to prototype fires of brush, pine forests, slash, etc. [1, 4, 6, 18, 22, 24, 28, 33, 34, 48, 53, 56, 60, 62, 64, 67, 72, 74]. In each of these analyses assumptions were made such that the heat fluxes on the right hand side of Eq. 2-3 which were regarded as important for the particular fuel bed under consideration could be evaluated in a manner independent of the volatile liberation rate whose combustion generates the heat which is transferred. With this limitation in mind a brief look at some of the heat transfer mechanisms described in Eq. 2-3 will be considered.

Radiative Transfer from the Burning Zone

Radiative transfer from the burning zone to the unburned fuel should always be considered as potentially important. It can probably only be ignored entirely for very thin fuelbeds such as a few layers of shredded newspaper (34, 60) or a fire spreading over a surface of relatively volatile material (16, 38) which is a different problem than the one considered here. The importance of this mechanism of radiative transfer has been confirmed experimentally for several types of different fuelbeds (2, 23, 42, 58, 67, 72).

The direct interchange area between a vertical black plane of infinite length and height L and an element of absorbing material at a distance x separated by absorbing material is presented in Appendix II (B-5). This function gives the distribution of radiation from the burning zone ahead of a vertical fire front in a fuelbed of randomly oriented particles. The absorption coefficient of the fuelbed for radiation can be calculated from Eq. (1-46) which states that it is equal to one fourth the fuel particle surface area per unit volume.

$$k = \frac{a'c}{4} \tag{2-8}$$

where a' is the surface area of a single particle, l^2
   c is the concentration of particles per unit volume, l^{-3}

The integral of this distribution from the fire front to infinity gives the total rate of heat transferred by this mechanism to the potential fuel, $Q/L)_R$.

The distribution presented in Appendix II (B-5) indicates that an exponential function is a reasonable approximation and this has often been used for analytical purposes (21, 22, 48, 62, 67, 75).

$$Q/L)_R = \sigma \, \varepsilon_B \, T_B^4 \int_0^\infty \bar{F}_{12} \, dx \tag{2-9}$$

assuming

$$F_{12} = e^{-Bkx} \tag{2-10}$$

$$Q/L)_R = \sigma \epsilon_B T_B^4 \int_0^\infty e^{-Bkx} \, dx \tag{2-11}$$

$$Q/L)_R = \frac{\sigma \epsilon_B T_B^4}{B} \tag{2-12}$$

This relation indicates that the radiative heat transfer through the bed is nearly independent of the absorption coefficient (B is a slight function of absorption coefficient) and thus of the voidage. Since a reduced voidage means more fuel to bring to ignition per unit volume of bed the strong dependence of the rate of spread on fuelbed voidage is at least partly explained by this mechanism of heat transfer. That is the same radiative transfer rate to the potential fuel must be spread over more fuel at a lower voidage.

A radiative analysis of this type could also be a reasonably good approximation for a fuelbed of ordered arrangement. However, a more detailed consideration of the particular geometry similar to that of Emmons and Shen (20) for linear paper arrays may be more appropriate for a well defined geometry.

Radiative Transfer from the Overhead Flame

There is considerable debate as to the importance of overhead flame radiation in the fire spread problem particularly in the absence of wind. Some investigators claim it is an important factor (1, 48, 64) while others claim not (26, 42, 62, 67). The evidence would seem to indicate that its relative importance depends on the type of fuelbed and the wind velocity and direction.

A rigorous analysis of radiation emitted by such an ill defined flame as that produced by the combustion of volatile material from the pyrolysis of solid material is out of question at this time.

Assuming that the flame can be represented by a uniform temperature body of emitting gray gas the distribution of radiation on the surface of the fuelbed can be obtained from an interchange factor such as that presented in Appendix II (B-3 or B-4). These factors have not been evaluated for other than vertical geometries, however.

If it is assumed that the flame can be represented by a plane of infinite length and height H with an emissive power $\epsilon_F \sigma T_F^4$ the distribution of radiant energy on the <u>surface</u> of the fuelbed can be obtained from the direct interchange area given in Appendix I (A-3). The radiation from the flame, however will penetrate the fuel bed in depth as accounted for in the previous analysis. This calculation has not yet been carried out and would be a useful piece of information to contribute to the overall picture.

Assuming that all the radiation which strikes the surface of the fuelbed is absorbed the total heat transferred by a flame represented by a uniformly emitting plane would be

$$Q/L)_F = \varepsilon_F \, \sigma T_F^{\,4} \int_0^\infty F_{12} \, dx \qquad (2\text{-}13)$$

$$Q/L)_F = \varepsilon_F \, \sigma T_F^{\,4} \int_0^\infty 1/2 \left[\frac{1 - \frac{x}{H}\cos\theta}{1 + (\frac{x}{H}) - 2\,\frac{x}{H}\cos\theta} \right] dx \qquad (2\text{-}14)$$

$$Q/L)_F = \frac{\varepsilon_F \sigma T_F^{\,4} \, H}{2} \, (1 + \cos\theta) \qquad (2\text{-}15)$$

The fact that back fires spread at substantially the same rate as fires in still air, Fig. (2-14) would seem to indicate that this mechanism is of secondary importance in these cases. This has been substantiated by experimental measurements for excelsior (23) and wood crib (26, 42) fires.

The increase in the rate of fire spread with an increase in head wind velocity predicted by Eq. (2-15) is probably partially due to this mechanism of heat transfer as the flame leans out over the potential fuel.

It is also observed that some types of fuelbeds burn with the leading edge of the fire front at the surface with a slightly curved front sweeping backward through the bed (22, 28, 63). As the wind velocity increases the fire front becomes less vertical within the bed. This is undoubtedly due to the high rate of heat transfer at the fuelbed surface provided by flame radiation and also probably increased convection. Frandsen (28) and Fang (22) have pointed out that the sloping fire front within the fuelbed provides additional transfer by radiation from the burning zone, a vertical radiative component, to produce a stable uniformly moving fire front.

Convective Transfer at the Fire Front

The convection heat transfer at the fire front is the most difficult to describe mathematically because it is the most difficult to describe physically. Many of the mathematical models mentioned above (2, 6, 34, 48, 62, 64, 74, 75) attempt to account for this mechanism of heat transfer in one way or another. However, in all cases the description of the detailed mechanism is rather vague.

Hottel et al (34) describe a process of turbulent diffusion produced by the flame. Albini (1) postulates a flame attachment mechanism in which the flame leaps from one particle to another. Fang and Steward (23) describe a mechanism in which small flames run along individual particles and jump across gaps between them. Steward (62), Fig. 2-5a,b presented single frames of motion pictures

which show small individual flames attached to one or more
fuel particles (pine needles) both at the surface and within the bed. In the absence of wind these small fires are
spreading along the particles in a "haphazard" fashion
due to the irregular arrangement of the fuel.

Figure 2-5 c and d show similar single frames of motion pictures at the fire front in the presence of a light
and moderate wind at the surface of the fuelbed. It is
apparent from the photographs that the situation is drastically altered. The wind increases the activity of the
small fires and presumably by so doing increase convective
transfer to the particle to which the flame is attached as
well as neighboring particles. As the wind velocity increases the overhead flame is swept forward and produces
direct contact between the main portion of the flame and
the fuel particle on or near the surface, Fig. 2-5d.
This is an entirely different convection mechanism and is
only present after a certain wind velocity is achieved.

Photographs taken through a glass which reveal the
inside of the bed give essentially the same picture in the
presence or absence of wind, Fig. 2-5b, except the slope
of the burning front sweeps back at a more acute angle.
This would increase the radiative transfer within the bed
by the radiative mechanism discussed above. A reliable
mathematical expression has yet to be written for this
process, but it is apparent from these photographs as well
as others from different types of fuelbeds that this
mechanism is strongly influenced by the fuelbed particles
and structure.

It should also be mentioned here that the small fires
spreading along the fuel undoubtedly produce very high
temperature gradients within the fuel particles and thus
conduction within the particle itself cannot be ruled out
as a mechanism of transfer. Although this mechanism by
nature must occur over very short distances it could be
contributing significantly.

Fang and Steward (23) and Fang (22) have found that
an increase in the length of the fuel particles in a
bed of randomly oriented particles gives an increase in
the rate of fire spread other conditions being similar,
Fig. 2-13. This could be explained by the end of a fuel
particle preventing conduction and therefore more ends reduce conductive transfer.

The geometric difficulties encountered with randomly
oriented particles within the fuelbed can be overcome by
using geometrically uniform beds. Steward and Waibel (64)
have recently studied uniform matrices of poplar match
splints where the fuel particles are all vertical and
evenly spaced. The convective transfer is more easily
described for such a geometry and can be developed on the
basis of transfer from one or more neighboring particles
to another by gas phase conduction as postulated by Vogel
and Williams (74).

Assuming there is a uniform heat transfer coefficient over the surface of the particle facing the burning fuel particles the convection heat transfer to a single particle from another burning particle can be written as

$$q_p = h \ (T_g - T_s) = \frac{k}{\Delta y} \ (T_g - T_s) \qquad (2\text{-}16)$$

where h is the heat transfer coefficient, h/tl^2T
T_g is the temperature of the gas at the surface of the flame, T
T_s is the temperature of the surface, T
k is the thermal conductivity of the gas, h/tlT
Δy is the distance between the flame surface and the particle surface, l.

In the matrix arrangement studied by Steward and Waibel (64) the gas phase conductive transfer would be from the three closest particles only. If it is assumed that the transfer is made primarily on one side of the approximately square cross sectional match splints and equally over this entire side.

$$q_p = kdL \ (\frac{1}{\Delta y_1} \ + \ \frac{2}{\Delta y_2}) \ (T_g - T_o) \qquad (2\text{-}17)$$

where   d is the distance along one side of the fuel particle, l
L is the length of the particle, L
Δy_1 is the distance between the flame surface and the particle surface for the nearest particle, l
Δy_2 is the distance between the flame surface and the particle surface for the two neighboring particles, l

If C is the center to center distance between the particles in the matrix the convective heat flux per unit length of fire front can be obtained

$$Q/L)_c = \frac{kdL}{C} \ (\frac{1}{\Delta y_1} \ + \ \frac{2}{\Delta y_2}) \ (T_g - T_o) \qquad (2\text{-}18)$$

where for this particular geometry

$$\Delta y_1 = (C - d) - y^* \qquad (2\text{-}19)$$

$$\Delta y_2 = \sqrt{2} \ (C - d) - y^* \qquad (2\text{-}20)$$

where y* = the distance between the flame surface and the burning fuel particle surface, l.
The quantity y* has been termed the flame stand off distance.

Using reasonable values for the quantities in equation 2-18, k = 10^{-4} cals/cm-sec°K, $T_G - T_s$ = 1500°K and the actual distances for a particular geometry, d = 0.9 inches, L = 1.5 inches, C = 40 inches and estimating Δy_1

and Δy_2 as .16 inches and .28 inches respectively gives
a heat flux per unit length of fire front of 260 Btu/ft.hr.
(0.6 cal/cm-sec.). This corresponds to a heat flux to the
fuel particle surface of 9,300 Btu/hr ft^2 (0.7 cals/cm^2-
sec) which indicates that this mechanism of heat transfer
would predominate for the matchsplint matrices. This conclusion is substantiated by thermocouple measurements (64)
which indicate that the major part of the heat is transferred to the heating fuel particle after the row just
previous has ignited. The thermocouple measurements suggest that somewhere between 5 and 10% of the enthalpy required for ignition is transferred further than one row,
i.e. radiative transfer.

In this same paper it is also demonstrated that as
the center to center distance between the particles is
reduced below a certain amount (0.135 inches) the fire
will not propagate. It is observed that at these close
spacings the flames on the individual particles are not
able to propagate down the particles into the fuelbed.
This of course makes it impossible to transfer heat by
this convective mechanism.

It can therefore be concluded that convection heat
transfer at the fire front is an important mechanism in
nearly all types of fuelbeds. It is also apparent that
it will be a strong function of the fuel particle and fuel
bed geometry which makes it difficult to formulate for
naturally occurring fuelbeds.

Convective Heat Loss from the Heating Fuel Particles

As the fuel receives heat by the various mechanisms
discussed above it will immediately begin to lose heat to
the surroundings since the ambient air will normally be at
a lower temperature than the particle. This heat transfer
would be primarily by convection although reradiation will
also be occurring. If it is assumed that there is a mean
particle heat transfer coefficient h the heat loss from a
particle with a surface temperature T_s to ambient air at
T_o will be given by

$$Q_p = h \, a' \, (T_s - T_o) \qquad (2\text{-}21)$$

Let the surface area of the particles per unit volume
be a_v' and the convection heat transfer per unit volume of
bed will be

$$q_v = h \, a_v' \, (T_s - T_o) \qquad (2\text{-}22)$$

The surface temperature of the particles and the heat
transfer coefficient will vary with position ahead of the
fire front and the distance below the fuel bed surface.
Neglecting the variation in depth one can write the total
heat loss to the point where the fuel enters the fire
front as

$$Q/L)_c = \int_0^\infty h \, a_v' \, (T_s - T_o) \, 1 \, dx \qquad (2\text{-}23)$$

This relation requires a knowledge of the surface temperature of the fuel particles as a function of position ahead of the fire front which is one of the unknowns in the problem. Various procedures can be used to obtain this distribution as it is a common type of integral equation. Steward (62) obtained this distribution by first solving a simplified differential equation which was then substituted into the general integral equation (2-3). Numerical procedures are of course always possible.

The particle surface heat transfer coefficient in Eq. (2-23) also presents certain difficulties. Ambient air circulation within the fuelbed can arise from three sources, air drawn into the fire front by the flame and burning zone, external wind, and bouyancy effects produced by the heating particles themselves. Any of these are difficult to evaluate for the complicated fuelbed geometries normally encountered and only rough estimates can be made from engineering data. Steward (62) concluded that in the absence of wind the circulation produced by the fire front would predominate over buoyancy induced flow. This conclusion was based on some visual observations with smoke tracer (58). He proceeded to evaluate the heat transfer coefficient on the basis of flow in packed beds often used in the chemical industry. When significant wind is present, however, the circulation produced by the fire is undoubtedly overwhelmed by it.

The previous heat flux analyses are presented as examples and were kept as general as possible. It seems that each type of fuel bed must be treated separately in this regard with one type of mechanism important in one case and another type in another. It must be admitted that the lack of generality is somewhat discouraging but a difficult problem cannot be simplified by a wish no matter how well intended.

Enthalpy Required to Bring the Fuel to Ignition

The only quantity on the right hand side of the energy balance on the potential fuel which requires further comment is h_{ig}, the enthalpy required to bring the fuel from ambient conditions to the point where it enters the conflagration. This quantity is in general quite complicated and would depend on the type of fuel particle, particle geometry, fuelbed geometry and rate of heating. However, it has been found that the fire front consumes a particle when the surface temperature of the particle is somewhere between 300 - 375°C for a considerable variety of cellulosic fuels (20, 21, 23, 55, 64, 74). This corresponds to a similar criteria for piloted ignition of cellulosic fuels as discussed in Chapter 3. If the fuel particles are receiving heat at a moderate rate, less than 2 cals/cm^2-sec. (26,500 Btu/ft^2hr), which is the case in this type of fire spread, the passage of the fuel into the fire front at a fixed surface temperature in the

vicinity of 350°C seems a good assumption and a useful
simplification.

A fixed surface temperature ignition reduces the problem to analysing for the heat flux within the particle
which is primarily by conduction. Solutions for the conduction equation for various shapes and conditions can be
obtained from the well known standard source Carslaw and
Jaeger (52). An example of how to apply this type of solution to a specific problem can be seen in reference to
fire spread in vertical evenly spaced matchsplints (64,74).

Rothernal (54) has used a combination of measured
values and theoretical calculations to determine a factor
which is multiplied times the enthalpy required to heat
the entire solid material to a fixed temperature. He
found .92 for excelsior, .50 for 1/4" sticks, and .22
for 1/2" sticks. This reflects the necessity of only
surface heating as the size of the fuel particle increases.
This type of analysis reduces the evaluation of h_{ig} to a
reasonably accurate and manageable level.

The Energy Balance on the Burning Fuel

The quantities necessary for the evaluation of Eq.
(2-4) have received considerably less attention because of the tempting simplification of postulating the
overhead flame and burning zone as a given boundary condition. However, a fire spread model which does not
account for changes in the burning characteristics of the
fire front with the variables which define the fuelbed
and external conditions cannot claim to be an entirely
representative picture.

The evolution of the heat fluxes on the right hand
side can be carried out in a fashion similar to that described for the rate of spread, Equation 2-3. Assuming
that a given flame can be assumed the interchange areas
presented in Appendix II, B-3 and B-4, would be useful
in the absence of wind. No evaluations of this type for
non-vertical flames due to horizontal wind are presently
available. Convection heat transfer from the flame and
within the bed itself is rather ill defined but an analysis similar to that of Block (10) for a stationary fire
in wood cribs would also probably be useful for spreading
fires.

The most difficult quantity to determine however, is
the enthalpy required to liberate a unit mass of volatile
material from the solid fuel. For a single fuel particle
this quantity is known to vary with temperature, rate of
heating, size of particle, shape of particle and physical
properties (8, 47, 48, 52, 70). When the particle is
located in an array even if the conditions external to
the particle do not alter the chemical kinetics of the
pyrolysis reactions within the particle the rate of heat
transfer to the surface of the particle will be controlled

by the bed structure. Since the pyrolysis within the
solid fuel is a function of the rate of heating this
quantity must in general be a function of the position of
the particle within the bed.

There are three methods by which it is possible to
determine this qunatity or the rate of formation of volatiles directly, direct weight measurements of a fuelbed
through which a fire is spreading, weight measurements of
individual particles in a controlled heating environment,
or the determination of the kinetics of the pyrolysis of
wood to be used in a theoretical heat transfer analysis.
In many respects these three methods compliment each other
in that the information gained in one can be useful in a
study of the others.

Direct measurements on the weight of the fuel bed as
a fire spreads through it have been carried out by Anderson and Rothermal (4, 55) and Fang (22). Some of the fire
intensities given in Fig. 2-10, 2-12, 2-16 and 2-17 were
obtained in this manner. Some of the data of Fang (22)
are presented in Fig. 2-19 where the measured steady
burning rate is plotted vs the wind velocity for several
types of fuel. In all cases the burning rate increases
with wind velocity for head fires and shows partically no
change for back fires.

A more detailed method of obtaining burning rates has
been proposed by Frandsen and Rothermel (29) and Berlad,
Rothermel, and Frandsen (7) in which a small section of
the fuelbed is weighed so that the local burning rate can
be determined as the fire passes. The local burning
intensities plotted in Fig. 2-2 were obtained in this
fashion. It is seen in this case that even though the
fire intensity and thus the overall burning rate is greater for one case the local burning intensity reaches a
higher value for a short period of time in the other.

A number of measurements have been carried out on
the weight loss of a single burning fuel particle (8, 36,
42, 48, 52, 70). The most useful of these studies for the
purposes here are the investigation by Phillips and Becker
(48) and Roberts (52) where the heat flux to the surface
of the fuel particle was controlled and measured.

Roberts found the maximum rate of weight loss of oven
dry wood squares to be essentially directly proportional
to the level or irradiation and proposed the equation

$$m_v = /L = 4.5 \times 10^{-4} I \qquad (2-24)$$

where $m_v = g/m^2$-sec and $I =$ watts/m^2

This empirical relation was found to agree reasonably
well with a theoretical analysis based on the conduction
equation with constant thermal conductivity, no internal
convective heat transfer and a volatile generation rate
with first order kinetics.

Phillips and Becker (48) carried out an extensive study on the rate of weight loss of single sticks of white pine exposed to gas streams of various velocities and temperatures. The temperature of the wall of the enclosure was always maintained at the same temperature as the gas stream. They were able to correlate their data by an empirical relation.

$$\frac{m_{o/L}}{q_p \pi d_o} \propto t_{1/2}^{a} \qquad (2\text{-}25)$$

where $m_{o/L}$ is the initial mass of the specimen per unit length g/m

q_p is the total flux density at the surface of the specimen based on the initial diameter, watts/m^2

d_o is the initial diameter of the specimen,

$t_{1/2}$ is the time required for the specimen to reach one half the original mass, sec.

a is an experimentally determined exponent which varies from 0.66 to 1.01 depending on temperature.

The correlation shows a definite temperature effect as demonstrated by the variation of a and the proportionality constant. The line for 895°C is approximately in the middle of the data and is given by the relation

$$\frac{m_{o/L}}{q_p \pi d_o} = .89 \times 10^{-3} \, t_{1/2} \, 0.85 \qquad (2\text{-}26)$$

where the units are given above.

Gross (31) carried out a number of experiments in which he measured the weight loss of burning wood cribs as a function of time. He found that above a certain bed porosity the burning rate expressed in percent consumed per unit time times the individual stick thickness to the 1.6 power was a constant.

$$R_o^{1.6} = constant \qquad (2\text{-}27)$$

If this relation is expressed in terms of volatile liberation rate per initial unit area of the stick one obtains

$$m_v \propto \frac{1}{d_o^{0.6}} \qquad (2\text{-}28)$$

An equivalent expression for Phillips and Becker's formula yields

$$m_v \propto \frac{q_p^{1.127}}{d_o^{0.127}} \qquad (2\text{-}29)$$

It is also interesting to compare the actual volatile liberation rates measured by the various workers. For a heating rate of 1 cal/cm-sec (4.18 x 10^4 watts/m^2, 13,250 Btu/ft^{2}hr), which is probably a reasonable level of heat

transfer to the fuel in a burning bed. Roberts formula
gives 18.8 g/m^2-sec. For a 3 cm diameter white pine stick
and the same heating rate Phillips and Beckers formula for
895°C gives 8.75 g/m^2-sec. Gross reported a volatile
liberation rate per unit area on a Douglas Fire crib fire
of 4.63 g/m^2-sec. From data by Fang (22) on the measured
weight loss and burn depth as a fire spreads through a
fuelbed in the absence of wind the rate of volatile liberation per unit initial fuel area gives 2.66 g/m^2-sec for
poplar wood shavings (excelsior).

The above numbers are actually quite close when it is
considered that the rather high value from Roberts formula
is based on the maximum burning rate of oven dry wood
whereas for the other results the wood contained significant amounts of moisture. Also the low values obtained
from Fang's data are based on initial fuel particle area
and a substantial part of the fuel has been consumed as
the fire passes (about 70%).

The above data indicate that it may be reasonable to
postulate a burning law on the basis of a volatilization
rate per unit area for a specified heat flux per unit
area, Eq. 2-24 or Eq. 2-29. Robert's relation,
2-24, also implied a specific amount of energy is required
to generate a unit mass of volatile over a wide range of
surface heating rates encountered in fire problems. He
gives a value of 2.2 KJ/g (525 cals/g) as the quantity
desired for Eq. 2-4.

Chapter 3 presents a detailed analysis of the heat
transfer within a solid material undergoing pyrolysis
so this problem will not be discussed here. However, it
should be pointed out that some of the conclusion reached
in Chapter 3 are based on heating to ignition and not for
heating during sustained burning. For instance, it has
been shown for burning cellulosic materials that internal
convection by volatile movement is an important mechanism
of heat transfer (46, 52), the change of physical properties with pyrolysis can substantially alter the internal
temperature distribution of the material (37, 49, 52) and
the heat of reactions resulting from pyrolysis becomes
significant (37, 44, 49, 52).

A volatile liberation rate per unit area of fuel based
on a combination of theoretical analysis of heat transfer
within the material and at the fuel surface in conjunction
with experimental measurements seems within reach at this
time.

Relations Governing the Overhead Flame

The heat flux terms in Eqs. (2-3) and (2-4) must
be determined from the temperature distribution in the
burning zone and overhead flame. The analysis of radiation
from the latter also requires a knowledge of the concentration distribution of the radiating gases and particles

in the flame. At the present time it is only possible to make very rough approximations of these distributions although this may be adequate for a number of applications.

The size of the overhead flame is primarily determined by the rate of mass liberation of volatile combustible material by the fuel bed and the distance over which it is liberated. Steward (61) developed a mathematical model on the basis of a given amount of combustible fuel entering the atmosphere through a source of specified width and infinite length at a uniform velocity. It was assumed that the fuel burnt to stoichiometric completion as it mixed with the   surrounding   air. The entrainment of the fuel jet was also assumed to obey the jet mixing laws derived for isothermal or nearly isothermal systems (40, 43, 57). A solution based on the equation of the conservation of energy, a force balance, and continuity indicated that the height at which a stoichiometric amount of air was entrained or a specified per cent excess air was entrained was given by

$$\frac{H}{D_B} \propto \left[\frac{(Q/L)^2 (s + w\rho_a/\rho_o)^2 \, w}{k_e^2 D_B^3 \, g\Delta H_c^2 \, \rho_a^2 \, (1-w)^3} \right]^{\frac{1}{3}} \qquad (2-26)$$

where H is the height in question, l
    w is the inverse coefficient of expansion due to combustion
    D_B is the width of the base, l
    Q/L is the energy liberation per unit length by combustion, h/lt
    s is the mass air to fuel ratio for stoichiometic combustion
    k_e is the jet entrainment coefficient
    g is the gravitational constant, $1/t^2$
    ΔH_c is the heat of combustion of the fuel per unit mass, h/m
    ρ_a is the density of air, m/l^3
    ρ_o is the density of the inlet fuel, m/l^3

Data for visible flame heights of a number of gaseous fuels were found to correlate with the quantity in the brackets, the empirical form being

$$\frac{H}{D_B} = 20 \left[\frac{(Q/L)^2 (s + w\,\rho_a/\rho_o)^2 \, w}{D_B^3 \, g\Delta H_c^2 \, \rho_a^2 \, (1-w)^3} \right]^{0.363} \qquad (2-27)$$

This relation is similar to the one postulated by Thomas et al (66, 69) on dimensional arguments

$$\frac{H}{D_B} \, \mathbf{\delta} \left[\frac{(Q/L)^2}{g D_B} \right]^{\frac{2}{3}} \qquad (2-28)$$

Equation (2-27) as well as other empirical correlations for various types of line fires, wood cribs (27) and long strip sources [66] are presented in Fig. 2-20. The relations are from three sets of data by different workers.

It is apparent that there is reasonable agreement between the three curves although Thomas's relation gives somewhat higher values than the other two.

These relations give the visible flame height essentially in terms of the mass volatile liberation rate per unit length, Q/L / ΔH_c, and the depth of the burning zone D_B, the other quantities being essentially the same for all cellulosic base fuels. However, this is only a measure of the size of the overhead flame and gives no indication with respect to the temperature distribution within it. Presently the best approach at this point for radiative analysis from the flame is to use a mean radiating temperature and one of the gray gas interchange areas presented in Appendix II (B-3 or B-4) based on the size of the flame calculated by relation 2-27 or Fig. 2-20.

It is conceivable that a more sophisticated analysis might be carried out to obtain the flame radiation similar to that of Masliyah and Steward (41) for a fire above a circular source. The model of the line fire given by Steward (61) predicts the variation of the temperature with height in the flame, but horizontal variations have been ignored since "top hat" profiles were assumed within the fire. A numerical solution incorporating finite difference techniques similar to that used in combustion chambers (30) is also feasible at this stage.

The analysis immediately above applies only to flames in the absence of wind. Any significant wind will greatly alter the nature of the flame. If the wind is normal to the fire front it will bend the flame in the direction in which it is blowing. Pipkin and Sliepcevich (51) carried out an analysis on the flame bending angle of hydrocarbon flames fed from cylindrical and channel sources in a cross wind. The analysis involved writing a force balance on the entire flame.

In a more detailed analysis Fang (22) solved a set of differential equations based on force, mass, and energy balances within the flame and a simplified burning law as the fuel and air mixes. He obtained an expression for the flame bending angle

$$\frac{tan\ \theta}{cos\ \theta} = \frac{C_F\ u^2}{2(1-\frac{\rho_f}{\rho_a})\ g D_B} \tag{2-29}$$

where C_F is the flame drag coefficient assumed constant over the flame

U is the wind velocity, l/t

ρ_f is the density of the gas in the flame, m/l^3

ρ_a is the density of the surrounding air, m/l^3

D_B is the width of the fuel source, l

g is the gravitational constant, l/t

Figure 2-21 presents a plot of some experimental flame data based on this equation. It is seen that a substantial amount of the data seem to be well correlated although there are some wide variations.

The interaction of the wind with the flame will also affect its temperature distribution by modifying the rate of mixing. The simplification of considering overall flame properties as for the still air flame is probably the best approach at present.

Thomas (68) has correlated data on visible flame lengths for lines fires in a uniform wind with the relation

$$\frac{L}{D_B} = 70 \left[\frac{(m/L)^2}{3\rho_o^2 D_B^3} \right]^{0.43} \left[\frac{U^2}{g D_B} \right]^{-0.11} \tag{2-30}$$

where m/L is the mass rate of fuel entering the flame per
  unit length, m/lt
  U is the wind velocity, l t
  L is the length of the flame, l

This relation can be used in the same manner as those presented above for still air.

The radiative analysis from a bent flame would be similar to that of a vertical one as presented above. However, there are at present no interchange areas calculated for this geometry and one is forced to use the black body view factor Appendix I (A-3) with a flame emissivity.

CONCLUSIONS

This chapter has attempted to outline the most important aspects of fire spread through a fuel bed. Considerable data has been presented to support the various contentions made. A particular emphasis was placed on the heat transfer mechanisms occurring within a spreading fire.

The mathematical analysis presented in the final section was kept in as general terms as possible. It was also intended that it should be primarily concerned with a physical understanding of the problem so that mathematical detail was kept to a minimum. It is hoped that the model presented is a comprehensive one in that it can be used as the starting point for nearly all types of fuelbeds. However, it should be emphasized that any detailed fire spread analysis depends very heavily on the type of fuelbed under consideration. With the data that has been collected it is quite apparent that the various heat flux terms must be tailored to the particular problem of interest.

REFERENCES

1. Albini, F.A., "A Physical Model for Fire Spread in
    Brush", Eleventh Symposium (International) on
    Combustion, p. 553, The Combustion Institute,
    Pittsburgh (1967).

2. Anderson, H.E., "Heat Transfer in Fire Spread",
    USDA, Forest Service, Research Paper INT-69,
    Intermountain Forest and Range Experiment Station,
    Ogden, Utah (1969).

3. Anderson, H.E., Brackebusch, A.P., Mutch, R.W. and
    Rothermel, R.C., "Mechanisms of Fire Spread
    Research", Progress Report No. 2, U.S. Forest
    Service, Research Paper INT-28 (1966).

4. Anderson, H.E. and Rothermel, R.C., "Influence of
    Moisture and Wind Upon the Characteristics of
    Free Burning Fires", Tenth Symposium
    (International) on Combustion, p. 1009, The
    Combustion Institute, Pittsburgh (1965).

5. Arseneau, D.F., "The Differential Thermal Analysis
    of Wood", Canadian J. of Chem., $\underline{39}$, 1915 (1961).

6. Berlad, A.L., "Fire Spread in Solid Fuel Arrays,
    Combustion and Flame, $\underline{14}$, 123 (1970).

7. Berlad, A.L., Rothermel, R.C. and Frandsen, W.,
    "The Structure of Some Quasi-Steady Fire-Spread
    Waves", Thirteenth Symposium (International)
    on Combustion, p. 927, The Combustion Institute,
    Pittsburgh, (1971).

8. Blackshear, P.L. and Murty, K.A., "Heat and Mass
    Transfer to, from, and within Cellulosic Solids
    Burning in Air", Tenth Symposium (International)
    on Combustion, p. 911, The Combustion Institute,
    Pittsburgh (1965).

9. Blackshear, P.L. and Murty Kanury, A., "On the
    Combustion of Wood I, A Scale Effect in
    Pyrolysis of Solids", Combustion Science and
    Technology, $\underline{2}$, 1 (1970).

10. Block, J.A., "A Theoretical and Experimental Study
    of Nonpropagating Free Burning Fires",
    Thirteenth Symposium (International) on
    Combustion, p. 971, The Combustion Institute,
    Pittsburgh (1971).

11. Browne, F.L., "Theories of the Combustion of Wood
        and its Control, A Survey of the Literature",
        USDA, Forest Service, Report No. 2136 (1958).

12. Byram, G.M., Clements, H.B., Bishop, M.E., and
        Nelson, R.M., "Final Report, Project Fire Model,
        An Experimental Study of Model Fires", USDA,
        Forest Service, Southern Forest Fire Laboratory,
        Macon, Georgia (1966).

13. Carslaw, H.S. and Jaeger, J.C., "Conduction of Heat
        in Solids", Second Edition, Oxford University
        Press, London (1959).

14. Curry, J.R. and Fons, W.L., "Forest Fire Behavior
        Studies", Mechanical Engineering, 62, 219 (1940).

15. Davis, K.P., "Forest Fire, Control and Use", McGraw-
        Hill Book Company,   New York (1959).

16. deRis, J.N., "Spread of a Laminar Diffusion Flame",
        Twelfth Symposium (International) on Combustion,
        p. 241, The Combustion Institute, Pittsburgh
        (1969).

17. Dunlap, F., "The Specific Heat of Wood" Forest
        Service Bulletin No. 110, U.S. Department of
        Agriculture, Washington, D.C. (1912).

18. Emmons, H.W., "Fire in a Forest", Fire Research
        Abstracts and Reviews, 5, 163 (1963).

19. Emmons, H.W., "Fundamental Problems of the Free
        Burning Fire", Tenth Symposium (International)
        on Combustion, p. 951, The Combustion Institute,
        Pittsburgh (1965).

20. Emmons, H.W. and Shen, T., "Fire Spread in Paper
        Arrays", Thirteenth Symposium (International)
        on Combustion, p. 917, The Combustion Institute,
        Pittsburgh (1971).

21. Fang, J.B., "Fire Spreading Studies", M.Sc. Thesis,
        Chemical Engineering Department, University of
        New Brunswick, Fredericton, N.B., Canada (1966).

22. Fang, J.B., "An Investigation of the Effect of
        Controlled Wind on the Rate of Fire Spread",
        Ph.D. Thesis, Chemical Engineering Department,
        University of New Brunswick, Fredericton, N.B.,
        Canada (1969).

23. Fang, J.B. and Steward, F.R., "Flame Spread through
        Randomly Packed Fuel Particles", Combustion and
        Flame 13, 392 (1969).

24.   Fons, W.L., "Analysis of Fire Spread in Light Forest
      Fuels", Journal of Agricultural Research, $\underline{72}$,
      93 (1946).

25.   Fons, W.L., Bruce, H.D. and Pong, W.Y., "A Steady-
      State Technique for Studying the Properties of
      Free Burning Wood Fires", The Use of Models in
      Fire Research, p. 219, Publication 786, National
      Academy of Sciences, National Research Council,
      Washington, D.C. (1961).

26.   Fons, W.L., Clements, H.B., Elliott, E.R. and
      George, P.M., "Project Fire Model, Summary
      Progress Report II", U.S.D.A. Forest Service,
      Southeastern Forest Experimental Station, Macon,
      Georgia (1962).

27.   Fons, W.L., Clements, H.B. and George, P.M., "Scale
      Effects on Propagation Rate of Laboratory Crib
      Fires", Ninth Symposium (International) on
      Combustion, p. 860, Academic Press, New York
      (1963).

28.   Frandsen, W.H., "Fire Spread through Porous Fuels
      from the Conservation of Energy", Combustion and
      Flame, $\underline{16}$, 9 (1971).

29.   Frandsen, W.H. and Ruthermel, R.C., "Measuring the
      Energy-Release Rate of a Spreading Fire",
      Combustion and Flame, $\underline{19}$, 17 (1972).

30.   Gosman, A.D., Pun, W.M., Runchal, A.K., Spalding,
      D.B. and Wolfshtein, M., "Heat and Mass Transfer
      in Recirculating Flows", Academic Press (1969).

31.   Gross, D., "Experiments in the Burning of Cross Piles
      of Wood", Journal of Research, National Bureau
      of Standards C, $\underline{66C}$, 2, 99 (1962).

32.   Hearman, R.F.S. and Burcham, J.N., "Specific Heat
      and Heat of Wetting of Wood", Nature, $\underline{176}$,
      978 (1955).

33.   Hottel, H.C., Williams, G.C. and Kwentus, G.C.,
      "Fuel Preheating in Free Burning Fires"
      Thirteenth Symposium (International) on
      Combustion, p. 963, The Combustion Institute,
      Pittsburgh (1971).

34.   Hottel, H.C., Williams, G.C. and Steward, F.R.,
      "The Modeling of Fire Spread through a Fuel Bed",
      Tenth Symposium (International) on Combustion,
      p. 997, The Combustion Institute, Pittsburgh
      (1965).

35. King, N.K. and Vines, R.G., "Variation in the
        Flammability of the Leaves of Some Australian
        Forest Species", CSIRO Report, Australia (1969).

36. Kosdon, F.J., Williams, F.A. and Buman, C.,
        "Combustion of Vertical Cellulosic Cylinders in
        Air", Twelfth Symposium (International) on
        Combustion, p. 253, The Combustion Institute,
        Pittsburgh (1969).

37. Kung, H.C. and Kalelkar, A.S., "On the Heat of
        Reaction in Wood Pyrolysis", Combustion and
        Flame, 20, 91 (1973).

38. Lastrina, F.A., Magee, R.S., and McAlevy, R.F.,
        "Flame Spread Over Fuel Beds: Solid Phase
        Energy Considerations", Thirteenth Symposium
        (International) on Combustion, p. 935, The
        Combustion Institute, Pittsburgh (1971).

39. Lawlor, B.J., "Analysis of Fire Spread in Fuel Beds",
        B.Sc. Thesis, Department of Chemical Engineering,
        University of New Brunswick, Fredericton, N.B.
        (1972).

40. Lee, S.L. and Emmons, H.W., "A Study of Natural
        Convection Above a Line Fire", Journal of Fluid
        Mechanics, 11, 353 (1961).

41. Masliyah, J.H. and Steward, F.R., "Radiative
        Transfer from a Turbulent Diffusion Buoyant
        Flame with Mixing Controlled Combustion",
        Combustion and Flame, 13, 613 (1969).

42. McCarter, R.J. and Broido, A., "Radiative and
        Convective Energy from Wood Crib Fires",
        Pyrodynamics, 2, 65 (1965).

43. Morton, B.R., Taylor, G.I. and Turner, J.G.,
        "Turbulent Gravitational Convection from
        Maintained and Instantaneous Sources",
        Proceedings of the Royal Society, A, 234, 1
        (1956).

44. Murty, K. A.   and Blackshear, P.L., "Pyrolysis
        Effect in the Transfer of Heat and Mass in
        Thermally Decomposing Organic Solids", Eleventh
        Symposium (International) on Combustion, p. 517,
        The Combustion Institute, Pittsburgh (1967).

45. Murty Kanury, A. and Blackshear, P.L., "Some
        Consideration Pertaining to the Problem of Wood
        Burning", Combustion Science and Technology, 1,
        339 (1970).

46. Murty Kanury, A. and Blackshear, P.L., "On the
    Combustion of Wood II, the Influence of Internal
    Convection on the Transient Pyrolysis of
    Cellulose", Combustion Science and Technology,
    2, 5 (1970).

47. Nolan, P.F., Brown, D.J. and Rothwell, E., "Gama-
    Radiographic Study of Wood Combustion",
    Fourteenth Symposium (International) on
    Combustion, to be published.

48. Pagni, P.J. and Peterson, T.G., "Flame Spread Through
    Porous Fuels", Fourteenth Symposium (Inter-
    national) on Combustion, to be published.

49. Panton, R.L. and Rittmann, J.G., "Pyrolysis of a
    Slab of Porous Material", Thirteenth Symposium
    (International) on Combustion, p. 881, The
    Combustion Institute, Pittsburgh (1971).

50. Phillips, A.M. and Becker, H.A., "Pyrolysis and
    Burning of Single Sticks of Wood under Uniform
    Ambient Conditions", 22nd Canadian Chemical
    Engineering Conference, Toronto, (Sept. 1972).

51. Pipkin, O.A. and Sliepcevich, C.M., "Effect of
    Wind on Buoyant Diffusion Flames", I and E.C.
    Fundamentals, 3, 147 (1964).

52. Roberts, A.F., "Problems Associated with the
    Theoretical Analysis of the Burning of Wood",
    Thirteenth Symposium (International) on
    Combustion, p. 893, The Combustion Institute,
    Pittsburgh (1971).

53. Rothermel, R.C., "Tailoring the Fire Spread Rate
    Model to the Field", Central States Section
    Spring Meeting, The Combustion Institute,
    Minneapolis, Minnesota (1969).

54. Rothermel, R.C., "A Mathematical Model for
    Predicting Fire Spread in Wildland Fuels",
    USDA, Forest Service, Research Paper INT 115
    (1972).

55. Rothermel, R.C. and Anderson, H.E., "Fire Spread
    Characteristics Determined in the Laboratory",
    U.S. Forest Service, Research Paper INT-30
    Ogden, Utah, (1966).

56. Rothermel, R.C., Frandsen, W.H. and Philpot, C.W.,
    "Recent Advances on the Fire Spread Model",
    Western States Section Meeting of the Combustion
    Institute, Paper No. 71-7,(April 1971)

57. Rouse, H., Yih, C.S. and Humphreys, H.W.,
        "Gravitational Convection from a Boundary
        Source", Tellus, 4, 201 (1952).

58. Sandhu, S.S., "Mechanisms of Heat Transfer to
        Fibrous Fuel Ahead of an Advancing Flame",
        M.Sc. Thesis, Department of Chemical
        Engineering, University of New Brunswick,
        Fredericton, N.B., Canada. (1970)

59. Smith, P.G. and Thomas, P.H., "The Rate of Burning
        of Wood Cribs", Fire Technology, 6, 29 (1970).

60. Steward, F.R., "Flame Spread Through a Solid Fuel",
        D.Sc. Thesis, Chemical Engineering Department,
        MIT, Cambridge, Mass. (1962).

61. Steward, F.R., "Linear Flame Heights for Various
        Fuels", Combustion and Flame, 8, 171 (1964).

62. Steward, F.R., "A Mechanistic Fire Spread Model",
        Combustion Science and Technology, 4, 177 (1971).

63. Steward, F.R., "Some Observations on Convective Heat
        Transfer During Fire Spread Through a High
        Voidage Fuel Bed", Eastern Section Meeting
        Combustion Institute (1972).

64. Steward, F.R. and Waibel, R.T., "Fire Spread
        Through Uniform Fuel Matrices", Western States
        Section Meeting, The Combustion Institute,
        WSC1 73-3, Tempe, Arizona (1973).

65. Tarifa, C.S., Notario, P.P.D., and Moreno, F.G.,
        "On the Flight Paths and Lifetimes of Burning
        Particles of Wood," Tenth Symposium
        (International) on Combustion, p. 1021, The
        Combustion Institute, Pittsburgh (1965).

66. Thomas, P.H., "The Size of Flames from Natural Fires",
        Ninth Symposium (International) on Combustion,
        p. 844, Academic Press, New York (1963).

67. Thomas, P.H., "Some Aspects of the Growth and Spread
        of Fire in the Open", Forestry, 40, 139 (1967).

68. Thomas, P.H. "Rates of Spread in Some Wind
        Driven Fires", Forestry, 44, No. 2, 156 (1971).

69. Thomas, P.H., Webster, C.T. and Raftery, M.M., "Some
        Experiments on Buoyant Diffusion Flames",
        Combustion and Flame, 5, 359 (1961).

70. Tinney, E.R., "The Combustion of Wooden Dowels in
        Heated Air", Tenth Symposium (International)
        on Combustion, p. 925, The Combustion Institute
        Pittsburgh (1965).

71. Van Wagner, C.E., "Three Experimental Fires in
        Jack Pine Slash", Canadian Department of
        Forestry, Publication No. 1146 (1966).

72. Van Wagner, C.E., "Fire Behavior Mechanisms in a Red
        Pine Plantation; Field and Laboratory Evidence",
        Canadian Department of Forestry, Publication No.
        1229 (1968).

73. Vines, R.G., "Heat Transfer Through Bark and the
        Resistance of Trees to Fire", Australian
        Journal of Botany, 16, 499 (1968).

74. Vogel, M. and Williams, F.A., "Flame Propagation
        Along Matchstick Arrays", Combustion Science and
        Technology, 1, 429 (1970).

75. Woods Hole Study Group, "A Study of Fire Problems",
        pp 102-12, National Academy of Sciences, National
        Research Council Publication No. 940,
        Washington, D.C. (1961)

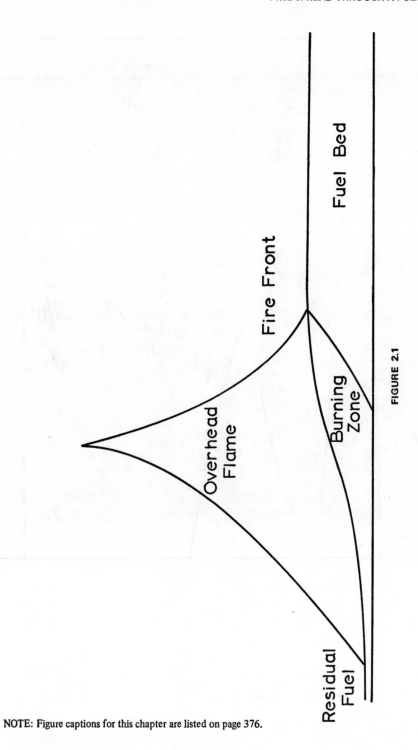

FIGURE 2.1

NOTE: Figure captions for this chapter are listed on page 376.

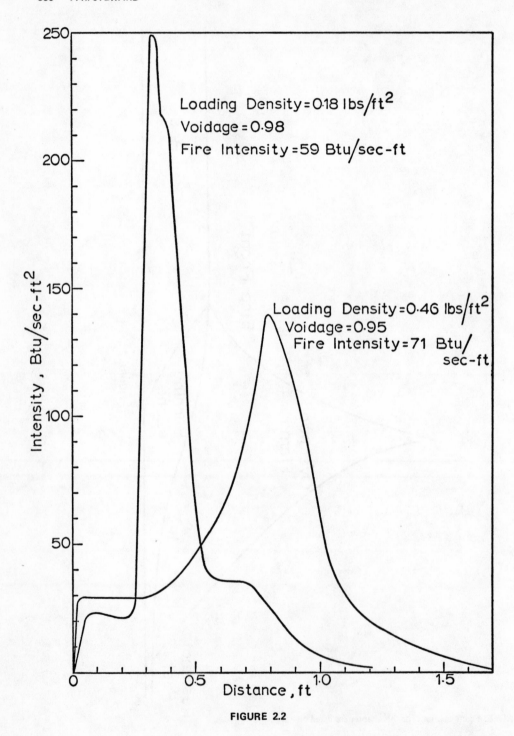

FIGURE 2.2

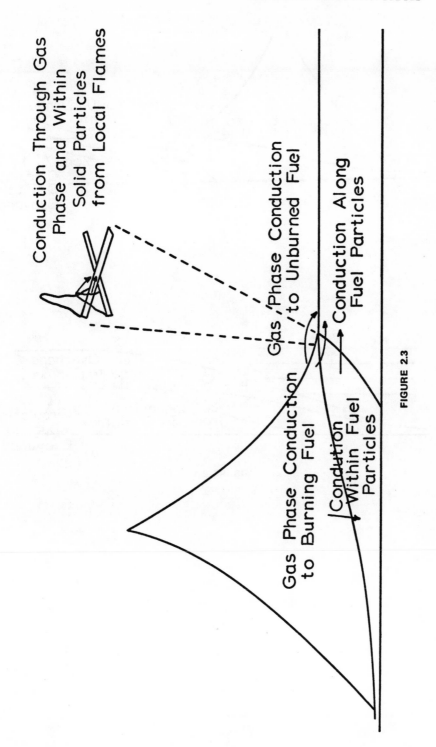

Conduction Through Gas Phase and Within Solid Particles from Local Flames

Gas Phase Conduction to Unburned Fuel

Conduction Along Fuel Particles

Gas Phase Conduction to Burning Fuel

Conduction Within Fuel Particles

FIGURE 2.3

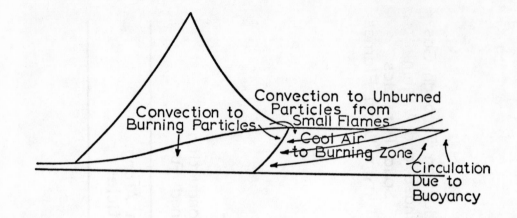

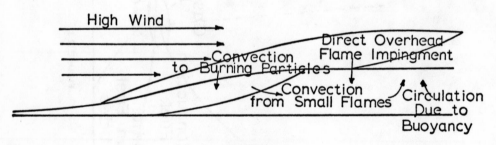

FIGURE 2.4

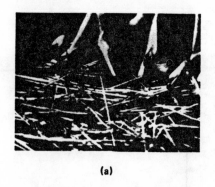

(a)

(b)

(c)

(d)

FIGURE 2.5

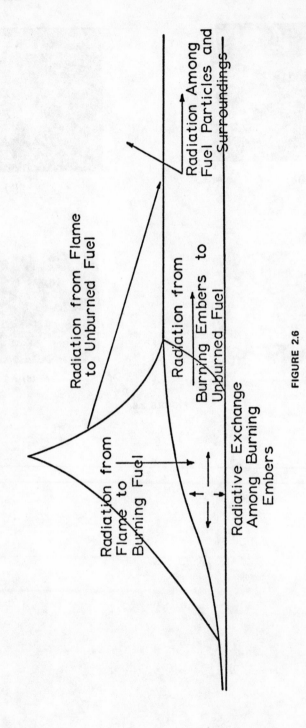

FIGURE 2.6

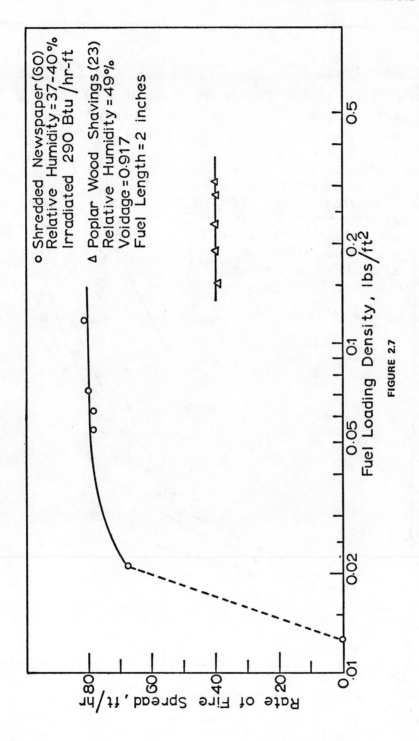

FIGURE 2.7

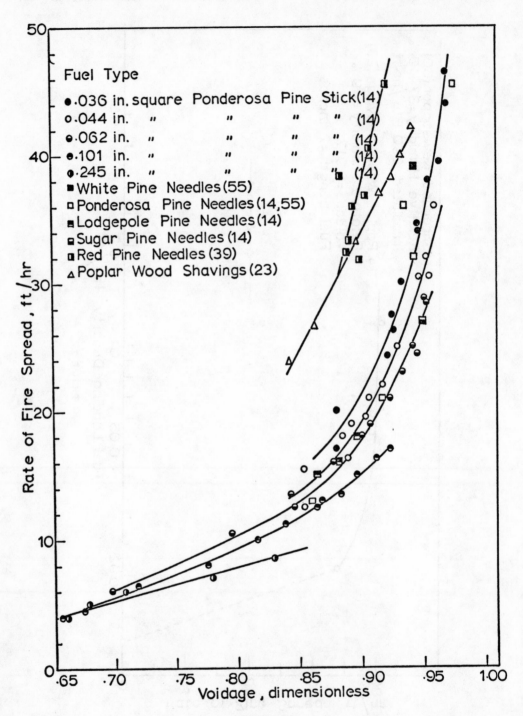

FIGURE 2.8

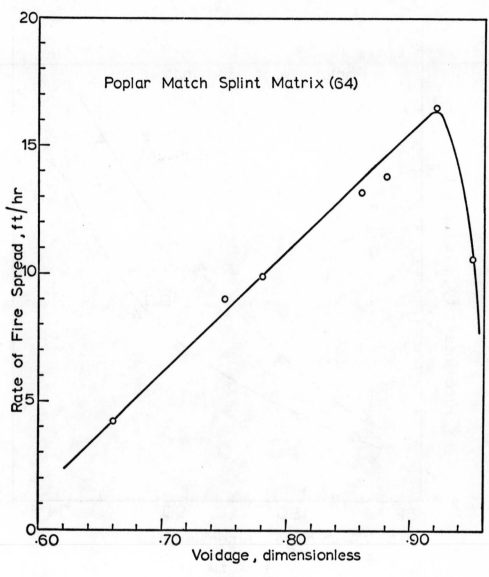

FIGURE 2.9

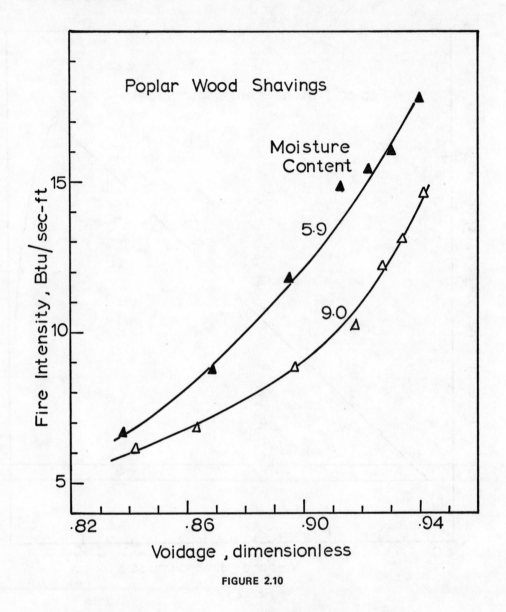

FIGURE 2.10

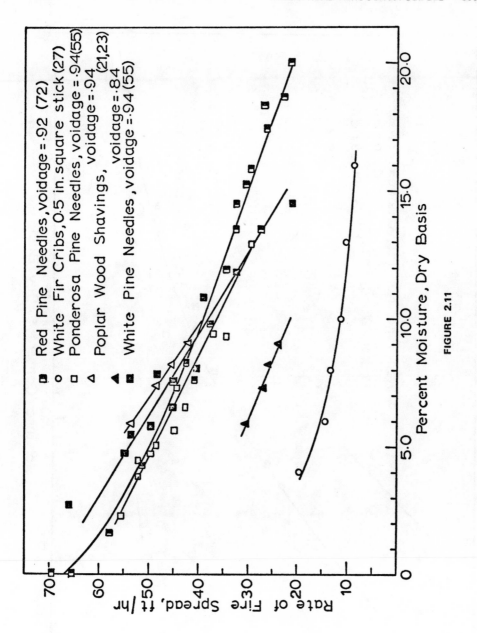

Red Pine Needles, voidage = .92 (72)
White Fir Cribs, 0.5 in. square stick (27)
Ponderosa Pine Needles, voidage = .94(55)
Poplar Wood Shavings, voidage = .94 (21,23)
White Pine Needles, voidage = .84
White Pine Needles, voidage = .94 (55)

Rate of Fire Spread, ft/hr

Percent Moisture, Dry Basis

FIGURE 2.11

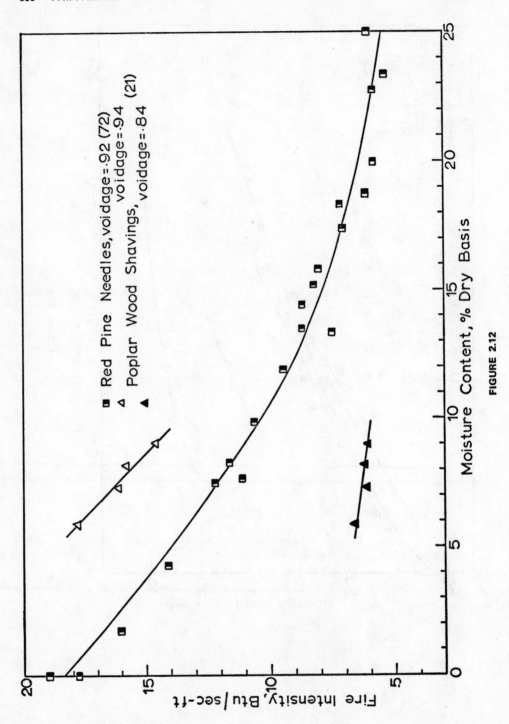

FIGURE 2.12

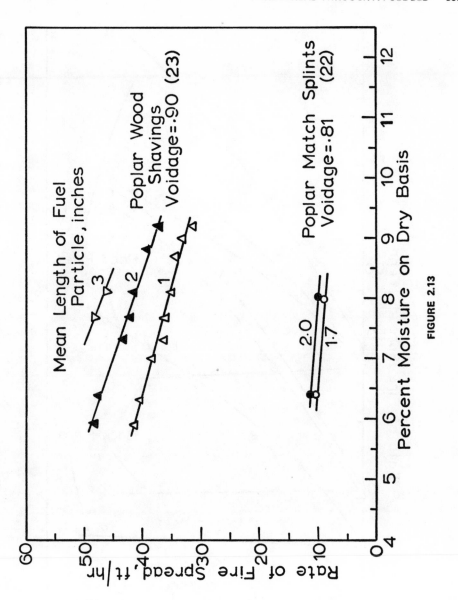

FIGURE 2.13

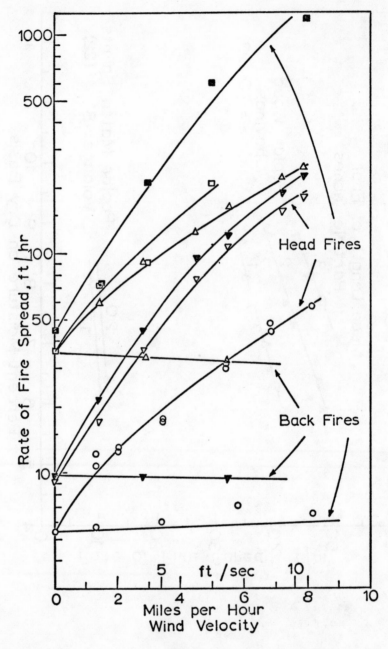

FIGURE 2.14

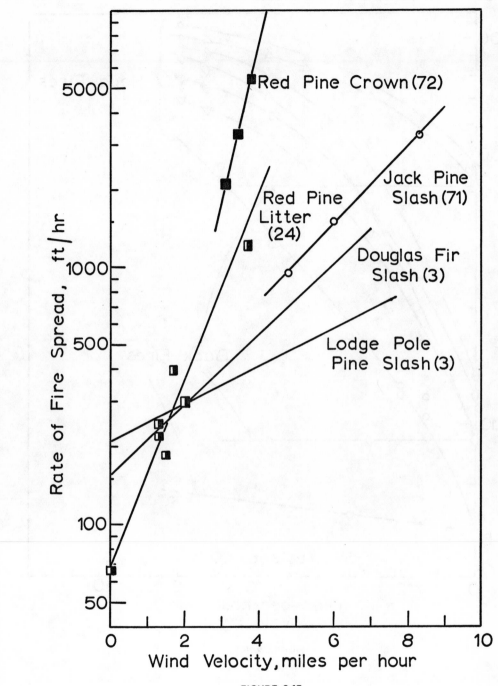

FIGURE 2.15

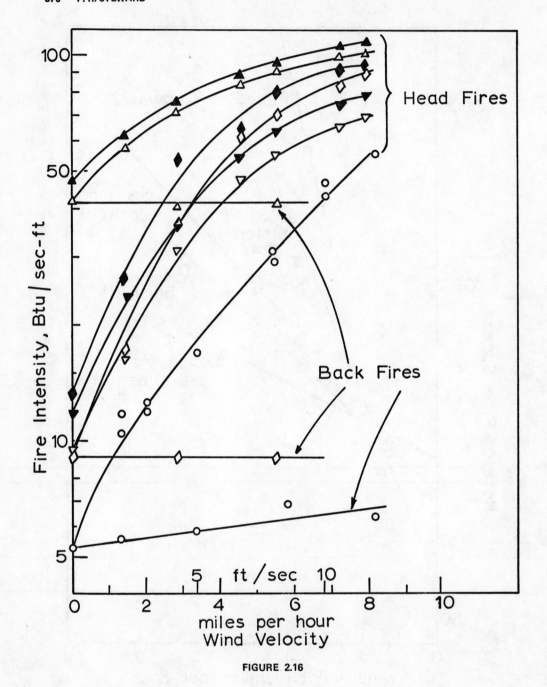

FIGURE 2.16

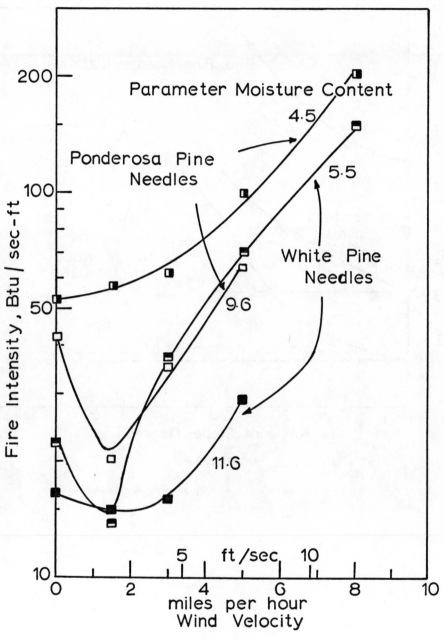

FIGURE 2.17

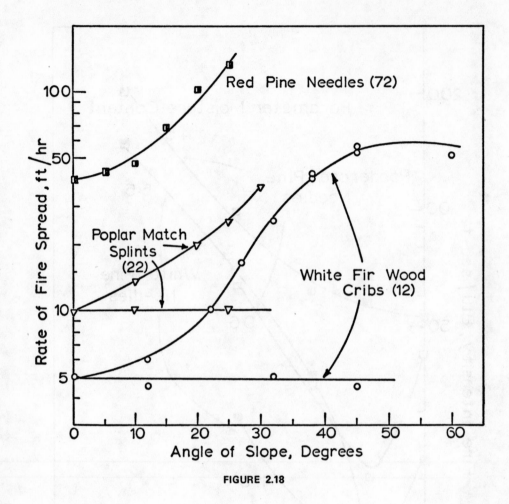

FIGURE 2.18

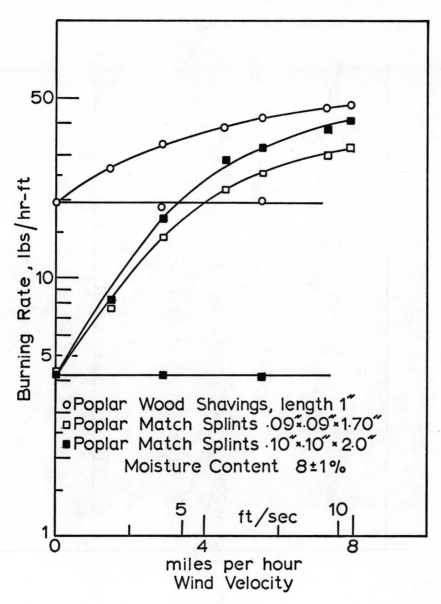

FIGURE 2.19

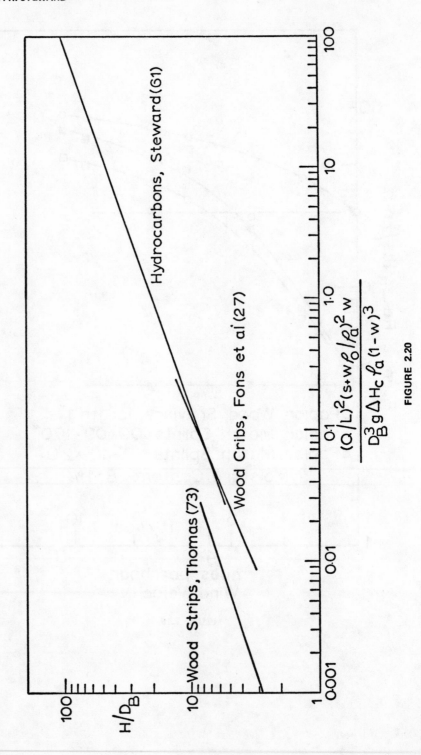

FIGURE 2.20

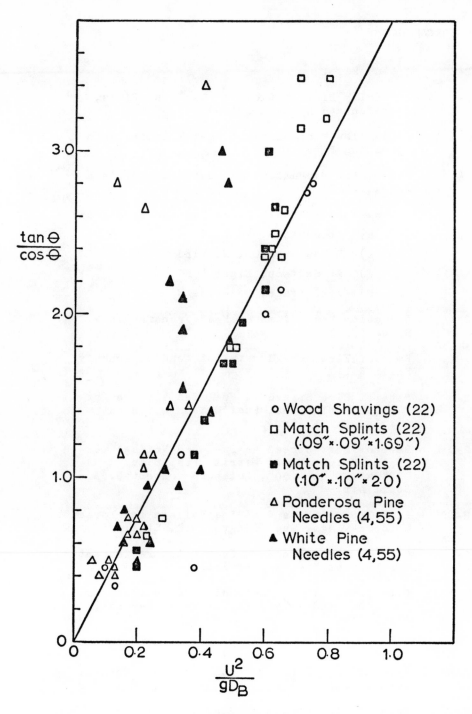

$$\frac{\tan \ominus}{\cos \ominus}$$

$$\frac{U^2}{gD_B}$$

○ Wood Shavings (22)
□ Match Splints (22)
   (.09″×.09″×1.69″)
▣ Match Splints (22)
   (.10″×.10″×2.0)
△ Ponderosa Pine
   Needles (4,55)
▲ White Pine
   Needles (4,55)

FIGURE 2.21

FIGURE CAPTIONS

2-1   A Line Fire.

2-2   Intensity Distribution for Two Line Fires,
         Frandsen and Rothermel (29).

2-3   Conduction Mechanisms of Heat Transfer in a
         Spreading Fire.

2-4   Convection Mechanisms of Heat Transfer in a
         Spreading Fire.

2-5   Photographs of Small Flames at the Fire Front

         a) At Surface, No Wind

         b) Below Surface, No Wind

         c) At Surface, Light Wind

         d) At Surface, Moderate Wind

2-6   Radiation Mechanisms of Heat Transfer in a
         Spreading Fire.

2-7   Rate of Fire Spread vs. Fuel Loading Density,
         Still Air.

2-8   Rate of Fire Spread vs Fuel Bed Voidage for
         Various Fuel Types, Fuel Moisture Content
         9 ± 1%, Still Air.

2-9   Rate of Fire Spread vs Fuel Bed Voidage for
         Poplar Match Splint Matrix, Fuel Size
         .09" x .09" x 1.50", Moisture Content 5.5%,
         Still Air.

2-10   Fire Intensity vs Fuel Bed Voidage for Poplar
         Wood Shavings, Fuel Length = 1", Still Air.

2-11   Rate of Fire Spread vs Fuel Moisture Content
         for Various Fuel Types, Still Air.

2-12   Fire Intensity vs Fuel Moisture Content for
         Red Pine Needles and Poplar Wood Shavings,
         Still Air.

2-13   Rate of Fire Spread vs Fuel Moisture Content
         for Various Fuel Particle Lengths, Still Air.

2-14   Rate of Fire Spread vs Wind Velocity for
         Various Fuel Types (Laboratory).

○ White Fir Sticks .256 in Square (12)
            1 in. spacing
            1.52 in. high.

▽ Poplar Match Splints .09 in. Square (22)
                1.7 in. length
▼   voidage .81
                .10 in. square
                2.0 in. length
                voidage .81

△ Poplar Wood Shavings Voidage .92 (22)
            mean length 1 in.

☐ Ponderosa Pine Needles Voidage .94 (4.55).

▨ White Pine Needles voidage .94 (4.55).

    All Fuels moisture content 9 ± 1%.

2-15   Rate of Fire Spread vs Wind Velocity for
        Various Fuel Types (Field).

2-16   Fire Intensity vs Wind Velocity for Various
        Fuel Types (Laboratory).

○ White Fir Sticks (12)
                Fuel Loading Density = .57 lbs/ft^2
                Spacing = 1.0 in.
                Height = 1.52 in.
                Cross Section = 0.256 in. square.

    Poplar Match Splints (22)
                Fuel Loading Density = .441 lbs/ft^2
                Voidage = 0.81
                Length = 2.0 in.
                Cross Section = 0.1 in. square

◇   Moisture Content = 8.3%
◆   Moisture Content = 6.5%

    Poplar Wood Shavings (22)
                Fuel Loading Density = .193 lbs/ft^2
                Voidage = 0.92
                Mean Length = 1.0 in.

△   Moisture Content = 8.1%
▲   Moisture Content = 6.5%.

Poplar Match Splints (22)

Fuel Loading Density = .441 lbs/ft^2
Voidage = 0.81
Length = 1.7 in.
Cross Section = 0.09 in. square

▽   Moisture Content = 8.3%

▼   Moisture Content = 6.5%

2-17   Fire Intensity vs Wind Velocity for Ponderous Pine and White Pine Needle Beds (4, 55).

2-18   Rate of Fire Spread vs Angle of Slope for Various Fuel Types, Still Air, Poplar Match splints (.09" x 109" x 1.00").

2-19   Burning Rate vs Wind Velocity for Various Types of Fuel Beds.

2-20   Correlation of Line Flame Heights with Flame Height to Base Ratio vs Dimensionless Combustion Number.

2-21   Correlation of Flame Bending Angle in a Cross Wind.

Chapter 3

IGNITION CHARACTERISTICS OF CELLULOSIC MATERIALS

F. R. STEWARD
Fire Science Centre, University of New Brunswick
Fredericton, N. B., Canada

3.1 INTRODUCTION

The ignition and burning of cellulosic materials have been of interest to mankind for a very long period of time. However, only in the past 30 years or so has it been subject to detailed and quantitative scientific scrutiny. The practical purposes of such studies have been to determine quantitatively the amount of heating required to ignite various materials under a number of typical conditions of exposure and to develop treatments for the material so as to suppress or prevent ignition. The cellulosic materials of most practical interest in this regard are wood and wood products and cellulose base fibers for obvious reasons.

This chapter will first discuss the general phenomena associated with ignition produced by external heating, but will concentrate on the physics associated with the problem which involves heat transfer, mass transfer, and fluid dynamics.

The basic elements in this chapter were originally prepared for a lecture presented at the Eastern Section Meeting of the Combustion Institute (58). Two other reviews on this subject have also recently appeared in the literature (44, 65).

3.2 TYPES OF IGNITION

Although there are a number of variations, three types of ignition have been generally accepted in the literature;

NOTE: All figures quoted in text are at the end of the chapter.

spontaneous ignition with external heating, pilot ignition, and spontaneous ignition by self heating.

The last of these is brought about by the generation of heat within the material by means of an exothermic chemical reaction which gradually raises the temperature of the substance until it bursts into flames. This process normally occurs over a relatively long period of time and is basically different in mechanism from the other two types of ignition for which the major portion of the heat is supplied externally.

Spontaneous ignition with external heating is defined as the appearance of a flame at the surface of a material which has been exposed to some type of external heating.

Pilot ignition is defined as the appearance of a flame at the surface of a material which has been exposed to some type of external heating with the presence of an ignition source in the volatile stream formed as the material is heated.

In both spontaneous ignition with external heating and pilot ignition two types of flaming have been observed. The flame appears at the surface only momentarily. This is often called "transient ignition". The flame is maintained at the surface and the sample continues to burn in the absence of additional heating. This is often called "sustained or persistent ignition".

Another phenomenon should also be mentioned here which has been referred to as "glowing ignition". Glowing ignition is defined as the appearance of red hot or glowing areas at the surface of the material when exposed to external heating. However, no flame appears at the surface. This phenomenon can be observed when an ignition source is either present or absent but is not considered to constitute ignition according to the above definitions of spontaneous and pilot ignition.

The fire spread equation in Chapter 2, Eq. (2-3) involves the enthalpy required to heat the potential fuel from ambient conditions to the point at which it enters the general conflagration. Although this process is not identical to the normal experimental ignition tests it can not be far different from the pilot ignition above.

This chapter is only concerned with ignition, spontaneous or pilot when external heating is involved as the mechanism of spontaneous ignition produced by self heating is entirely different.

3.3 DESCRIPTION OF THE SYSTEM

The experimental apparatus for studying ignition phenomena have varied widely using many types of materials, a number of external heat sources and several types of ignition sources for pilot ignition studies. The materials include various types of wood (1, 9, 11, 15, 21, 22, 27, 29, 32, 43, 52, 54, 56, 62, 63), various types of cloth (6, 30, 52, 66), α cellulose (3, 16, 38, 39, 40, 41, 63, 64, 66, 68), and paper (11, 52). The external heating has consisted of irradiation by a hot surface (15, 32, 43, 52, 53, 55, 56), a carbon arc (3, 16, 38, 39, 40, 41, 52, 68, a tungsten lamp (6, 52, 53, 66), a flame (29, 66) nuclear weapons (11, 68), hot gases forced pass the surface of the sample (1, 63, 64) and a combination of irradiation by a hot surface and the flow of hot gases past the surface of the sample (21). The most normal type of pilot ignition source has been a small flame, (1, 15, 29, 32, 43, 55, 56), but an electrical spark (1) or a hot surface (1, 63, 64), have also been used.

Whatever the cellulosic material, means of heating or source of ignition is present, the overall process is generally the same and is shown schematically in Fig. 3-1. The external heat source transfers heat to the material in some fashion. The material increases in temperature as it receives this heat and simultaneously redistributes the heat by conduction. As certain temperature levels are reached within the material volatiles are formed and proceed to the surface where they are expelled into the surrounding air. These volatiles will consist mainly of water vapor in the early stages resulting from hygroscopic moisture but as the temperature of the material increases decomposition reactions occur which release combustible products. The volatile stream is hotter and thus less dense than the surrounding air and therefore rises as it mixes with the air. If sufficient heat is supplied a combustible mixture is eventually formed at some point in this volatile stream and ignition occurs at this spot and flashes back to the surface where it may or may not be stabilized.

Pilot ignition will occur when a mixture of volatiles and air within the combustible limits at the temperature of the source reaches the vicinity of the ignition source.

Spontaneous ignition will occur when some point in the volatile stream attains a combustible mixture. This of course involves both a temperature and concentration criteria and is thus more difficult to achieve.

In either case the entire process involves a number of interacting chemical and physical processes which are presently only understood in a gross fashion. Although the chemical and physical processes are interdependent it is useful to look at them first separately to simplify the problem.

3.4 CHEMICAL CONSIDERATIONS

The three major constituents of wood have been determined as hemicellulose, cellulose and lignin (1, 10, 14, 36). Although other compounds are known to be

present in small amounts (8). Because of the complex
structure of a mixture of such compounds it is not
difficult to understand the problems in attempting to
investigate the chemistry of decomposition of wood. In
the past ten years this problem has lent itself to a
certain amount of clarification by the use of thermogravimetric analysis (TGA) and differential thermal
analysis (DTA), but this field of study is really only in
its infancy. A detailed review of the literature on these
methods has recently been prepared by Beall and Eickner
(10). A much older but useful review of the literature
on the pyrolysis of wood in general is given by Browne
(14).

One method of simplifying a study of the chemical
reactions taking place as wood is heated and undergoes
pyrolysis is to study the pyrolysis of major chemical
constituents individually. This method appears to be
justified by Arseneau (8) who found that the DTA of
balsam fir is similar to the sum of the DTA's of its
constituents, hemicellulose, cellulose, lignin, and other
minor extracts. However, care should be exercised in this
approach as a recent study by Roberts (48) indicated that
the DTA of wood more closely resembles that of lignin than
the other two major constituents. It is also known that
small concentrations of inorganic impurities greatly
effect the pyrolysis of cellulose alone (13).

Although the differential thermal analysis of wood
varies from species to species the findings of a number of
workers, presented by Beall and Eickner (10) give the same
general characteristics; an endothermic region starting
just below 100°C which continues to between 250 - 300°C
where it reaches a maximum, followed by an exothermic
region which persists to about 450 - 550°C.

Various thermogravimetric analysis (TGA) of wood
(10) indicate that, after the weight loss due to the
evaporation of hygroscopic water, significant weight loss
of the wood itself begins to occur just about 200°C and
is nearly completed by 400°C.

It has been stated that the major constituents of
wood decompose on heating in the order hemicellulose,
cellulose, and lignin (14, 20, 46) although there is
disagreement on the order of cellulose and lignin (1). Of
these three major constituents the decomposition of
cellulose has received far greater attention than the
other two probably because its structure is known making
its chemistry more amenable to theoretical consideration
and it is the primary constituent of ordinary paper.

A number of differential thermal analysis of various
workers are presented by Beall and Eickner (10) which show
the same general characteristics, although somewhat
different in detail; an endothermic region around 100°C

where hygroscopic water is evaporated, a region up to about 275°C which has been characterized as a shallow endotherm, a shallow exotherm or a plateau by different workers, a deep endotherm around 275 to 325°C which reverses to a strong exotherm about 350°C and finally disappears at approximately 400 - 500°C.

TGA curves for cellulose indicates that significant weight loss of the material itself does not begin until about 275°C but proceeds rapidly and is nearly complete at about 375°C.

In addition to differential thermal analysis a number of studies have determined the various products of pyrolysis (14, 17, 20, 23, 24, 25, 27, 36, 39, 42, 50, 51, 60). The complexity of the reaction is demonstrated by the large number of products identified. For instance Schwenker and Beck (50) detected 37 volatile compounds alone although only 18 were identified. However, once again the results are generally similar as pyrolysis of the cellulose produces volatiles, tar, and char of varying amounts and composition depending on the rate of heating, composition of surrounding atmosphere, and the starting material particularly with respect to the trace impurities.

The most important observation from this pyrolysis work seems to be the close relationship between the amount of levoglucosan formed in the tar and the ease of ignition (12, 17, 24, 28, 36, 51). This has led to the interpretation of the DTA curve for cellulose in the following manner (28). The nearly neutral reaction in the region below 275°C decomposes the cellulose chain without forming levoglucosan while the strongly endothermic reaction in the region of 300°C produces a high levoglucosan content tar. These two reactions presumably compete for the original cellulose. The exothermic reaction at higher temperature is due to further decomposition of the products. However, it seems the decomposition of the material formed by the non tar producing endothermic reaction gives less combustible volatiles than the decomposition of the tar containing a high concentration of levoglucosan. This is supported by the observations that some flame retardants inhibit the formation of levoglucosan in particular (12, 17, 51, 60) as well as tar in general (14, 20, 24, 51).

3.5 PHYSICAL CONSIDERATIONS

The physical aspects of the problem are associated with the heat transfer, mass transfer, and fluid dynamics which arise as the material is heated. The external heat is primarily transferred to the surface of the material although a certain amount of radiation may penetrate cellulosic materials (25, 52, 53). The heat will be transferred into the material by conduction which is the only possible internal mechanism of heat transfer in the early stages of heating. As the temperature at any particular

point rises first hygroscopic water will be released
followed by the numerous chemical reactions described in
the previous section. As indicated these reactions are
first endothermic and eventually exothermic.

The water vapor and gaseous products of these
reactions "diffuse" to the surface of the material
although diffusion further into the sample probably occurs
to a limited extent. This internal flow provides an
additional mechanism of heat transfer.

Depending on the mechanism of external heating
energy will be transferred to or form the surface receiving heat by convection and radiation. If the sample
is thin enough so that heat penetrates to the surface not
being heated externally radiation and convection or conduction will transfer heat from this surface also.

The volatiles leaving the surface of the sample
flow upwards due to buoyancy and mix with the surrounding
air. Ignition has been shown to occur within this stream
a considerable distance from the surface, approximately
one inch (52, 53, 58, 62).

3.6 INTERACTION OF CHEMICAL AND PHYSICAL PHENOMENA

The interaction of the chemical and physical
phenomena is considerable. Within the material itself
the heat transfer determines the termperature distribution
within the solid material during the early stages of
heating. However, once temperatures are attained at
which chemical reactions commence the temperature distribution is affected by the heat of reaction, the flow
of the volatiles to the surface, and the change in
physical properties during decomposition. The temperature
distribution in turn determines the relative rates of the
various competing reactions. In addition the volatiles
flowing to the surface are in close contact with the
solid material and therefore undergoing secondary reactions.

The rate and chemical composition of the volatiles
reaching the surface determine the ability of the volatile
stream to mix with the surrounding air to form a combustible mixture.

3.7 MATHEMATICAL ANALYSIS

A mathematical description of the heat transfer
within the cellulosic material is not difficult in
principle and has been considered for a number of
boundary conditions and various assumptions (1, 7, 9, 25,
44, 52, 54, 55, 56, 58, 61, 63). The problem is usually
considered to be one dimensional with all heat transfer
occurring normal to the surface since experimental
samples are normally either thin relative to their
surface area or the level of penetration of heat is small
relative to the size of the sample. The energy balance

on a differential element dx per unit area located within the cellulosic material is given by

Conduction   *Radiation*   *Chemical Reaction*

$$\frac{\partial}{\partial x}\left[k\,\frac{\partial T}{\partial x}\right] \quad + \quad a\,I\,e^{-ax} \quad + \quad \sum_i Q_i\,r_i$$

Convection   *Thermal Inertia*

$$-\frac{\partial}{\partial x}\,(m_v C_p T) \qquad = \qquad \rho C_p\,\frac{\partial T}{\partial t} \qquad\qquad (3-1)$$

All nomenclature is given in the Appendix VI.

The absorption of irradiation within the solid is assumed to obey Beer's Law and the temperature of the volatiles passing through the element are assumed to have the same temperature as the solid at all levels.

If the chemical reactions are assumed to obey the Arhenius expression

$$r_i = A_i e^{-E_i/RT}\,m_s{}^{n_i} \qquad\qquad (3-2)$$

Where w is the mass of reacting material per unit volume and n_i is the order of reaction i.

The mass flow rate of volatiles through the element must be given as the rate at which they are produced deeper in the material. If inertial effects on mass flow are neglected

$$m_v = \int_0^x \sum_i A_i e^{-E_i/RT}\,m_s{}^{n_i}\,\gamma_i\,dx \qquad (3-3)$$

where γ_i is the mass of volatiles formed per unit mass of material reacted according to reaction i.

The above equation in this general form cannot be solved analytically. However, several cases have been solved in certain simpler forms using particular boundary conditions by the use of finite difference methods (9, 61, 64). With present computational facilities these equations are not difficult to solve numerically by finite difference methods, however, as with many physical problems of this type the available physical data does not warrant a great amount of mathematical sophistication. The physical properties of cellulosic materials are only known accurately for a few substances and are highly dependent on moisture content, temperature and fraction of decomposition (45). The chemical reaction rate constants have been determined by a number of workers (1, 2, 8, 34, 35, 37, 49, 57, 59) but many different values have been obtained (45). A recent critical review of the kinetic data on wood pyrolysis has been prepared by Roberts (47).

The complexity of the chemical kinetics makes it attractive to at least consider some of the solutions obtained with a number of simplifying assumptions. These are obtained by neglecting radiative transfer within the material, heat transfer by the flow of volatiles within the material, the heat absorbed or generated by chemical reaction, and the variation of the physical properties with temperature. The material is thus an inert slab with constant physical properties. Using such simplifications the following solutions for external heating supplied by irradiation at one face are of interest.

3-7-1   An Inert Slab with Constant Physical Properties
        Originally at a Uniform Temperature Absorbing
        Irradiation at a Uniform and Constant Rate at One
        Surface. Heat Loss at this Same Surface is at a
        Rate Proportional to the Temperature Difference
        Between the Surface and the Initial Temperature.
        The Heat Flux at the Other Surface is zero.

The general partial differential equation (3-1) for the energy balance in the solid reduces to

$$k \ \frac{\partial^2 T}{\partial x^2} \ = \ \rho \ C_p \ \frac{\partial T}{\partial t} \tag{3-4}$$

with the boundary conditions

$$0 < x < 1 \quad t = 0 \qquad T = T_0 \tag{3-5i}$$

$$x = 0 \qquad\quad t > 0 \qquad \frac{\partial T}{\partial x} = 0 \tag{3-5ii}$$

$$x = 1 \qquad\quad t > 0 \qquad I_0 - h \ (T - T_0) = k \ \frac{\partial T}{\partial x} \tag{3-5iii}$$

The solution for the temperature distribution in the solid for these conditions is given by Carslaw and Jaeger (18) as

$$T - T_0 = \frac{I_0}{h} \left[1 - \sum_{j=1}^{\infty} \frac{2 \ \frac{h1}{k} \ \cos \ (\beta_j x/1) \sec \beta_j e^{-\beta_j \alpha t / 1^2}}{\frac{h1}{k} \left[\frac{h1}{k} + 1 \right] \ + \ \beta_j^2} \right] \tag{3-6}$$

where β_j is given by the positive roots of the equation

$$\beta_j \ tan \ \beta_j = \frac{hl}{k} \qquad (3-7)$$

An interesting simplification occurs if the heat loss from the irradiated surface is assumedd negligible. The boundary condition (3-5iii) reduces to

$$x = 1 \qquad I_0 = k \ \frac{\partial T}{\partial x} \qquad (3-8)$$

The solution for the temperature distribution in the solid with this boundary condition is also given by Carslaw and Jaeger (18) as

$$T - T_0 = \left[\frac{2 \ I_0 \ (\alpha t)^{\frac{1}{2}}}{k} \right] \sum_{j = 0}^{\infty} \left\{ ierfc \left[\frac{(2j+1) \ 1-x}{2 \ (\alpha t)^{\frac{1}{2}}} \right] \right.$$

$$\left. + ierfc \left[\frac{(2j + 1) \ 1 + x}{2 \ (\alpha t)^{\frac{1}{2}}} \right] \right\} \qquad (3-9)$$

3-7-2   An Inert Slab of Thickness 1 with Constant Physical
        Properties and an Infinite Thermal Conductivity
        Originally at a Uniform Temperature Absorbing
        Irradiation at a Uniform and Constant Rate at One
        Surface. Heat Loss at this Same Surface is at a
        Rate Proportional to the Temperature Difference
        Between the Surface and the Initial Temperature.
        The Heat Loss at the Other Surface is zero.

An energy balance on the entire material is now pertinent. This is given by the total differential equation

$$I_0 - h \ (T - T_0) = \rho C_p \ \frac{dT}{dt} \qquad (3-10)$$

The solution to this equation is straight forward. The temperature of the entire material is given as a function of time by

$$T - T_0 = \frac{I_0}{h} \left[1 - e^{\frac{-ht}{\rho C_p l}} \right] \qquad (3-11)$$

Once again the simplification of no heat loss at the irradiated surface is of interest. In this case the differential equation (10) reduces to

$$I_o = \rho C_p \frac{dT}{dt} \qquad (3\text{-}12)$$

the solution of which is given by

$$T - T_o = \frac{I_o t}{\rho C_p l} \qquad (3\text{-}13)$$

3-7-3 An Inert Slab of Infinite Thickness with Constant
        Physical Properties Originally at a Uniform
        Temperature Irradiated at a Uniform and Constant
        Rate at its Surface. Heat Loss at the Surface
        is at a Rate Proportional to the Temperature
        Difference Between the Surface and the Initial
        Temperature.

The general partial differential equation for the
energy balance in the solid (3-1) reduces to the same
equation as for Case I above

$$k \frac{\partial^2 T}{\partial x^2} = \rho C_p \frac{T}{t} \qquad (3\text{-}14)$$

However, the boundary conditions are given by

$$0 < x < \infty \qquad t = 0 \qquad T = T_o \qquad\qquad (3\text{-}14i)$$

$$x = \infty \qquad t > 0 \qquad T = T_o \text{ or } \quad \frac{\partial T}{\partial x} = 0 \qquad (3\text{-}14ii)$$

$$x = 0 \qquad t > 0 \qquad I_o - h\,(T - T_o) = k\,\frac{\partial T}{\partial x} \qquad (3\text{-}14iii)$$

The solution for the temperature distribution within
the solid material is given by Varma [61] as

$$T - T_o = \frac{I_o}{h} \left[\text{erfc} \left[\frac{x}{2\sqrt{\alpha t}} \right] \right.$$

$$\left. - \exp\left[\frac{hx}{k} + \frac{h^2 \alpha t}{k} \right] \text{erfc} \left[\frac{x}{2\sqrt{kt}} + \frac{h\,(\alpha t)^{\frac{1}{2}}}{k} \right] \right] \qquad (3\text{-}15)$$

For this problem it is also useful to consider the simplified case of negligible heat loss at the irradiated surface. The boundary condition (3-14iii) becomes

$$x = 0 \qquad\qquad t > 0 \qquad\qquad I_0 = k \frac{dT}{dx} \qquad\qquad (3\text{-}16)$$

The temperature distribution within the solid material with this boundary condition is given by Carslaw and Jaeger (18) as

$$T - T_0 = \frac{2 I_0}{k} \left[\left[\frac{\alpha t}{h} \right]^{\frac{1}{2}} \quad e^{-\alpha t} \quad - \frac{x}{2} \, erfc \left[\frac{x}{2\sqrt{\alpha t}} \right] \right] \qquad (3\text{-}17)$$

Solutions for the conduction equation are possible for other simple boundary conditions. Some of these which involve convective rather than radiative heating can be found in the recent review (44). The above solutions are given as examples and will be used to interpret ignition criteria when heating of the material is by irradiation.

3.8 CRITERIA FOR IGNITION

In order to make use of the mathematical analyses based on energy balances within the cellulosic material various criteria for ignition have been postulated. These include:

1. a) The surface temperature of the material attains a certain fixed value. This has been suggested for spontaneous ignition of thin materials, Martin (39), spontaneous ignition of semi-infinite solids, Lawson and Simms (32), Simms (52), Alvares, Blackshear and Murty Kanury (4) and pilot ignition of semi infinite solids Simms (55), Simms and Law (56).

   b) The mean temperature of the material reaches a certain fixed value. This has been suggested for spontaneous ignition of thin materials by Simms (46).

2. The attainment of a particular rate of evolution of volatiles at the surface, Bamford, Crank, and Malan (9) for spontaneous ignition.

3. A combination of the two former depending on circumstances, Akita (1) for spontaneous and piloted ignition.

4. A thermal feed back criteria for the sustained spontaneous ignition of thin materials, Weatherford and Sheppard (63).

5. The surface decomposition reaction becomes self
sustaining, Anderson (7) for spontaneous ignition.

6. The reversal of the temperature gradient at the
surface of the solid for spontaneous ignition by
irradiation of semi infinite solids due to exothermic
reaction in the gas phase, Deverall and Lai (19).

There are certain bases and arguements given for all
of the above criteria for ignition but it is important to
remember two points. First, none of the above conditions
are mutually exclusive and two or more could give similar
times of heating to ignition for the same conditions.
Secondly, all of the above criteria are based on an indirect quantity which is assumed closely related to
ignition. This of course need not be the case or at
least may only apply over a limited range of heating
rates.

3.9 THE USE OF THE MATHEMATICAL ANALYSIS TO INTERPRET IGNITION CONDITIONS

All of the above criteria for ignition are related
in some fashion to the energy balance within the solid
material. Therefore it is useful to attempt an interpretation of ignition phenomena on the basis of the
simplified solutions to this equation for an inert
material given earlier.

Spontaneous Ignition of Thin Materials by Irradiation

The temperature distribution for the uniform and
constant irradiation of a thin inert material originally
at a uniform temperature is given by equation (6). From
this solution it is apparent that there exists a
functional relationship among the pertinent dimensionless
groups.

$$\frac{(T - T_o)h}{I_o} = \phi \left[\frac{\alpha t}{l^2}, \frac{hl}{k}, \frac{x}{l} \right] \tag{3-18}$$

If it is inferred that ignition is related to this
distribution or more precisely to the temperature at a
particular point; the time of irradiation to ignition,
t_{ig}, would be given by

$$\frac{\alpha t_{ig}}{l^2} = \phi \left[\frac{(T - T_o)h}{I_o} \quad \frac{hl}{k} \right] \tag{3-19}$$

Rearrangement yields

$$\frac{I_o \, t_{ig}}{\rho \, C_p l (T - T_o)} = \phi \left[\frac{I_o l \, (T - T_o)}{k}, \frac{hl}{k} \right] \tag{3-20}$$

Without choosing any particular temperature rise or Newtonian cooling coefficient, h, both of which are difficult to determine precisely, a plot of $\dfrac{I_o t_{ig}}{\rho C_p l}$ vs

$\dfrac{I_o l}{k}$ for different values of l should yield a unique diagram if the thermal conductivity of the material k, does not vary greatly. Martin (39) first presented this diagram for α cellulose as shown in Fig. 3-2. It indicates that over a wide range of irradiation intensities the regions of ignition, persistent and transient, and non ignition are reasonably well defined. This is of course an important simplification since the time of irradiation to ignition can be determined uniquely in terms of the level of irradiation, the thickness of the sample and the physical properties of the material.

If one takes the first criterion of spontaneous ignition listed above, that is the attainment of a certain surface temperature, the solution (3-6) can be evaluated at $x = 1$ to give the same diagram for a given surface temperature. This evaluation is given in Fig. (3-3) by the solid curves for $T - T_0 = 575°C$. The lowest curve on the diagram is the limiting case of no heat loss at the surface given by Eq. (3-9) at $x = 1$.

Now consider the solution for the uniform and constant irradiation of a thin inert material originally at a uniform temperature but with an infinite thermal conductivity given by Eq. (3-11).

The evaluation of this equation taking the temperature rise of the material as 575°C is presented in Figure 3-3 by the dotted curves. The straight line represents the limiting case of no surface heat loss.

The similar but not identical nature of Figs. 3-2 and 3-3 is evident. It appears that at high intensities of irradiation the level of the surface temperature is important to produce transient spontaneous ignition but the entire sample must be heated to a certain level to give sufficient volatile evolution for sustained ignition. At low levels of irradiation both criteria are satisfied simultaneously. It also appears that no single temperature level or Newtonian cooling coefficient suffices to correlate the data over the entire range. (Ablation at the high levels of irradiation could be at least partially accounted for by additional.1 Newtonian cooling). It should also not be forgotten that numerous simplifying assumptions were made to obtain the solutions presented in Fig. 3-3. Thus even though the analysis based on the heating of an inert slab cannot predict ignition phenomena in detail it suggests the most useful method of presenting the data and explains the resulting diagram in a general fashion.

Spontaneous and Pilot Ignition of Semi-Infinite Materials by Irradiation

The temperature distribution for the uniform and constant irradiation of a semi-infinite material originally at a uniform temperature is given by Eq. (3-15). From this solution it is apparent that there is a functional relation among the dimensionless groups given by

$$\frac{(T - T_0)h}{I_0} = \phi\left[\frac{h^2\alpha t}{k}, \frac{x}{(\alpha t)^{\frac{1}{2}}}\right] \qquad (3\text{-}21)$$

As above for thin materials if it is inferred that ignition is related to this distribution or more precisely the temperature at a particular point; the time of irradiation to ignition, t_{ig}, would be given by

$$\frac{h^2\alpha t}{k} = \phi\left[\frac{(T - T_0)\,h}{I_0}\right] \qquad (3\text{-}22)$$

Rearrangement gives

$$\frac{I_0\,t_{ig}^{\frac{1}{2}}}{(k\rho C_p)^{\frac{1}{2}}\,(T - T_0)} = \phi\left[\frac{I_0}{h\,(T - T_0)}\right] \qquad (3\text{-}23)$$

As above in order to avoid a particular temperature rise or a Newtonian cooling coefficient, h, a plot of $\dfrac{I_0\,t_{ig}^{\frac{1}{2}}}{(k\rho C_p)^{\frac{1}{2}}}$ vs I_0 suggests itself.

Figure *3-4* gives such a plot for data on spontaneous ignition from Simms (52) and Simms and Law (56) while Fig. 3-5 gives the same plot for pilot ignition data from Simms (55).

The curves in both figures are obtained from evaluating Eq. (3-5) at x 0 which corresponds to the surface of the materials.

It is apparent that in both cases the correlation is reasonable although a particular temperature rise or Newtonian cooling coefficient is not sufficient for a good representation of the data by the mathematical solution.

Variability of Ignition Time

Two recent studies on the ignition characteristics of high voidage fuels of the type often encountered in

forests (22, 62) have emphasized the additional important
variable of sample voidage, but more important the
variability of ignition times for seemingly identical
experimental conditions. Varma and Steward (62) report
a variation of ± 50% for time or irradiation to spontaneous ignition while
Garg and Steward (22) report ± 15% variations for time of irratiation to
ignition for pilot ignition for seemingly identical experimental conditions.

Recent data from our laboratory on the time of irradiation to pilot
ignition for a number of samples with identical moisture content using the
same intensity of irradiation is given in Figure 3-6. The diagram indicates
that the time of irradiation to ignition varies within
much greater limits than can be explained by experimental
error and is thus not a unique value. A possible explanation of this variation may be in the gas phase
ignition. Figure 3-7 shows the flame flashing back to
the surface of a cedar wood sample from the pilot flame.
The sample did not ignite during this particular flash
back. High speed motion pictures indicate that the number
of flash backs before ignition can vary from one to four.

Figure 3-8 shows a sample of wood shavings
igniting spontaneously which is reproduced from an
earlier paper (62). It is apparent that the ignition
occurs in the volatile stream a substantial distance from
the surface of the sample. Earlier photographs of
spontaneous ignition by Simms (53) are similar. This
strongly indicates that the mixing in the volatile stream
is also playing an important role in forming a combustible mixture.

Two recent studies (4, 19) have attempted to account
for the energy and mass transport in the gas phase, but
much remains to be done in this difficult area.

3.10 CONCLUSIONS

The quantitative assessment of the time and energy
required to produce either spontaneous or pilot ignition
of cellulosic materials has greatly improved in recent
years due to the significant effort applied to obtaining
experimental data.

A beginning has been made in the understanding of the
chemistry of the pyrolysis of cellulose and how this is
related to ignition phenomena. However, the pyrolysis of
the other two major constituents of wood, hemicellulose
and lignin, are only beginning to be extensively investigated.

Mathematical analyses based on simplified energy
balances in the solid material have suggested useful
methods of correlating data and can be used to explain
ignition characteristics in a qualitative manner.
Additional analysis by finite difference techniques may be
useful but should be employed sparingly in view of the lack
of precise physical data.

Since ignition occurs in the gas phase it seems that
this aspect of the problem should receive more attention
than it was given in the past.

REFERENCES

1. Akita, K., "Studies on the Mechanism of Ignition of
        Wood", Report of the Fire Research Institute of
        Japan, 9, No. 1-2 (1959) (Fire Research
        Abstracts and Reviews, 4, 109 (1962)).

2. Akita, K. and Kase, M., "Determination of Kinetic
        Parameters for Pyrolysis of Cellulose and
        Cellulose Treated with Ammonium Phosphate by
        Differential Thermal Analysis and Thermal
        Gravimetric Analysis", J. Poly. Sci. (A-1), 5,
        833 (1967).

3. Alvares, N.J., "Measurement of the Temperature of the
        Thermal Irradiated Surface of Alpha Cellulose",
        USNRDL-TR-735, Defense Atomic Support Agency
        Contact MIDR 526-64 (1964).

4. Alvares, N.J., Blackshear, P.L. and Murty Kanury, A.,
        "The Influence of Free Convection on the Ignition
        of Vertical Cellulosic Panels by Thermal
        Radiation", Combustion Science and Technology,
        1, 407 (1970).

5. Alvares, N.J. and Martin, S., "Mechanisms of Ignition
        of Thermally Irradiated Cellulose", Thirteenth
        Symposium (International) on Combustion, p. 905,
        The Combustion Institute, Pittsburgh (1971).

6. Alvares, N.J. and Wiltshire, L.L., "Ignition and Fire
        Spread in a Thermal Radiation Field", USNRDL-
        TR-68-56, DASA, NWFR A-8 Subtask RLD 5032
        (1970).

7. Anderson, W.H., "Theory of Surface Ignition with
        Application to Cellulose, Explosives and
        Propellants", Combustion Science and Technology,
        2, 213 (1970).

8. Arseneau, D.F., "The Differential Thermal Analysis of
        Wood", Canadian J. of Chem., 39, 1915 (1961).

9. Bamford, C.H., Crank, J. and Malan, D.H., "The
        Combustion of Wood, Part I", Proceedings of
        Cambridge Philosophical Society, 42, 166, (1946).

10. Beall, F.C. and Eickner, H.W., "Thermal Degradation
        of Wood Components: a Review of the Literature",
        USDA, Forest Service Research Paper, FPL 130
        (1970).

11. Bracciaventi, J., Heilferty, R. and Derksen, W., "Radiant Exposures for Ignition of Timber by Thermal Radiation from Nuclear Weapons", Final Report under Defense Atomic Support Agency Sub-task 12, 009 (1966).

12. Broido, A. and Kilzer, F.J., "A Critique of the Present State of Knowledge of the Mechanism of Cellulose Pyrolysis", Fire Research Abstracts and Reviews, 5, 157 (1963).

13. Broido, A. and Martin, S.B., "Effect of Potassium Bicarbonate on the Ignition of Cellulose by Radiation", Fire Research Abstracts and Reviews, 3, 193 (1961).

14. Browne, F.L., "Theories of the Combustion of Wood and its Control, A Survey of the Literature", USDA, Forest Service, Report No. 2136 (1958).

15. Buschman, A.J., "Ignition of Some Woods Exposed to Low Level Thermal Radiation", Technical Report No. 1 ONR Contract N Aopr 24-60 and OCDM Contract CDM-SR-60-22 (1961).

16. Butler, G.P., Martin, S. and Lai, W., "Thermal Radiation Damage to Cellulosic Materials, Part II, Ignition of Alpha Cellulose by Square Wave Exposure", USNRDL-TR-135 (AFSWP-906), (1956).

17. Byrne, G.A., Gardiner, D. and Holmes, F.H., "The Pyrolysis of Cellulose and the Action of Flame-Retardants", J. of App. Chem., 16, 81 (1966).

18. Carslaw, H.S. and Jaeger, J.C., "Conduction of Heat in Solids", Second Edition, Oxford University Press, London (1959).

19. Deverall, L.T. and Lai, W., "A Criterion for Thermal Ignition of Cellulosic Materials", Combustion and Flame, 13, 8 (1969).

20. Eickner, H.W., "Basic Research on the Pyrolysis and Combustion of Wood", Forest Product Journal, 12, 194 (1962).

21. Fons, W., "Heating and Ignition of Small Wood Cylinders", Ind. Eng. Chem., 42, 2130 (1950).

22. Garg, D.R. and Steward, F.R., "Pilot Ignition of Cellulosic Materials Containing High Void Spaces", Combustion and Flame, 17, 287 (1971).

23. Greenwood, C.T., Knox, J.H. and Milne, E., "Analysis

of the Thermal Decomposition Products of Carbohydrates by Gas Chromatography", Chem. and Ind. (Lond.), 1878 (1961).

24.   Holmes, F.H. and Shaw, C.J.G., "The Pyrolysis of Cellulose and the Action of Flame - Retardants, I Significance and Analysis of the Tar", J. Of Appl. Chem., 11, 210 (1961).

25.   Hottel, H.C. and Williams, C.C., "Transient Heat Flow in Organic Materials Exposed to High Intensity Radiation", Ind. Eng. Chem., 47, 1136 (1955).

26.   Kato, K., "Pyrolysis of Cellulose, Part III Comparative Studies of the Volatile Compounds from Pyrolysates of Cellulose and its Related Compounds", Agricultural and Biological Chemistry, 31, 657 (1967).

27.   Keylwerth, R. and Christoph, N., "Basic Investigations Concerning the Thermal Decomposition and Ignition of Wood", Mitt. Deut. Ges. Holzforsch, 50, 125 (1963).
      (Fire Research Abstract and Reviews, 9, 176 (1967)).

28.   Kilzer, F.J. and Broido, A., "Speculations on the Nature of Cellulose Pyrolysis", Pyrodynamics, 2, 151 (1965).

29.   Koohyar, A.N., Welker, J.R. and Sliepcevich, C.M., "The Irradiation and Ignition of Wood by Flame", Fire Technology, 4, 284 (1968).

30.   Laible, R.C., "Recent Work on the Mechanism of the Thermal Degradation of Cellulose", American Dyestuff Reporter, 47, 173 (1958).

31.   Lawson, D.I., "Wood and Fire Research", J. of the Institute of Wood Science, 4, 3 (1959).

32.   Lawson, D.I. and Simms, D.L., "The Ignition of Wood by Radiation", Brit. J. of Appl. Phys. 3, 288 (1952).

33.   Lipska, A.E., "The Fire Retardance Effectiveness of High Molecular Weight, High Oxygen Containing Inorganic Additives in Cellulosis and Synthetic Materials", Western States Section Meeting, The Combustion Institute, paper WSCl 73-13, (1973).

34.   Lipska, A.E. and Parker, W.J., "Kinetics of the Pyrolysis of Cellulose Over the Temperature Range 250-300°C", J. Appl. Polymer Sci., 10, 1439 (1966).

35. Lipska, A.E. and Wodley, F.A., "Isothermal Pryrolysis
     of Cellulose: Kinetics and Gas Chromatographic
     Mass Spectrometric Analysis of the Degradation
     Products", J. of Applied Polymer Science, 13,
     851 (1969).

36. MacKay, G.D.M., "Mechanism of Thermal Degradation of
     Cellulose: A Review of Literature", "Wood, Fire
     Behaviour and Fire Retardant Treatment",
     Canadian Wood Council, Ottawa (1966).

37. Madorsky, S.L., Hart, V.E. and Straus, S., "Pyrolysis
     of Cellulose in a Vacuum", J. Res. Nat. Bur.
     Standards, 56, 343 (1956).

38. Martin, S., "Ignition - Ablation Responses of
     Cellulosic Materials to High Radiant Heat Loads",
     Pyrodynamics, 2, 145 (1965).

39. Martin, S., "Diffusion - Controlled Ignition of
     Cellulosic Materials by Intense Radiant Energy"
     Tenth Symposium (International) on Combustion,
     p. 877, The Combustion Institute, Pittsburgh,
     (1965).

40. Martin, S.B. and Lai, W., "Thermal Radiation Damage
     to Cellulosic Materials Part III, Ignition of
     Alpha Cellulose by Pulses Simulating Nuclear
     Weapon Air Bursts", USNRDL-TR-252-AFSWP-1082
     (1958).

41. Martin, S., Lincoln, K.A. and Ramstad, R.W., "Thermal
     Radiation Damage to Cellulosic Materials, Part IV
     Influence of Moisture Content and the Radiant
     Absorptivity of Cellulosic Materials on their
     Ignition Behavior", USNRDL-TR-295, AFSWP 1117
     (1958).

42. Martin, S.B. and Ramstead, R.W., "Compact Two-Stage
     Gas Chromatograph for Flash Pyrolysis Studies",
     Anal. Chem., 33, 982 (1961).

43. Mosey, E.B. and Muir, W.E., "Pilot Ignition of
     Building Materials by Radiation", Fire
     Technology, 4, 46 (1968).

44. Murty Kanury, A., "Ignition of Cellulosic Solids,
     A Review", Fire Research Abstracts and Reviews,
     14, 24 (1972).

45. Murty Kanury, A. and Blackshear, P.L., "Some Consid-
     ations Pertaining to the Problem of Wood Burning",
     Combustion Science and Technology, 1, 339 (1970).

46. Ramiah, M.V., "Thermogravimetric and Differential
     Thermal Analysis of Cellulose, Hemicellulose, and
     Lignin", J. of Applied Polymer Science, 14,
     1323 (1970).

47. Roberts, A.F., "A Review of Kinetics Data for the
     Pyrolysis of Wood and Related Substances",
     Combustion and Flame, 14, 261 (1970).

48. Roberts, A.F., "The Heat of Reaction During the
     Pyrolysis of Wood", Combustion and Flame, 17,
     79 (1971).

49. Roberts, A.F. and Clough, G., "Thermal Decomposition
     of Wood in an Inert Atmosphere", Ninth Symposium
     (International) on Combustion, p. 158, The
     Combustion Institute, Pittsburgh, (1963).

50. Schwenker, R.F. and Beck, L.R. Jr., "Study of the
     Pyrolytic Decomposition of Cellulose by Gas
     Chromatography", J. Polymer Sci. Part C, 1,
     No. 2, 331 (1963).

51. Schwenker, R.F. and Pacsu, E., "Chemically Modifying
     Cellulose for Flame Resistance", Ind. Eng. Chem.,
     50, 91 (1958).

52. Simms, D.L., "Ignition of Cellulosic Materials by
     Radiation", Combustion and Flame, 4, 293 (1960).

53. Simms, D.L., "Experiments on the Ignition of
     Cellulosic Materials by Thermal Radiation",
     Combustion and Flame, 5, 369 (1961).

54. Simms, D.L., "Damage to Cellulosic Solids by Thermal
     Radiation", Combustion and Flame, 6, 303 (1962).

55. Simms, D.L., "On the Pilot Ignition of Wood by
     Radiation", Combustion and Flame, 7, 253 (1963).

56. Simms, D.L. and Law, M., "The Ignition of Wet and
     Dry Wood by Radiation", Combustion and Flame, 11,
     377 (1967).

57. Stamm, A.J., "Thermal Degradation of Wood and
     Cellulose", Ind. Eng. Chem., 48, 413 (1956).

58. Steward, F.R., "Ignition Characteristics of
     Cellulosic Materials", Review Paper, Eastern
     Section Meeting of the Combustion Institute
     (1971).

59. Tang, W.K. "Effect of Inorganic Salts on Pyrolysis of
     Wood, Alpha - Cellulose, and Lignin Determined
     by Dynamic Thermogravimetry", USFS Research
     Paper, FPL 71 (1967).

60.   Tsuchiya, Y. and Sumi, K., "Thermal Decomposition
      Products of Cellulose", J. of Appl. Polymer
      Science, $\underline{14}$, 2003 (1970).

61.   Varma, A., "Spontaneous Ignition of High Voidage
      Fuels", M.Sc. Thesis, Chem. Eng. Dept.,
      University of New Brunswick, Fredericton, N.B.
      (1968).

62.   Varma, A. and Steward, F.R., "Spontaneous Ignition of
      High Voidage Cellulosic Materials", The Journal
      of Fire and Flammability, $\underline{1}$, 154 (1970).

63.   Weatherford, W.D. and Sheppard, D.M., "Basic
      Studies of the Mechanism of Ignition of
      Cellulosic Materials", Tenth Symposium
      (International) on Combustion, p. 897, The
      Combustion Institute, Pittsburgh (1965).

64.   Weatherford, W.D. and Valtierra, M.L., "Pilot
      Ignition of Convection-heated Cellulose Slabs",
      Combustion and Flame, $\underline{10}$, 279 (1966).

65.   Welker, J.R., "The Pyrolysis and Ignition of
      Cellulosic Materials, A Literature Review", J.
      of Fire and Flammability, $\underline{1}$, 12 (1970).

66.   Welker, J.R., Wesson, H.R. and Sliepcevich, C.M.,
      "Ignition of Alpha-Cellulose and Cotton Fabric
      by Flame Radiation", Fire Technology, $\underline{5}$, 59
      (1969).

67.   Wesson, H.R., Welker, J.R. and Sliepcevich, C.M.,
      "The Piloted Ignition of Wood by Thermal
      Radiation", Combustion and Flame, $\underline{16}$, 303 (1971).

68.   Wiltshire, L.L. and Parker, W.J., "Ignition of
      Retardant - Treated Cloth by Nuclear Weapon
      Thermal Pulses", USNRDL-TR-68-139, OCD Work
      Unit 2542A (1969).

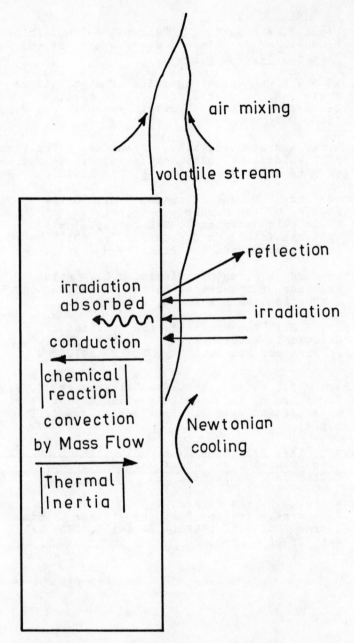

FIGURE 3.1

NOTE: Figure captions for this chapter are listed on page 407.

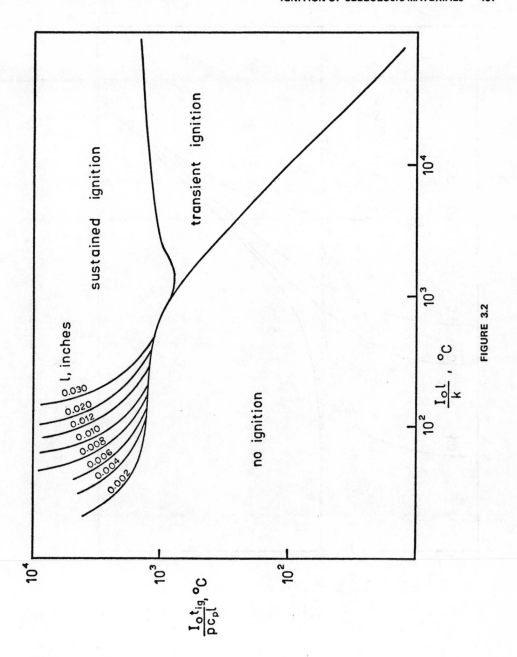

FIGURE 3.2

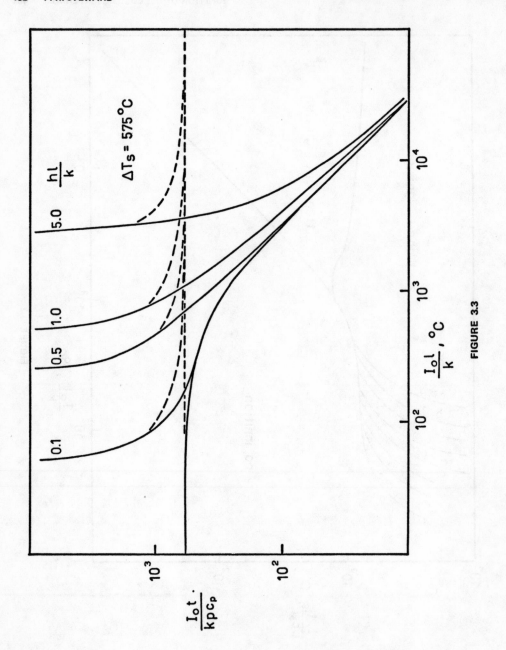

FIGURE 3.3

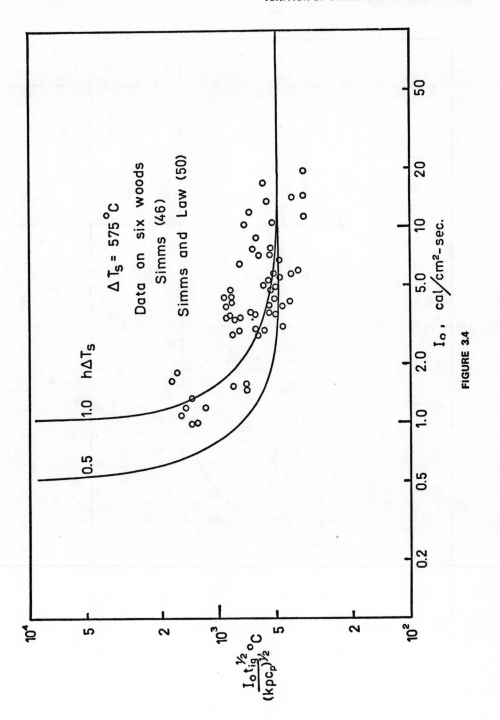

FIGURE 3.4

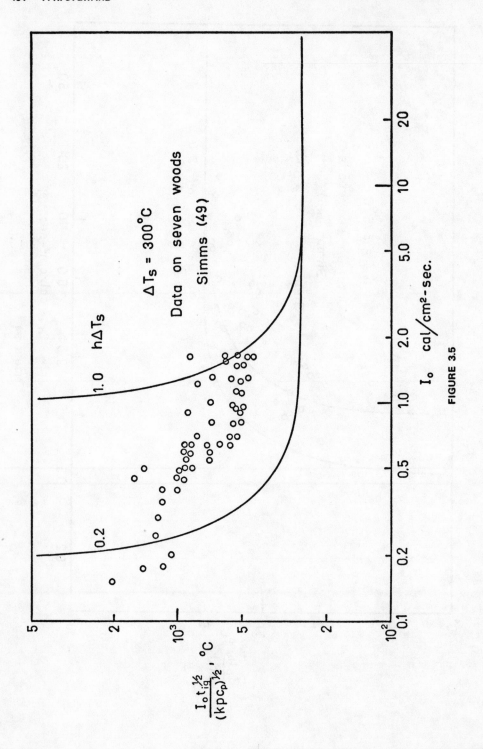

FIGURE 3.5

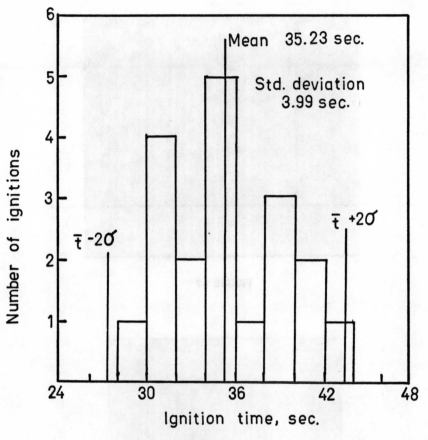

FIGURE 3.6

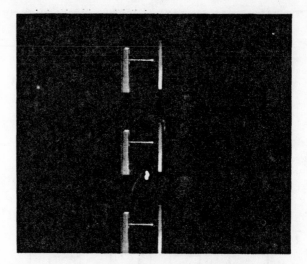

FIGURE 3.7

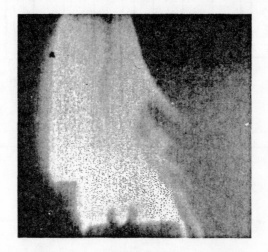

FIGURE 3.8

FIGURE CAPTIONS

3-1   Mechanisms of Heat Transfer Associated with Irradiation of Cellulosic Material.

3-2   Diagram for the Determination of the Time of Irradiation to Spontanwous Ignition for Thin Sample of α Cellulose, Martin (37).

3-3   Curves for Fixed Surface or Sample Temperature Rise Due to Irradiation, Analytical Solutions.

3-4   Diagram for the Determination of the Time of Irradiation to Spontaneous Ignition for Semi-Infinite Samples of Various Woods. Curves Represent Analytical Solution for a Fixed Surface Temperature Rise.

3-5   Diagram for the Determination of the Time of Irradiation to Pilot Ignition for Semi-Infinite Samples of Various Woods. Curves Represent Analytical Solution for a Fixed Surface Temperature Rise.

3-6   Histogram for the Time of Irradiation to Pilot Ignition of Cedar, Intensity of Radiation 0.738 cals/cm^2 sec., Pilot Flame 1" above Sample. Sample Size 3 1/4" x 3 1/4" x 3/8", Moisture Content. 8.8%, Irradiated Against Grain.

3-7   Flash Back from Pilot Flame.

3-8   Instant of Spontaneous Ignition.

Appendix I

BLACK BODY VIEW FACTORS

Nearly all radiative transfer analysis between surfaces depends on a knowledge of the direct interchange areas on view factors of the surfaces involved. A number of these have been tabulated here which are often encountered in fire problems. Additional ones may be found in more extensive tabulations (2) (4) and (5).

The first group, Section A, contains those which have one dimension infinite and are useful in considering line fire problems. Of particular interest here are those which give the distribution of radiation over the surface of a plane receiving radiation from another defined plane. These view factors were obtained by the cross string method described in the text or the appropriate differentiation thereof.

The second group, Section B, contains those which are often encountered in enclosures or fire fronts of finite length.

The third group, Section C, contains those which are associated with fires in circular symmetry and yield the radiative distribution around such fires if assumed black.

Each geometry is defined by a diagram and a plot of the view factors vs. the defining parameters is given. When possible the analytical expression is also given.

NOTE: All figures quoted in text are at the end of the chapter.

REFERENCES

1. Masliyah, J.H. and Steward, F. R. , "The Distribution
    of Radiation on a Horizontal Surface Surrounding
    the Base of a Black Frustum", Second Western
    Heat Transfer Conference (Canadian), Vancouver
    (1968).

2. Siegal, R. and Howell, J. R. , "Thermal Radiation
    Heat Transfer", McGraw-Hill Book Company, New
    York (1972).

3. Sparrow, E. M. and Cess, R. D. , "Radiation Heat
    Transfer", Brooks/Cole Publishing Company,
    Belmont, California (1966).

4. Stevenson, J. A. and Grafton, J. C. , "Radiation
    Heat Transfer Analysis for Space Vehicles",
    Rep.'SID 61-91, North American Aviation (AFASD
    TR 61-119 pt 1) (1961).

5. Wiebelt, J. A., "Engineering Radiation Heat Trans-
    fer", Holt, Rinehart and Winston, New York
    (1966).

A. SYSTEMS WITH ONE DIMENSION

$$F_{12} = \frac{1 + (\frac{X}{H}) - \sqrt{1 + (\frac{X}{H})^2 - 2 (\frac{X}{H}) \cos \theta}}{2}$$

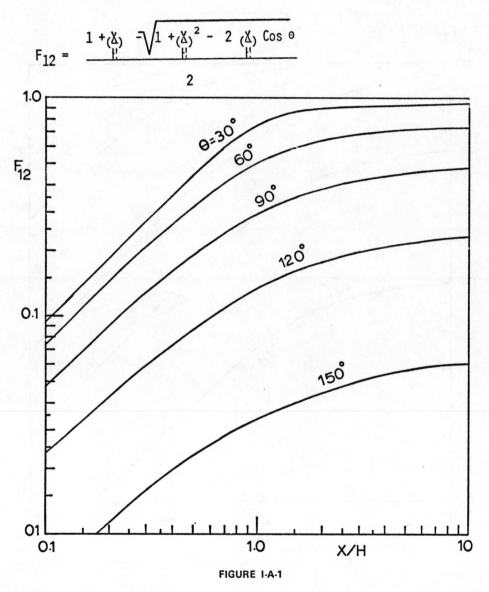

FIGURE I-A-1

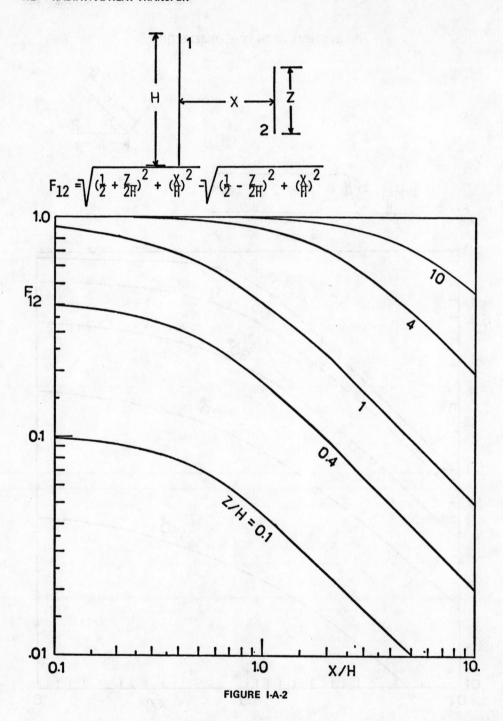

$$F_{12} = \sqrt{\left(\frac{1}{2} + \frac{Z}{2H}\right)^2 + \left(\frac{X}{H}\right)^2} - \sqrt{\left(\frac{1}{2} - \frac{Z}{2H}\right)^2 + \left(\frac{X}{H}\right)^2}$$

FIGURE I-A-2

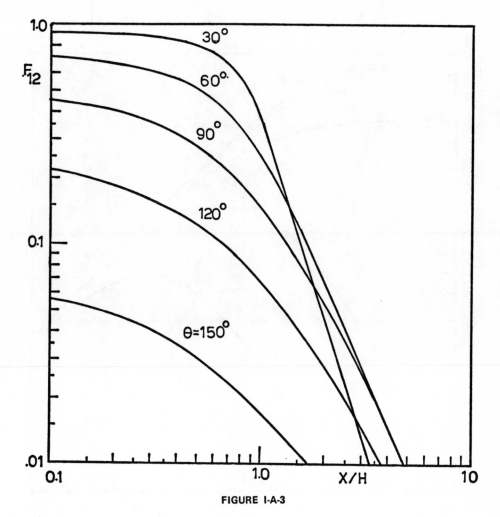

$$F_{12} = \frac{1}{2}\left[1 - \frac{(\frac{X}{H} - \cos\theta)}{\sqrt{1 + (\frac{X}{H})^2 - 2(\frac{X}{H})\cos\theta}}\right]$$

FIGURE I-A-3

$$F_{12} = \frac{1}{4} \left\{ \frac{(1 + \frac{Z}{H})}{\sqrt{(\frac{1}{2} + \frac{Z}{2H})^2 + (\frac{X}{H})^2}} + \frac{(1 - \frac{Z}{H})}{\sqrt{(\frac{1}{2} - \frac{Z}{2H})^2 + (\frac{X}{H})^2}} \right\}$$

FIGURE I-A-4

B.   RECTANGULAR SYSTEMS

$$F_{12} \left(\frac{\pi XY}{2}\right) = \text{LN} \left[\frac{(1+X^2)(1+Y^2)}{(1+X^2+Y^2)}\right]^{\frac{1}{2}} + Y\sqrt{1+X^2} \ \text{TAN}^{-1} \left(\frac{Y}{\sqrt{1+X^2}}\right)$$

$$+ X\sqrt{1+Y^2} \ \text{TAN}^{-1} \left(\frac{X}{\sqrt{1+Y^2}}\right) - Y \ \text{TAN}^{-1} Y - X \ \text{TAN}^{-1} X$$

$$X = \frac{A}{C}$$

$$Y = \frac{B}{C}$$

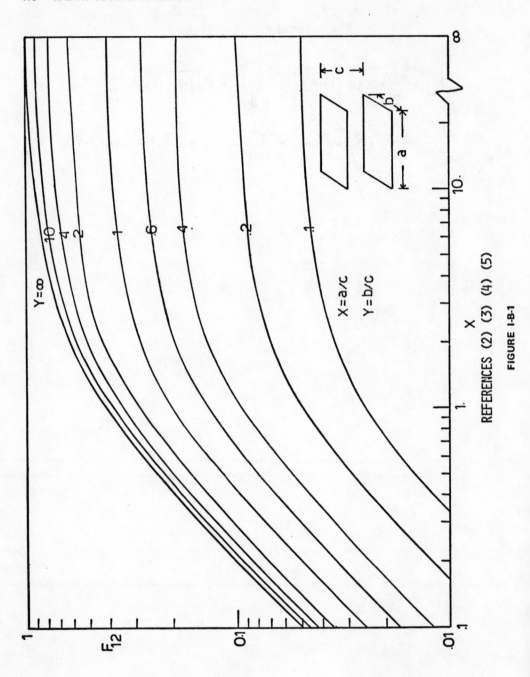

X = a/c
Y = b/c

REFERENCES (2) (3) (4) (5)

FIGURE I-B-1

$$F_{12} = \frac{2}{\pi} \left[\frac{X}{\sqrt{1+X^2}} \; TAN^{-1} \frac{Y}{\sqrt{1+Y^2}} + \frac{Y}{\sqrt{1+Y^2}} \; TAN^{-1} \frac{X}{\sqrt{1+Y^2}} \right]$$

$$X = \frac{A}{C}$$

$$Y = \frac{B}{C}$$

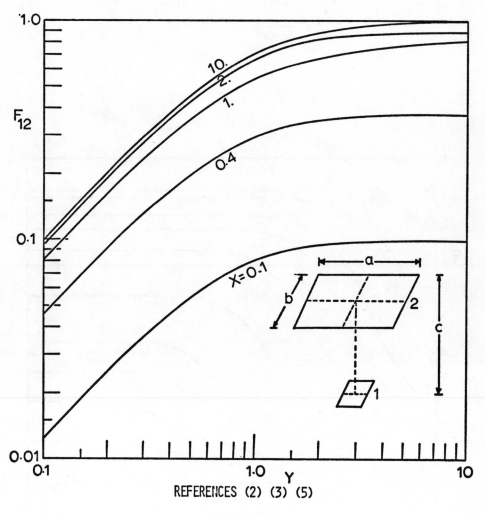

REFERENCES (2) (3) (5)

FIGURE I-B-2

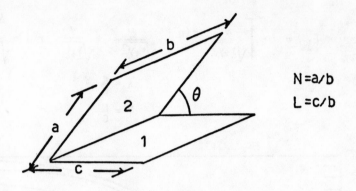

$N = a/b$
$L = c/b$

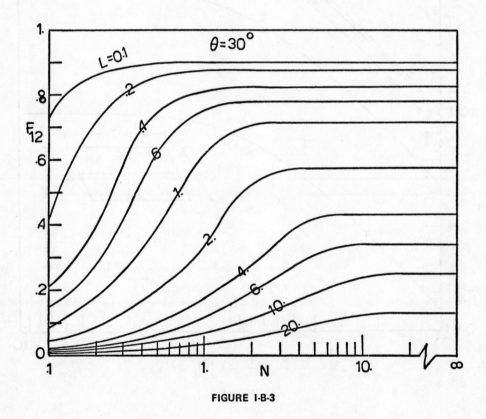

FIGURE I-B-3

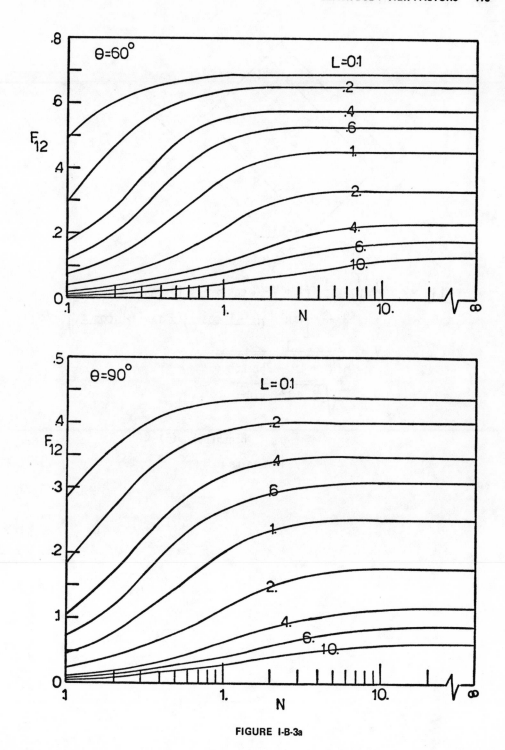

FIGURE I-B-3a

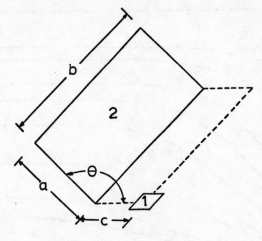

$$F_{12} = \frac{1}{2\pi}\left\{ \text{TAN}^{-1}\left(\frac{1}{L}\right) + V\left(N \cos \phi - \phi L\right) \text{TAN}^{-1} V \right.$$
$$\left. + \frac{\cos \phi}{W}\left[\text{TAN}^{-1}\left(\frac{N - L \cos \phi}{W}\right) + \text{TAN}^{-1}\left(\frac{L \cos \phi}{W}\right)\right]\right\}$$

$$V = \frac{1}{\sqrt{N^2 + L^2 - 2NL \cos \phi}}$$

$$W = \sqrt{1 + L^2 \sin^2 \phi}$$

REFERENCE (5)

FIGURE I-B-4

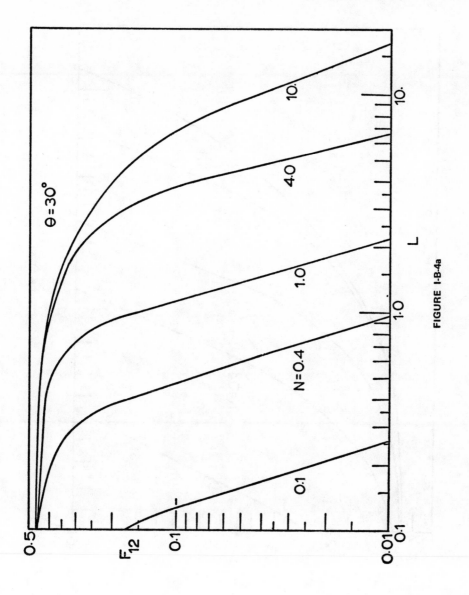

FIGURE I-B-4a

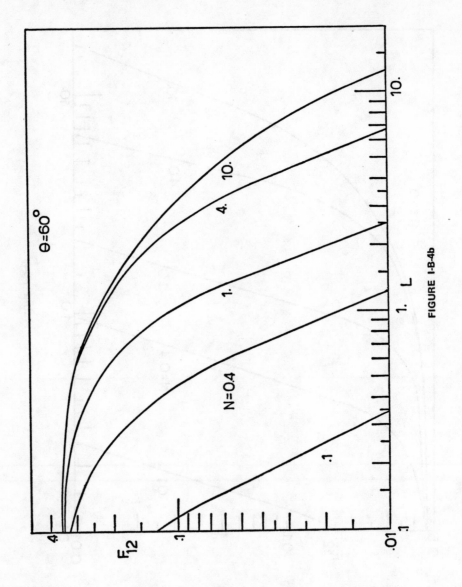

FIGURE I-B-4b

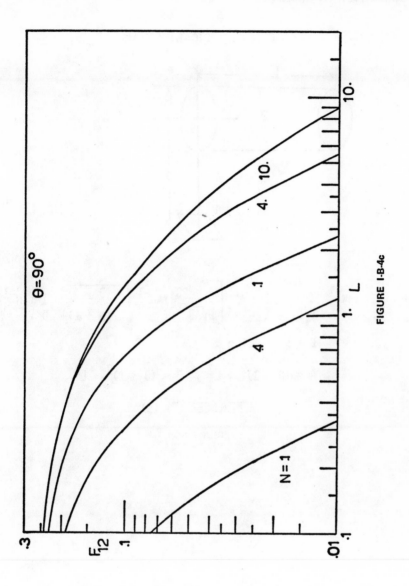

FIGURE I-B-4c

C. CYLINDRICAL SYSTEMS

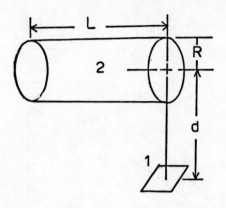

$$F_{12} = \frac{1}{\pi D} \, TAN^{-1} \, (\frac{L}{\sqrt{D^2 - 1}}) + \frac{L}{\pi} \left[\frac{A - 2D}{D\sqrt{AB}} \, TAN^{-1} \sqrt{\frac{A \, (D - 1)}{B \, (D + 1)}} - \frac{1}{D} \, TAN^{-1} \sqrt{\frac{D - 1}{D + 1}} \right]$$

$$D = \frac{d}{R} \, , \quad L = \frac{L}{R}$$

$$A = (D + 1)^2 + L^2 \, , \quad B = (D - 1)^2 + L^2$$

REFERENCES (3) (5)

FIGURE I-C-1

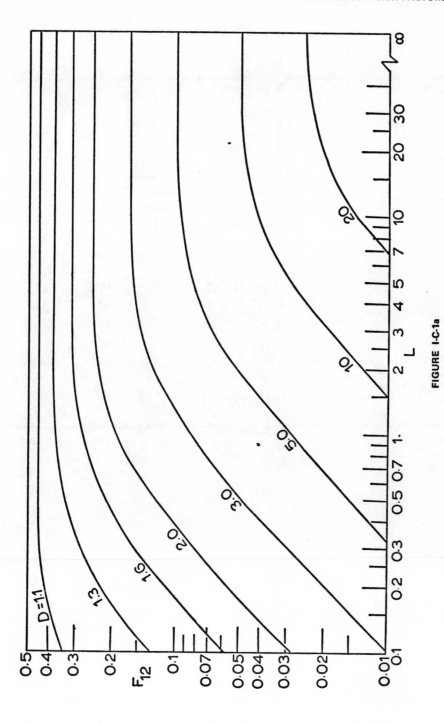

FIGURE I-C-1a

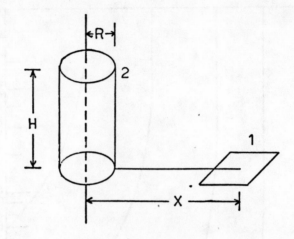

$$F_{12} = \frac{1}{\pi}\left[\text{TAN}^{-1}\sqrt{\frac{X' + R'}{X' - R'}} - \frac{(1.0 + X'^2 - R'^2)}{(1.0 + X'^2 + R'^2)} \text{TAN}^{-1}\sqrt{\frac{X' - R'}{X' + R'}} \right]$$

$$X' = \frac{X}{H}$$

$$R' = \frac{R}{H}$$

REFERENCE (1)

FIGURE I-C-2

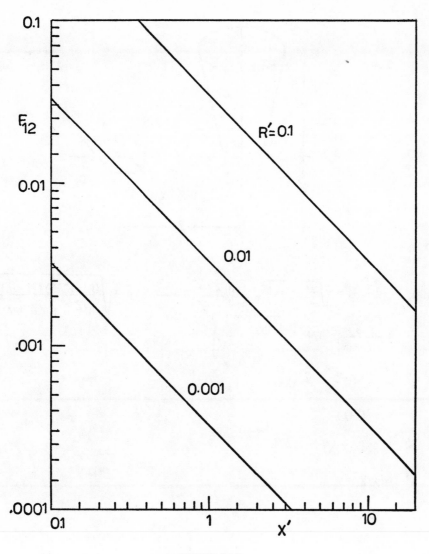

FIGURE I-C-2a

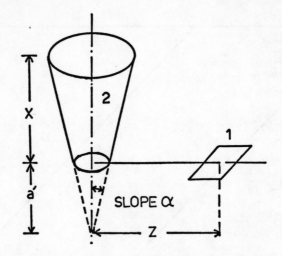

$$F_{12} = \frac{1}{\Pi}\left\{ \text{TAN}^{-1}\sqrt{\frac{Z' + \alpha A'}{Z' - \alpha A'}} - \frac{1.0 + Z'^2 - B'^2}{\sqrt{A'^4 - C'^4}}\ \text{TAN}^{-1}\sqrt{\frac{(A'^2 + C'^2)(Z' - \alpha A')}{(A'^2 - C'^2)(Z' + \alpha A')}} \right.$$

$$\left. + \frac{\alpha}{\sqrt{1 + \alpha^2}}\ \text{TAN}^{-1}\sqrt{\frac{1 + \alpha^2}{Z'^2 - \alpha^2 A'}} \right\}$$

$$A^2 = X^2 + Z^2 +\ ^2\ (X + A)$$

$$B = \alpha\ (X + A)$$

$$C^2 = 2Z\alpha\ (X + A)$$

$$A' = A/X$$

$$Z' = Z/X$$

$$A' = A/X$$
$$B' = B/X$$

$$C' = C/X$$

REFERENCE (1)

FIGURE I-C-3

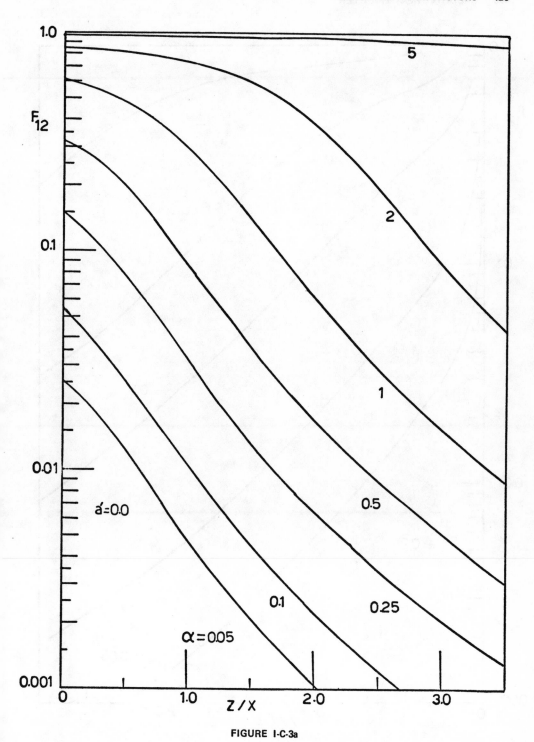

FIGURE I-C-3a

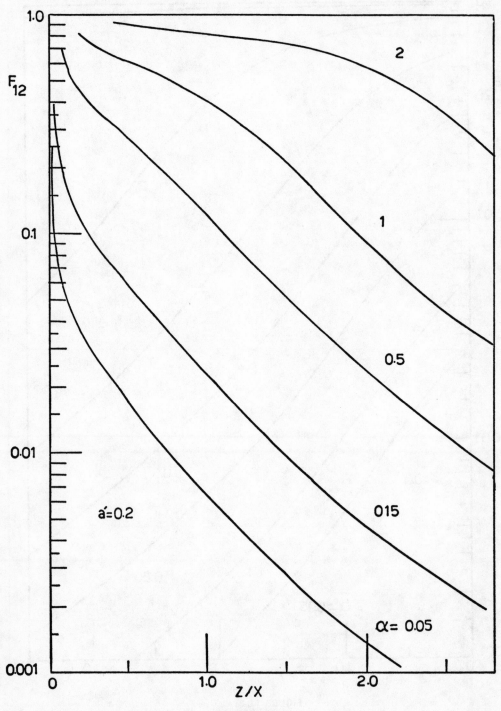

FIGURE I-C-3b

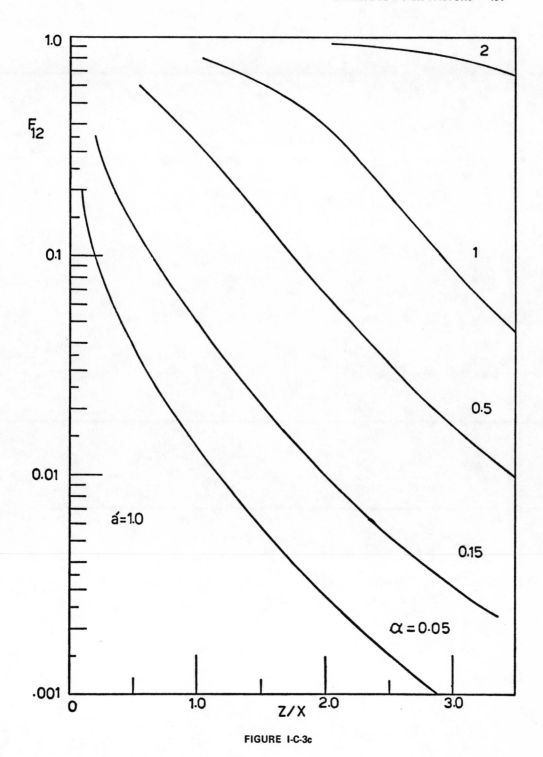

FIGURE I-C-3c

Appendix II

DIRECT INTERCHANGE AREAS WITH ABSORBING AND EMITTING MATERIAL PRESENT

Radiative transfer in systems which contain an absorbing and emitting material are often encountered in fire problems as any flame involves gas emission and absorption and in many cases emission and absorption by particulate matter. Fuelbeds with high voidage such as brush, leaves, needles, etc. can also be treated as an absorbing and emitting medium. A number of these have been tabulated here which are often encountered in fire problems. Additional ones may be found in other tabulations [1, 3, 5].

As indicated in the text a quantity termed the mean beam length is often useful in making more simplified calculations. These have been presented when appropriate.

The first group, Section A, contains the direct interchange areas associated with enclosures.

The second group, Section B, presents those involving distributions around line fires and mean beam lengths for systems of one dimension infinite.

The third group, Section C, gives the distribution around a circular fire and the mean beam lengths associated with such a system.

Each geometry is defined by a diagram and a plot of the direct interchange area or a modification thereof vs. the defining parameters as given.

The mean beam length is also presented when appropriate.

NOTE: All figures quoted in text are at the end of the chapter.

A. RECTANGULAR SYSTEMS

1. Self Interchange of an Emitting Cube of Side B(3,4)
    Self absorption of radiation given by

$$Q = 4kB^3(1- \phi')\sigma T^4 \qquad (II-1)$$

The remainder of the emission escapes from the cube.

2. Self Interchange of An Emitting Rectangular Parallele-
    piped 2B x B x B (3,4).

    Self absorption of radiation given by

$$Q = 4kB^3(1-\phi')\,\sigma T^4 \qquad (II-2)$$

The remainder of the emissive escapes from the parallele-
piped.

3. Opposed Squares (1, 3, 4, 5)
    The view factor is presented. Radiant interchange be-
    tween surfaces given by

$$Q_{1 \rightleftharpoons 2} = A_1 F_{12}\sigma(T_1^4 - T_2^4) \qquad (II-3)$$

4. Adjacent Squares [1, 3, 5]
    The view factor is presented. Radiant Interchange
    between surfaces given by

$$Q_{1 \rightleftharpoons 2} = A_1 F_{12}\sigma(T_1^4 - T_2^4) \qquad (II-4)$$

B. SYSTEMS WITH ONE DIMENSION INFINITE

1. Mean beam length for a volume of elemental cross section
    and length 2-Y separated from an elemental area by a
    minimum absorbing length s (4,5).

    Mean beam length given by

$$L_m = L_h \Delta \qquad (II-5)$$

This quantity corresponds to the mean transmissivity
of the system for radiation emitted by the volume of
elemental cross section headed for the elemental area.

2. Mean beam length for an surface of elemental thickness
    and length 2Y separated from an elemental area by a
    minimum absorbing length, s. (4,5)

    Mean beam length given by

$$L_m = L_h \Delta \qquad (II-6)$$

This quantity corresponds to the mean transmissivity
of the system for radiation emitted by the surface of
elemental thickness headed for the elemental area.

3. View factors and mean beam length giving radiative
    distribution around an infinite parallelepiped of
    emitting material (4,5).

    Radiant flux density distribution given by

    $$q = F_{12} \sigma T_2^{4} \qquad (II\text{-}7)$$

    Mean beam length given by

    $$L_m = L_{\hbar} Z \qquad (II\text{-}8)$$

which can be used to calculate the radiative flux distribution according to the relation

$$F_{12} = \frac{(1 - e^{-kL_m})}{2} \left[1 - \frac{X/Z}{\sqrt{1 - (X/Z)^2}} \right] \quad (II\text{-}9)$$

    All primed distances are made dimensionless by mul-
tiplication of the absorption coefficient of the material.

4. View factor and mean beam length giving the radiative
    distribution around an infinite emitting wedge of
    emitting material (4,5)

    Radiative flux distribution given by

    $$q = F_{12} \sigma T_2^{4} \qquad (II\text{-}10)$$

    Mean beam length given by

    $$L_m = L_{\hbar} Z \qquad (II\text{-}11)$$

which can be used to calculate the radiative flux distribution according to the relation

$$F_{12} = \frac{(1 - e^{-kL_m})}{2} \left[1 - \frac{(X/Z - \cos\theta)}{\sqrt{1 + (X/Z)^2 - 2(X/Z)\cos\theta}} \right]$$
$$(II\text{-}12)$$

    All primed distances are made dimensionless by mul-
tiplication of the absorption coefficient of the material.

5. View factor and coefficient for an exponential giving
    distribution of absorbed radiation in an infinite slab
    of absorbing material (2).

    Radiative flux density distribution given by

    $$q_{1\to2} = \frac{Q_{1\to2}}{\ell dx} = \overline{F_{12}} \, kL \, \sigma \, (T_1^{4} - T_2^{4}) \qquad (II\text{-}13)$$

where $\overline{F_{12}}$ is the fraction of radiation emitted by the plane
1 which passes a plane located at a distance x from the
plane.

The coefficient for an exponential relation can be defined by the equation

$$\int_0^\infty e^{-B\dot{x}}d\dot{x} = \int_0^\infty \overline{\frac{F_{12}}{kLdx'}}d\dot{x} \qquad (II\text{-}14)$$

This yields the exponential relation which gives the identical area under the curve as the rigorous relationship.

C. CYLINDRICAL SYSTEMS

1. View factor and mean beam length giving the radiative distribution around a cone of emitting material (7).

    Radiative flux density distribution given by

    $$q = \frac{Q}{dA} = F_{12}\ \sigma T_2^4 \qquad (II\text{-}15)$$

    Mean beam length given by

    $$L_m = L_n Z \qquad (II\text{-}16)$$

which can be used to calculate the radiative flux distribution according to the relation

$$F_{12} = (1 - e^{-kL_m})\ \frac{1}{\pi}\left[\frac{\pi}{4} + \frac{a}{\sqrt{1 + a^2}}\ tan^{-1}\left[\sqrt{\frac{1 + a^2}{x'}}\right]\right.$$

$$- \frac{(x'^2 + (1 - a^2))}{(x'^2 + 2(1-a^2)\ x'^2 + (1+a^2))^2)^{1/2}}\ tan^{-1}$$

$$\left.\sqrt{\frac{x'^2 + 2ax' + (1+a^2)}{x'^2 - 2ax' + (1+a^2)}}\ \right] \qquad (II\text{-}17)$$

where "a" is the slope of the cone.

2. Mean beam length for a wedge or cone of emitting material and an infinite plane surrounding its base

    The mean beam length is given by

    $$L_m = L_n Z \qquad (II\text{-}18)$$

which can be used to calculate the direct radiative interchange according to the relations

$$F_{12} = (1 - e^{-kL_m})\ 1/2\left[1 + \frac{a}{\sqrt{1 + a^2}}\right] \qquad (II\text{-}19)$$

REFERENCES

1. Dunkle, R. V. , "Geometive Mean Beam Lengths for
        Radiant Heat Transfer Calculations", J. Heat
        Transfer, <u>86</u>, 75 - 80 (1964).

2. Fang, J. B. and Steward, F. R. , "A Radiative Inter-
        change Factor for Fire Propagation in a Porous
        Medium", Combustion Science and Technology, <u>4</u>,
        187-90 (1971).

3. Hottel, H. C. and Cohen, E. S., "Radiant Heat
        Exchange in a Gas Filled Enclosure", AIChE
        Journal, <u>4</u>, 3-14 (1958).

4. Hottel, H. C. and Sarofim, A. F. , "Radiative Heat
        Transfer",McGraw-Hill Book Company, New York
        (1967).

5. Siegel, R. and Howell, J. R. , "Thermal Radiative
        Heat Transfer", McGraw-Hill Book Company,
        New York (1972).

6. Steward, F. R. , "Fire Spread through Solid Fuel",
        DSc Thesis, Chem. Eng. Dept., M.I.T. Cambridge,
        Mass., (1962).

7. Steward, F. R. ,"The Distribution of Radiation on a
        Horizontal Surface Surrounding a Uniformly
        Emitting Gray Gas Cone", 57th Annual Meeting
        AIChE, Boston, (1964).

A.   RECTANGULAR SYSTEMS

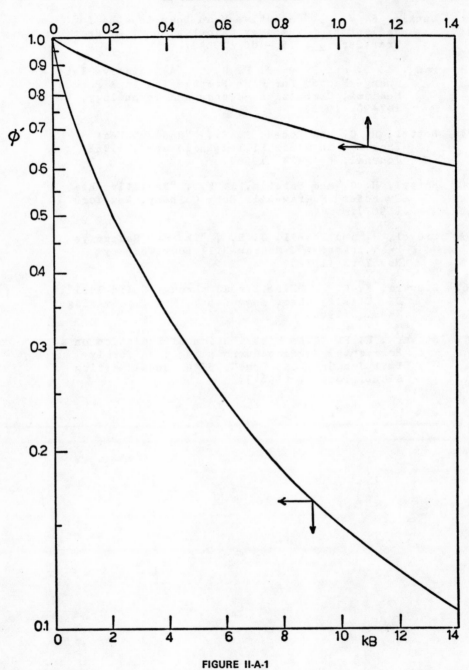

FIGURE II-A-1

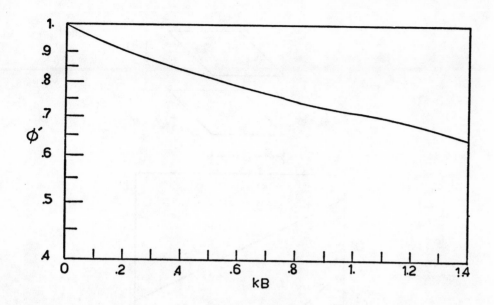

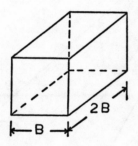

FIGURE II-A-2

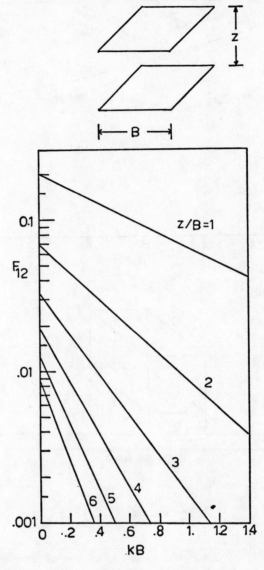

FIGURE II-A-3

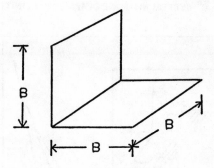

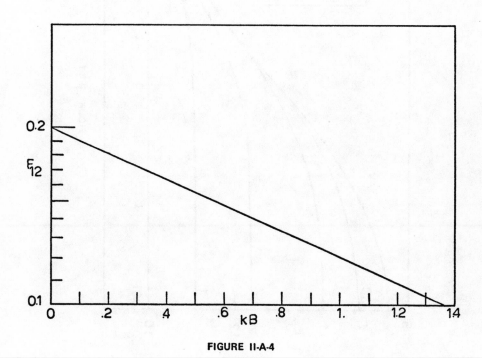

FIGURE II-A-4

B. SYSTEM WITH ONE DIMENSION INFINITE

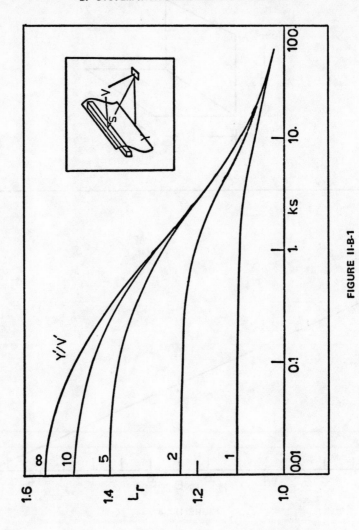

FIGURE II-B-1

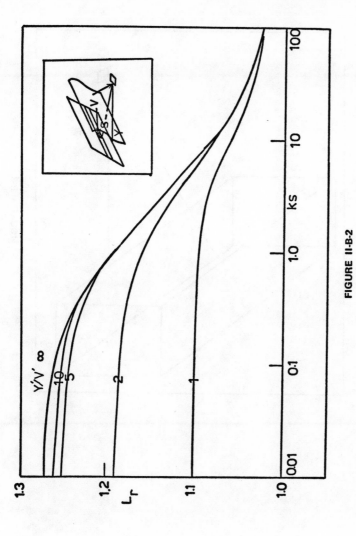

FIGURE II-B-2

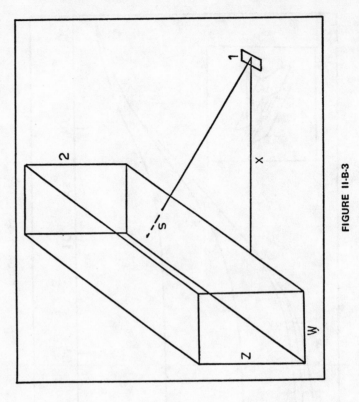

FIGURE II-B-3

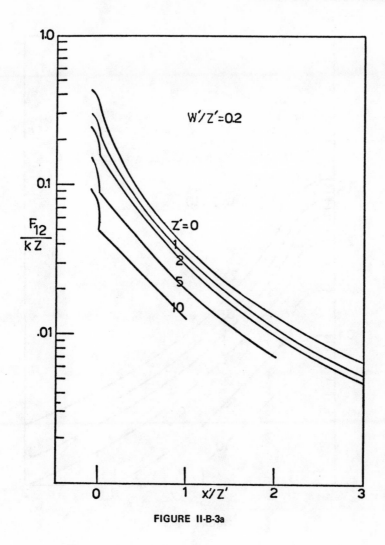

FIGURE II-B-3a

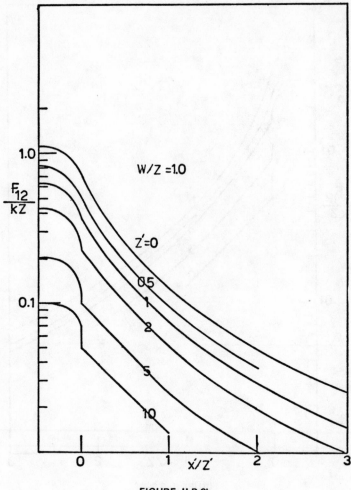

FIGURE II-B-3b

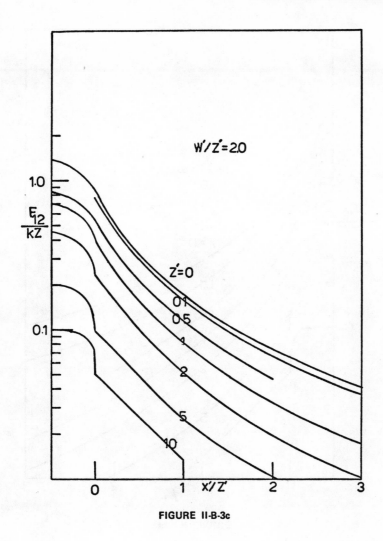

FIGURE II-B-3c

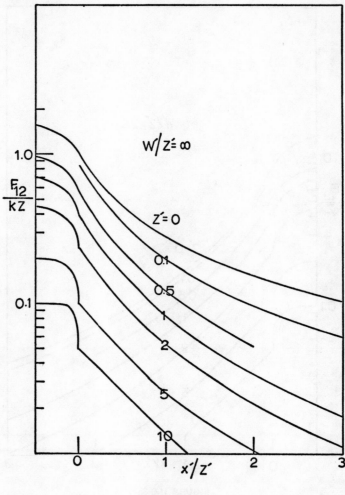

FIGURE II-B-3d

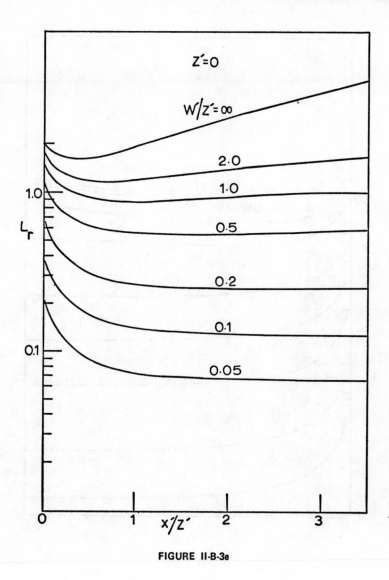

FIGURE II-B-3e

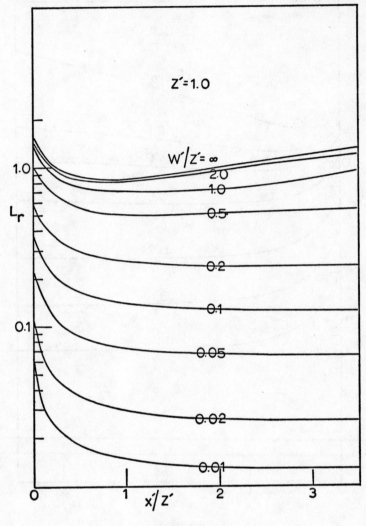

FIGURE II-B-3f

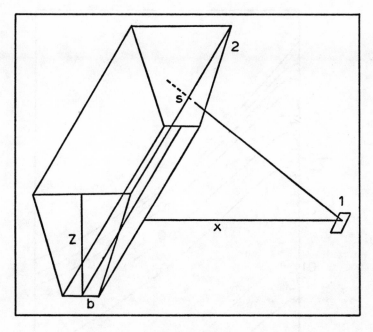

FIGURE II-B-4

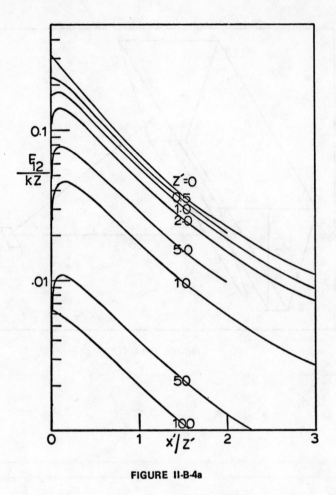

FIGURE II-B-4a

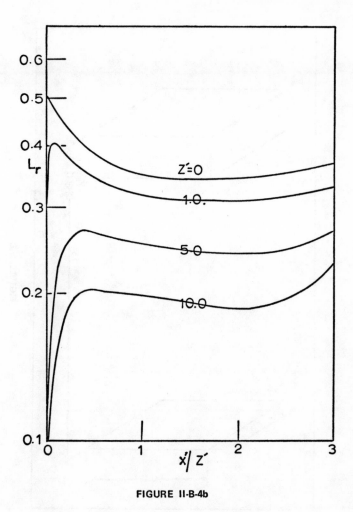

FIGURE II-B-4b

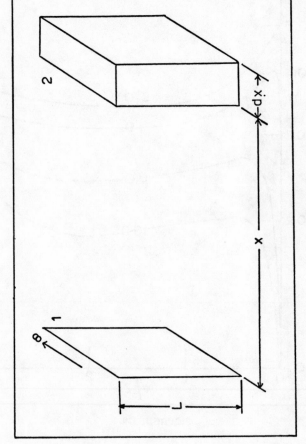

FIGURE II-B-5

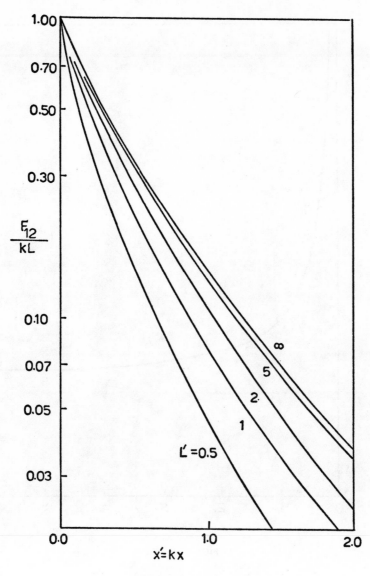

FIGURE II-B-5a

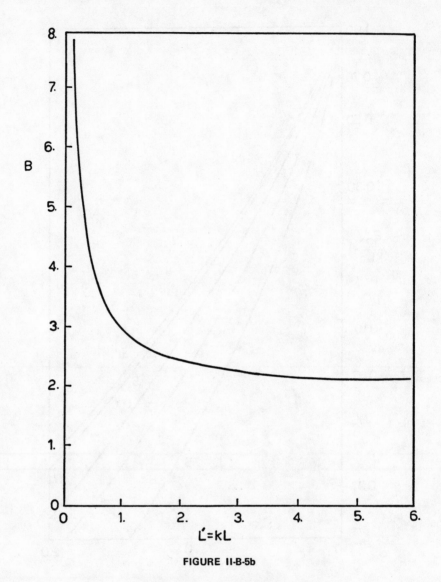

FIGURE II-B-5b

C. CYLINDRICAL SYSTEMS

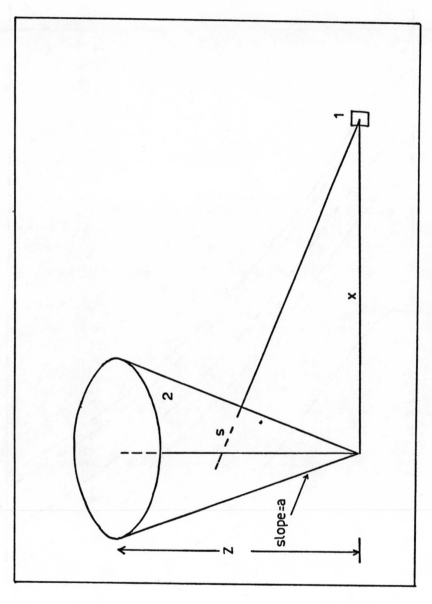

FIGURE II-C-1

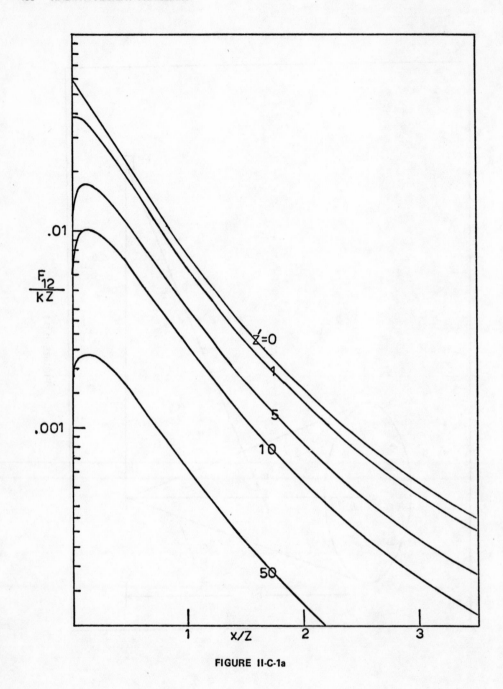

FIGURE II-C-1a

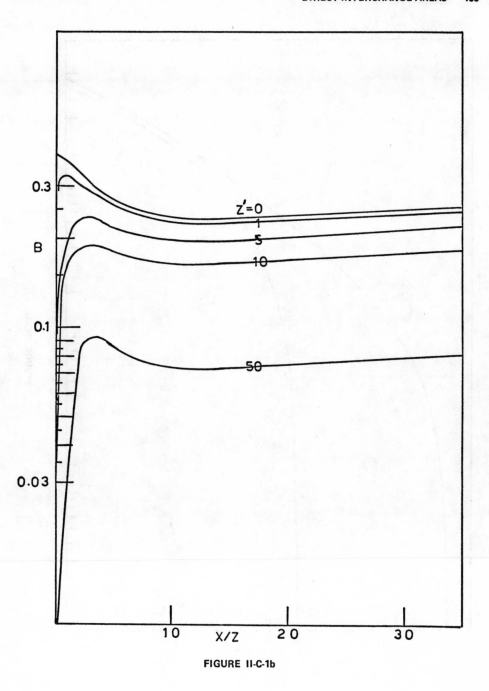

FIGURE II-C-1b

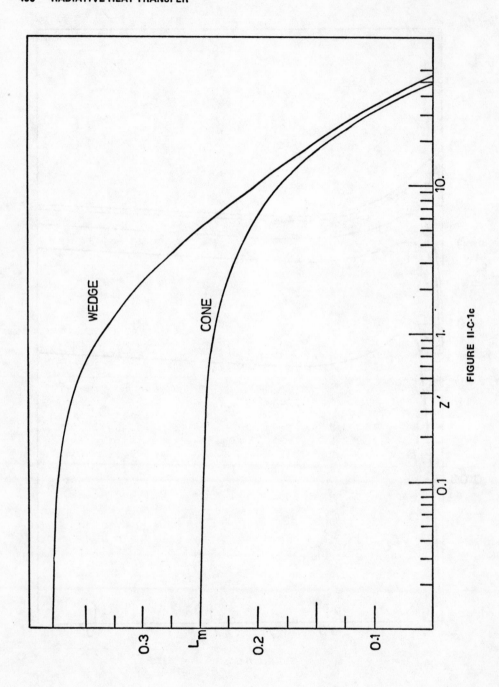

FIGURE II-C-1c

Appendix III

EXAMPLE OF TOTAL SURFACE TO SURFACE RADIATIVE INTERCHANGE IN AN ENCLOSURE

This example problem is to demonstrate the method for calculating the total radiative interchange from one surface to another surface within an enclosure.

The problem is shown in Fig. III-1 and can be considered to be a room of dimensions 16' x 8' x 8'. Surface 1 is assumed covered with flame, a burning curtain or drape, with an emissivity of 0.5 and temperature of 2000°F. The floor, surface 2, is covered with a heavy rug, emissivity 0.95, and the ceilingg has been painted white, emissivity 0.90. The remaining side walls, surfaces 4 and 5 are of walnut panel, emissivity 0.83. Surface 4 includes two walls as the system is sysmmetrical. All nonburning surfaces are taken at 60°F.

It is desired to find the radiative heat flux from the burning surface 1 to the other surfaces in the system. The solution to this problem can be obtained by the appropriate evaluation of Eq. 1-39. The evaluation of this equation requires the view factors for each of the surfaces in the system with all the other surfaces. These can be obtained from Appendix I B for the appropriate geometries and are listed.

View Factors for the Enclosure

$F_{11} = 0.0000$ $F_{21} = 0.1162$ $F_{31} = 0.1162$ $F_{41} = 0.1162$ $F_{51} = 0.0700$

$F_{12} = 0.2325$ $F_{22} = 0.0000$ $F_{32} = 0.2976$ $F_{42} = 0.2350$ $F_{52} = 0.2325$

$F_{13} = 0.2325$ $F_{23} = 0.2976$ $F_{33} = 0.0000$ $F_{43} = 0.2350$ $F_{53} = 0.2325$

NOTE: All figures quoted in text are at the end of the chapter.

$F_{14} = 0.4650$ $F_{24} = 0.4700$ $F_{34} = 0.4700$ $F_{44} = 0.2976$ $F_{54} = 0.4650$

$F_{15} = 0.0700$ $F_{25} = 0.1162$ $F_{35} = 0.1162$ $F_{45} = 0.1162$ $F_{55} = 0.0000$

The other quantities given by the definition of the problem are the surface areas and the emissivity and reflectivity of the surfaces.

Areas in Square Feet

$A_1 = 64$ $A_2 = 128$ $A_3 = 128$ $A_4 = 256$ $A_5 = 64$

$\varepsilon_1 = 0.5$ $\varepsilon_2 = 0.95$ $\varepsilon_3 = 0.90$ $\varepsilon_4 = 0.83$ $\varepsilon_5 = 0.83$

$\rho_1 = 0.5$ $\rho_2 = 0.05$ $\rho_3 = 0.10$ $\rho_4 = 0.17$ $\rho_5 = 0.17$

The substitution of the quantities into Eq. (1-39) yields the total interchange areas. This result was calculated with a short digital computer program which can handle any reasonable number of surfaces.

The total interchange areas for the enclosure given in square feet.

$A_1 \mathcal{F}_{11} = 0.26$

$A_1 \mathcal{F}_{12} = 8.12$

$A_1 \mathcal{F}_{13} = 7.51$

$A_1 \mathcal{F}_{14} = 13.86$

$A_1 \mathcal{F}_{15} = 2.70$

The net radiative heat flux between surface 1 and any of the other surfaces in the system is given by

$$q_{1 \rightleftharpoons k} = A_1 \mathcal{F}_{1k} \sigma \left(T_1^4 - T_k^4 \right) \qquad (III-1)$$

These fluxes are determined as

$q_{1 \rightleftharpoons 2} = 512,500 \ Btu/hr$

$q_{1 \rightleftharpoons 3} = 474,500 \ Btu/hr$

$q_{1 \rightleftharpoons 4} = 874,000 \ Btu/hr$

$q_{1 \rightleftharpoons 5} = 170,200 \ Btu/hr$

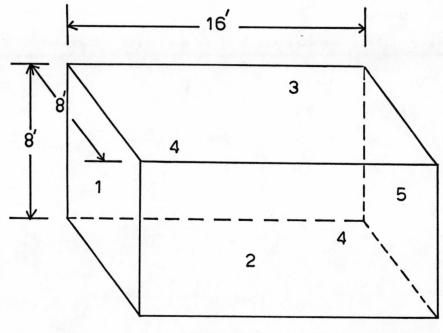

FIGURE III-1

NOTE: Figure captions for this chapter are listed on page 464.

FIGURE CAPTIONS

Figure III-1 Geometry for Surface Radiative Interchange in a Room.

Appendix IV

EMISSIVITIES OF COMBUSTION PRODUCT GASES

The gas emissivities associated with fire problems are restricted primarily to carbon dioxide, water vapor, and carbon monoxide. The charts and calculation procedures for determining the emissivity of mixtures of these gases, along with nonemitting ones, are presented here for convenience.

The emissivity of some other common gases have been determined and can be found in reference[2].

A. DATA FOR CARBON DIOXIDE

The emissivity of a given volume of gas containing carbon dioxide can be found from Figs. IV-1 and IV-2.

Figure IV-1 gives the emissivity of the carbon dioxide at one atmosphere total pressure as a function of the emitting gas temperature and the optical thickness in feet-atmospheres. The length used in this calculation would normally be the mean beam length for the particular geometry under consideration as discussed in the text.

A total pressure correction due to pressure broadening can be determined from Fig. IV-2. If the combustion gases are at one atmosphere total pressure no correction is required.

The absorptivity of a gas body, of carbon dioxide, α_G, for radiation originating at a temperature T_S can be determined from the relation

$$\alpha_G(T_G, T_S, pL) = \left[\frac{T_G}{T_S}\right]^{0.65} \varepsilon_G\left[T_S, pL\left(\frac{T_S}{T_G}\right)\right] \quad (IV-1)$$

Notice that ε_G should be evaluated at a temperature T_S and an optical thickness of $pL(T_S/T_G)$ for this calculation.

B. DATA FOR WATER VAPOR

The emissivity of a given volume of gas containing water vapor can be found from Figs. IV-3 and IV-4.

NOTE: All figures quoted in text are at the end of the chapter.

Figure IV-3 gives the emissivity of the water vapor at one atmosphere total pressure and a partial pressure of water vapor of zero as a function of emitting gas temperature and the optical thickness in feet-atmospheres. The length used in this calculation would normally be the mean beam length for the particular geometry under consideration as discussed in the text.

A total pressure correction which includes the correction for a finite partial pressure of water vapor can be obtained from Fig. IV-4.

The absorptivity of a gas body of water vapor, α_G, for radiation originating at a temperature T_S can be determined from the relation

$$\alpha_G(T_G, T_S, pL) \quad \left[\frac{T_G}{T_S}\right]^{0.45} \quad \varepsilon_G\left[T_S, pL\left(\frac{T_S}{T_G}\right)\right] \qquad (IV-2)$$

Notice that α_G should be evaluated at a temperature T_S and an optical thickness of $pL\,(T_S/T_G)$ for this calculation.

C. CORRECTION FOR CARBON DIOXIDE WATER VAPOR OVERLAP IN MIXTURES

The presence of carbon dioxide and water vapor together produces a certain amount of overlap in some of the bands. This correction can be made by the use of Fig. IV-5.

The procedure is the evaluate emissivities of carbon dioxide and water vapor independently as if the other gas was not present and subtract the value of $\Delta\varepsilon$ given by Fig. IV-5.

D. DATA FOR CARBON MONOXIDE

The emissivity of a given volume of carbon monoxide can be found from Fig. IV-6.

The emissivity is given in terms of the emitting gas temperature and the optical thickness in feet-atmospheres. The data is based on mixtures of carbon monoxide and nitrogen at one atmosphere total pressure·

Carbon monoxide and carbon dioxide overlap in the 4.3 μ region and it has been indicated that only 0.8 of the additional emissivity for carbon monoxide at $pL = 0.2$ ft-atm be added when the carbon dioxide partial pressure is high enough to yield $pL = 1.0$ ft-atm·

REFERENCES

1. Hottel, H. C. , "Radiant Heat Transmission", Chapter 4 of "Heat Transmission" third eddition, by W. H. McAdams, McGraw-Hill Book Company, Inc., New York (1954).

2. Hottel, H. C. and Sarofim, A. F. , "Radiative Transfer", McGraw-Hill Book Company, New York (1967).

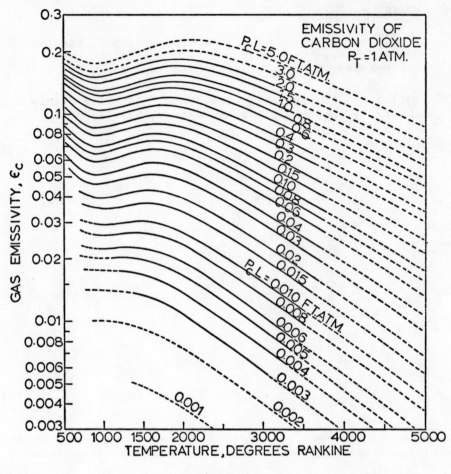

FIGURE IV-1

NOTE: Figure captions for this chapter are listed on page 474.

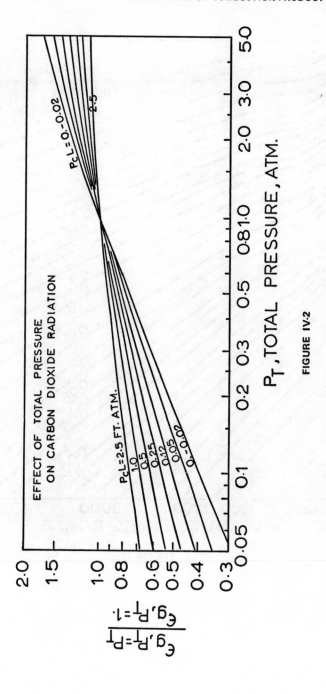

FIGURE IV-2

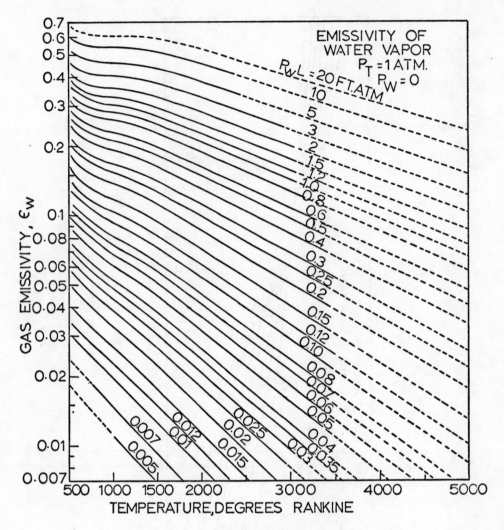

FIGURE IV-3

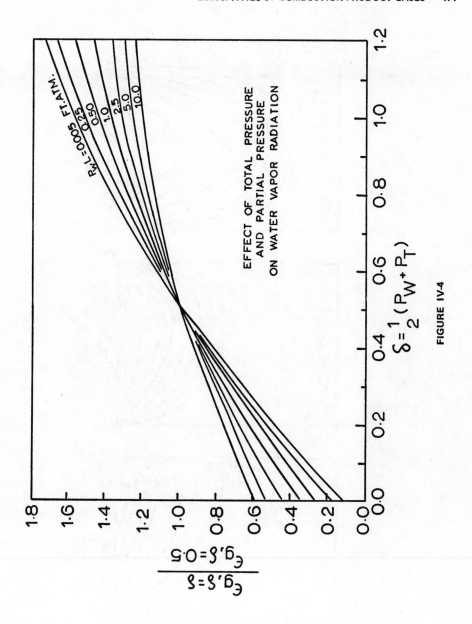

EFFECT OF TOTAL PRESSURE
AND PARTIAL PRESSURE
ON WATER VAPOR RADIATION

FIGURE IV-4

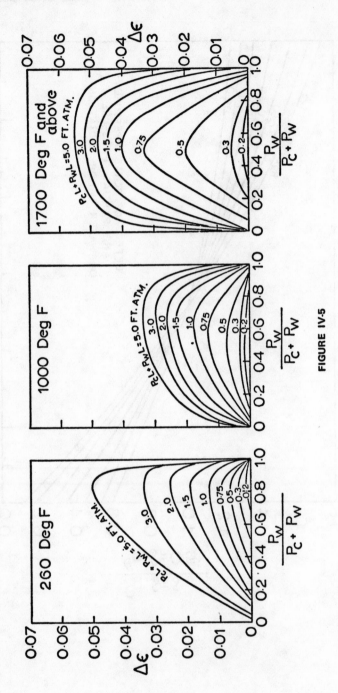

FIGURE IV-5

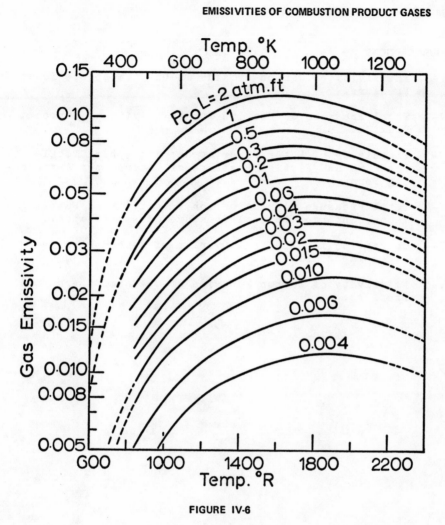

FIGURE IV-6

FIGURE CAPTIONS

IV-1   Emissivity of Carbon Dioxide at One Atmosphere Total
        Pressure.

IV-2   Total Pressure Correction for Emissivity of Carbon
        Dioxide.

IV-3   Emissivity of Water Vapor at One Atmosphere Total
        Pressure and Partial Pressure of Water Vapor Approach-
        ing Zero.

IV-4   Total Pressure Correction for Emissivity of Water
        Vapor, Includes Partial Pressure of Water Vapor.

IV-5   Correction for Spectral Overlap of Carbon Dioxide-
        Water Vapor Mixtures.

IV-6   Emissivity of Carbon Monoxide at One Atmosphere Total
        Pressure.

Appendix V

EXAMPLE OF TOTAL GAS TO SURFACE RADIATIVE INTERCHANGE IN AN ENCLOSURE

This example problem is to demonstrate the method for calculating the total radiative interchange from an emitting gas volume to a surface in an enclosure.

The problem is show in Fig. V-1 and can be considered to be a room of dimensions 16' x 8' x 8'. The room is filled with combustion products from burning a fuel of composition $(CH_2)_n$ with 20% excess air to form CO_2 and H_2O at 2000°F. The composition of the combustion product gas will be

$$CO_2 = 14.9\%$$
$$H_2O = 14.9\%$$
$$N_2 = 67.2\%$$
$$O_2 = 3.0\%$$

The floor of the room, surface 2, is covered with a rug, emissivity 0.95, and the ceiling, surface, has been painted white, emissivity 0.90. The walls, surfaces 1 and 4, are walnut panel, emissivity 0.83. The ends and sides can be taken as the same surfaces because of symmetry.

The mean beam length for an enclosure of pale emitting gas is given by

$$L_m = \frac{4V}{A} = \frac{4 \times 8 \times 8 \times 16}{(2 \times 8 \times 8 + 4 \times 8 \times 16)}$$

$$L_m = 6.4 \text{ ft.}$$

NOTE: All figures quoted in text are at the end of the chapter.

The partial pressures of carbon dioxide and water vapor are both 0.149 atmospheres. The mean optical thickness of the enclosure for the two radiating gases is given by

$$P_{CO_2} L_m = 0.955 \text{ ft. atm.}$$

$$P_{H_2O} L_m = 0.955 \text{ ft. atm.}$$

From the emissivity charts in Appendix IV the emissivity of each gas independently is given by

$$\varepsilon_{CO} = 0.140$$

$$\varepsilon_{H_2O} = 0.176$$

The connection for overlapping bands was found to be

$$\Delta\varepsilon = 0.040$$

therefore

$$\varepsilon_G = \varepsilon_{CO_2} + \varepsilon_{H_2O} - \Delta\varepsilon \qquad \text{(V-1)}$$

$$\varepsilon_G = 0.276$$

For the approximate calculation assume the radiating gas is gray and obtain an overall absorption coefficient for the gas.

$$\varepsilon_G = 1 - e^{-kL_m}$$

$$k = \frac{-1}{L_m} \ln(1 - \varepsilon_G) \qquad \text{(V-2)}$$

$$k = 0.050 \text{ ft}^{-1}$$

The total interchange factor between the gas volume and each surface is given by Eq. 1-77. This requires the view factors between all the surface in the system and the direct interchange area between the radiating gas and each of the surfaces in the system. These quantities were obtained from Appendix II, Series A.

View Factors

$$F_{11} = 0.030 \qquad F_{21} = 0.182 \qquad F_{31} = 0.182 \qquad F_{41} = 0.183$$

$$F_{12} = 0.182 \qquad F_{22} = 0.000 \qquad F_{32} = 0.179 \qquad F_{42} = 0.193$$

$$F_{13} = 0.182 \qquad F_{23} = 0.179 \qquad F_{33} = 0.000 \qquad F_{43} = 0.193$$

$F_{14} = 0.366$ $F_{24} = 0.386$ $F_{34} = 0.386$ $F_{44} = 0.179$

$F_{16} = 0.240$ $F_{26} = 0.253$ $F_{36} = 0.253$ $F_{46} = 0.252$

Direct Gas to Surface Interchange Areas in Square Feet

$$\overline{gs_1} = 30.72 \quad \overline{gs_2} = 32.36 \quad \overline{gs_3} = 32.36 \quad \overline{gs_4} = 64.72$$

The other quantities given by the definition of the problem are the surface areas and the emissivity and reflectivity of the surfaces.

Areas in Square Feet

$A_1 = 128$ $A_2 = 128$ $A_3 = 128$ $A_4 = 256$

$\varepsilon_1 = 0.83$ $\varepsilon_2 = 0.95$ $\varepsilon_3 = 0.90$ $\varepsilon_4 = 0.83$

$\rho_1 = 0.17$ $\rho_2 = 0.05$ $\rho_3 = 0.10$ $\rho_4 = 0.17$

The substitution of the quantities into Eq. 1-22 yields the total interchange areas. This result was calculated with a short digital computer program which can handle any reasonable number of volume and surface zones.

Total interchange areas for the enclosure in square feet

$$A_1\overline{\mathcal{K}}_{1G} = 28.30$$

$$A_2\overline{\mathcal{K}}_{2G} = 34.57$$

$$A_3\overline{\mathcal{K}}_{3G} = 32.47$$

$$A_4\overline{\mathcal{K}}_{4G} = 59.01$$

The net radiative heat flux between the combustion gas and any one of the surfaces in the system is given by

$$q_{G \rightleftarrows 1} = A_k\overline{\mathcal{K}}_{kG}\,\sigma(T_G^4 - T_k^4) \qquad \text{(V-3)}$$

These fluxes are determined as

$$q_{G \rightleftarrows 1} = 1{,}786{,}000 \text{ Btu/hr}$$

$$q_{G \rightleftarrows 2} = 2{,}180{,}000 \text{ Btu/hr}$$

$$q_{G \rightleftarrows 3} = 2{,}050{,}000 \text{ Btu/hr}$$

$$q_{G \rightleftarrows 4} = 3{,}720{,}000 \text{ Btu/hr}$$

These answers can be compared with a standard formula for radiative transfer in a furnace enclosure with a single surface.

$$A_{1\ 1G} = \frac{A_1}{\frac{1}{\varepsilon_1} + \frac{1}{\varepsilon_G} - 1} \qquad \text{(V-4)}$$

Taking an area average for the surface emissivity $\varepsilon_1 = 0.87$ and A_1 as the total surface area $640\ ft^2$ gives $A_1\overline{\mathcal{K}}_{1G} = 164\ ft^2$. This gives a total heat flux to all the surfaces.

$$q_{G\neq1} = 10,330,000\ Btu/hr$$

The sum of the heat fluxes for the previous more sophisticated analysis yields

$$q_{G\neq1+2+3+4} = 9,763,000\ Btu/hr$$

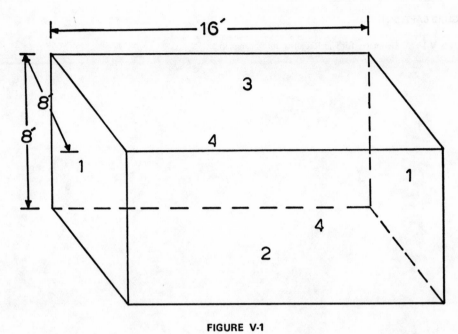

FIGURE V-1

NOTE: Figure captions for this chapter are listed on page 480.

FIGURE CAPTIONS

Figure V-1   Geometry for Gas Radiative Interchange in a
               Room.

Appendix VI

NOMENCLATURE

a	absorption coefficient of solid for radiation, ℓ^{-1}
a'	surface area of fuel particle, ℓ^2
a_p	mean projected area of a particle, ℓ^2
A	surface area, ℓ^2
A_i	frequency factor for the ith chemical reaction, $(\ell^3/m)^{n-1} t^{-1}$
$A_i F_{ij}$	direct black body interchange area between the surfaces i and j, ℓ^2
$A_i \overline{\mathcal{F}}_{ij}$	total interchange area between the surfaces i and j, ℓ^2
B	exponential coefficient for radiative interchange in a fuel bed, dimensionless
c	velocity of light in vacuum, ℓ/t
c	concentration of absorbing gas, m/ℓ^3
C	center to center distance between fuel particles in a matrix, ℓ
C'	proportionality constant for volume emission of a radiating gas, ℓ^{-1}
C_F	flame drag coefficient, dimensionless
C_p	heat capacity, h/mT

d	diameter of circular fuel particle or side of square particle, l
d_o	initial diameter of a fuel particle, l
D	determinant of coefficient matrix
D_{ij}	minor of determinat of coefficient matrix, less i-th row and j-th column
D_B	thickness of burning zone, l
E_i	activation energy of i-th reaction, h/MT
f	fraction of fuel consumed, dimensionless
F_{ij}	fraction of radiation leaving surface i interrupted by surface j, dimensionless
g	gravitational constant, l/t^2
$\overline{g_i s_j}$	direct interchange area between the volume i and the surface j, l^2
$\overline{g_i g_j}$	direct interchange area between the volume i and the volume j, l^2
$\overline{G_i S_j}$	total interchange area between the volume i and the surface j, l^2
$\overline{G_i G_j}$	total interchange area between the volume i and the volume j, l^2
h	Planck's constant, ht
h	surface convection heat transfer coefficient, $h/l^2 tT$
h_{ig}	enthalpy required to bring fuel from ambient conditions to the point where it enters the conflagration, h/m
h_v	enthalpy required to liberate a unit mass of volatile material from solid fuel, h/m
H	flame height, l
$_iH_j$	flux density of radiation on suface j which originated at surface i, $h/_t l^2$
AH_c	heat of combustion of fuel, h/m
I	intensity of irradiation, $h/_t l^2$
I_F	fire intensity, $h/_{lt}$
I_o	irradiation absorbed at the surface, $h/_t l^2$
I_s	intensity of irradiation at surface, $h/_t l^2$
k	thermal conductivity, $h/_t lT$
k	absorption coefficient of material, l^{-1}
k_c	absorption coefficient of material in concentration units, $l^2{}^{m-1}$

k_p	absorption coefficient of material in pressure units, $atm^{-1} \ell^{-1}$
k_e	jet entrainment coefficient, dimensionless
ℓ	thickness of fuel bed, ℓ
L	length, ℓ
L_m	mean beam length for a volume of radiating gas, ℓ
m	fuel loading density, m/ℓ^2
m_s	mass of solid undergoing pyrolysis per unit volume, m/ℓ^3
m_v	rate of volatile liberation per unit area of fuel surface, $m/\ell^2 t$
M	moisture content of fuel on a dry basis, mass of moisture per mass of dry fuel
$\dot{M}/L$	mass rate of volatile combustibles entering flame per unit length of fire front, $m/\ell t$
M_o/L	initial mass of fuel particle per unit length of fuel particle, m/ℓ
M_v/L	mass rate of volatiles liberated per unit length of fire front, $m/\ell t$
n_i	order of i-th chemical reaction, dimensionless
P	partial pressure of radiating gas, atm
q	heat flux density, $h/t\ell^2$
q_p	heat flux per unit area of fuel particle surface, $h/\ell^2 t$
q_v	heat flux per unit volume of fuel bed, $h/\ell^3 t$
Q	heat flux, h/t
Q_i	enthalpy of ith reaction, h/m
Q/L	heat transfer rate per unit length of fire front, $h/\ell t$
r	distance, ℓ
r_i	rate of ith chemical reaction, m/ℓ^3
R	gas constant, h/MT
R	rate of fire spread, ℓ/t
$_iR_j$	flux density of radiation leaving surface j which originated at surface i, $h/\ell^2 t$
s	mass air to fuel ratio required for stoichiometric combustion, mass air to mass fuel
$\overline{s_i s_j}$	direct interchange area between surface i and surface j, ℓ^2

$\overline{S_i S_j}$	total interchange area between surface i and surface j, ℓ^2
t	time, t
T	temperature, T
T_F	flame temperature, T
T_G	gas temperature, T
T_o	initial or surrounding temperature, T
T_s	surface temperature, T
U	wind velocity, ℓ/t
V	volume, ℓ^3
w	inverse volumetric coefficient of expansion due to combustion, dimensionless
W_B	hemispherical blank body emissive power, $h/\ell^2 t$
$W_{B\lambda}$	monochromatic emissive power of a black body, per unit wavelength, $h/\ell^3 t$
$W_{B\lambda}$	monochromatic emissive power of black body per unit frequency, h/ℓ^2
x	distance, ℓ
X	distance, ℓ
y	distance, ℓ
Y	distance, ℓ
z	distance, ℓ
Z	distance, ℓ
α	absorptivity, of a surface, dimensionless
α	thermal diffusivity, ℓ^2/t
α	slope of cone, dimensionless
γ	frequency, t^{-1}
γ_i	stoichiometric amount of volatiles released upon pyrolysis resulting from i-th chemical reaction, mass volatiles/mass pyrolysis
ϵ	emissivity of a surface, dimensionless
ϵ_B	emissivity of the burning zone, dimensionless
ϵ_F	emissivity of the flame, dimensionless
ϵ_p	emissivity of a volume of particles suspended, dimensionless
θ	planner angle
λ	wavelength, ℓ
ρ	reflectivity of a surface, dimensionless
ρ	density, m/ℓ^3

ρ_a	density of air, m/ℓ^3
ρ_f	density of gases in flame, m/ℓ^3
ρ_o	density of volatile fuel, m/ℓ^3
σ	Stefan-Boltzmann constant, $h/\ell^2 t T^4$
τ	transmissivity, dimensionless
ϕ	azumith angle
ϕ'	fraction of radiation leaving an emitting volume, dimensionless
ω	solid angle

Subscripts

a	air
c	convection
g	gas
i,j,k	surface and volume, integers
ℓ	loss
o	initial
p	particle
r	radiation
s	surface
v	volatile
γ	frequency
λ	wavelength

Units are given in

h	energy
ℓ	length
m	mass
M	mole
t	time
T	temperature

ACKNOWLEDGMENTS

I wish to express my appreciation to Mrs. Guniz Guruz and Mr. Kemal Guruz, graduate students in the Chemical Engineering Department at the University of New Brunswick, who helped me prepared the diagrams and the radiative interchange tabulations.

I also wish to thank Miss Charlotte Cogswell and Mrs. Darlene O'Donnell who typed the manuscript.

PART V

RADIATIVE TRANSFER PARAMETERS

Chapter 1

BAND MODELS OF INFRARED RADIATION

R. GOULARD
Purdue University, USA

The role of radiative exchange in fires has been shown in earlier presentations to be critical in the onset or propagation of a number of combustion processes. In some cases, a rough estimate of the radiation fluxes has proven to be adequate: hence the use of the well-known temperature-averaged, frequency-summed charts for the emissivity. If a finer understanding is required, one must seek more precise models of radiation transfer, without going to the extreme, of course, of a line-by-line account of molecular radiation. Such need for a compromise between computational speed and predictive accuracy has yielded over the last 15 years a successful set of band models which will be discussed in this section.

1. LINE EMISSION PROPERTIES

Given a medium of optical thickness τ_ν, it is convenient to characterize it over a certain frequency interval (that of a line for instance) by its equivalent width W:

$$W \equiv \int_0^{+\infty} (1 - e^{-\tau_\nu})\, d\nu \tag{1}$$

It can be seen that in the special case of an isothermal source, and since the Planck function B_ν does not vary much over the frequency interval of one line, W plays a similar role to that of the emissivity ε of the gas, as defined in previous presentations:

$$\int_0^{\infty} 1_\nu d\nu = \int_0^{+\infty} B_\nu (1 - e^{-\tau_\nu})\, d\nu = W \cdot B_\nu \tag{2}$$

NOTE: All figures quoted in text are at the end of the chapter.

In other words, W is a way to describe the absorption (or emission) characteristics corresponding to a line or a group of lines, whereas τ_ν describes it at one frequency only. It can be shown that it has also the convenient property of being measured and used directly, without the need for the knowledge of a slit function, if this function is symmetrical.

In Eq. 1 the optical thickness for a collision broadened line is:

$$\tau_\nu \equiv Q_\nu NL = NL \frac{\pi e^2}{m_e c} f \frac{1}{\pi} \frac{\frac{\gamma}{4\pi}}{(\nu - \nu)^2 + \frac{\gamma}{4\pi}^2} \tag{3}$$

or, in short:

$$\tau_\nu \equiv N'S \frac{1}{\pi} \frac{\alpha}{(\nu - \nu_0)^2 + \alpha^2}$$

where:

N' = NL is the product of the absorbing species density N and the path length L,

$S = \frac{\pi e^2}{m_e c} f$ is the absorption corss section of the line,

ν_0 is the frequency at the center of the line,

$\alpha = \frac{\gamma}{4\pi}$ is the reduced line half-width.

If one ignores the effect of Doppler broadening (which could be easily introduced by substituting the Voigt profile to the collision profile into Eq. 3), the dependence of the line thickness on thermodynamic properties is of the general form:

$$\alpha = \alpha_0 \frac{P}{P_0} \frac{T_0}{T}^n \tag{5}$$

It is therefore possible to condense all the absorption properties across the frequency range of a line into a single property W [Eq. 1], whose dependence on other properties is obtained from Eqs. 4 and 5. This integration of Eq. 1 gives the dependence of the equivalent width W on N' and α in the following form:

$$W = 2\pi\alpha L \frac{SN'}{2\pi\alpha} \tag{6}$$

where L, the "Ladenberg-Reiche function," is defined as:

$$L(x) \equiv xe^{-x} [1_0(x) + 1_1(x)] \tag{7}$$

where 1_0 and 1_1 are Bessel functions. This function L is tabulated in many texts [1].

It can be shown from Eq. 7 that for $x \ll 1$, there is a "weak line" approximation: $L(x) \simeq x$, whereas for $x \gg 1$, the "strong line" approximation: $L(x) \simeq \left(\frac{2x}{\pi}\right)^{1/2}$ applies. Fig. 1b illustrates these asymptotic trends of the curve of growth W.

The comparison between the familiar Fig. 1a where the radiation is emitted at one frequency and the curve of growth (Fig. 1b) where we consider the radiation from the whole frequency width of the line, is instructive. At low optical thicknesses (small $k_\nu L$ and SN'), doubling N' doubles the outgoing flux also: the medium is optically thin (points A and A'). When N' increases, the center of the line may give an optical thickness $Q_\nu N'$ at that frequency such that it approaches saturation (point B on Fig. 1) while it is still thin in the wings (point C). Hence a slower average rate of growth for W (point D). Since lines are so narrow and their wings so widespread, one cannot reach the black body limit in Fig. 1b as one does, ideally, in Fig. 1a. (This situation would be completely changed in the case of a soot-laden flame, since soot emission is continuous and not line-shaped: such flames often reach the black body limit.)

Goody [1] gives an experimental verification of Eq. 6 in the case of the rotational lines of the CO fundamental band.

2. BAND EMISSION PROPERTIES

In combustion problems, one is concerned with the radiation of many vibration-rotation lines and it would be an impossible task to calculate exactly their contributions to each spectral interval. In fact, the transition probabilities corresponding to certain bands such as H_2O are not all available yet. For these two reasons it is often chosen to invent a model which will duplicate roughly the overall spectral features of the molecular band of interest and to find experimentally the proper values to attach to the constants used in the model.

In practice, one defines an averaged absorption property based on all the lines within a frequency interval typical of current spectrometers (say 25 cm^{-1}). Such a set of values, every 25 cm^{-1} across the interval of the molecular band of interest describes its emission and absorption accurately. A recently published handbook [2] provides such information in the 300K - 3000K temperature range for H_2O, CO_2, CO, HF, HCl, CN, OH and NO (see Part II of these notes).

Two classes of line distributions seem to characterize most molecules found in combustion systems. For some diatomic molecules and in the wings of some triatomic molecules, there exists an even spacing of lines (CO, NO, some parts the CO_2, N_2O spectra). Elsasser derived a closed form of S_ν for this regular comb-like set of lines (Eq. 3) and integrated it in Eq. 1. Another common pattern is observed near the center of molecular bands where several branches overlap, and especially for complex molecules like H_2O: it is the random model which is widely used for such combustion products as CO_2 and H_2O. It will be discussed here in some detail.

First, it is convenient to consider one of the physical meanings of W, which is that it substitutes to the partial absorption and emission of a line across a large frequency span, a total absorption in a finite frequency interval W (Fig. 2a).

Hence the transmittance T in a spectral interval $\Delta \nu$ where a line of equivalent width W exists, is defined by the probability T that a passing photon will not be absorbed:

$$\bar{T} = 1 - \frac{W}{\Delta \nu} \tag{8}$$

If there exists n randomly distributed lines in this interval $\Delta \nu$ (Fig. 2b), the probability becomes a product:

$$\bar{T} = \left(1 - \frac{W_1}{\Delta \nu}\right) \left(1 - \frac{W_2}{\Delta \nu}\right) \cdots \left(1 - \frac{W_n}{\Delta \nu}\right) \tag{9}$$

If all lines were of equal strength $(W_1 = W_2 = \ldots = W_n)$, Eq. 9 would reduce to:

$$\bar{T} = \exp\left(-\frac{W}{d}\right) \tag{10}$$

where $d \equiv \frac{\Delta \nu}{n}$ is the average spacing of the lines in that interval.

3. THE RANDOM BAND MODEL

A more realistic model assumes that the population P(S) of the lines varies as a negative exponential function of their strength S:

$$P(S) = \frac{1}{S_0} \exp\left(-\frac{S}{S_0}\right) \tag{11}$$

This model, where both line strength and frequency are randomly distributed, yields, once Eqs. 11, 3 and 1 are integrated in Eq. 9, a transmittance T of the form:

$$\bar{T} = \exp\left(-\frac{kN'}{\sqrt{1 + \frac{kN'}{4a}}}\right) \tag{12}$$

where: $k \equiv \frac{S_0}{d}$ and $a \equiv \frac{\alpha}{d} = \frac{\gamma}{4\pi d}$

This model has been very successful in describing actual flames. Its application to a specific case is discussed in detail in the second part of these notes. T is an especially useful property because the flux expression uses the transmittance T directly:

$$1_\nu = \int_0^L B_\nu(T) \frac{d}{ds} (\bar{T}) \, ds \tag{13}$$

which can be put in numerical computational form:

$$l_v = \sum_{j=1}^{j=p} B_v(T_j)\,(\bar{T}_{v,j} - \bar{T}_{v,j-1})\,\Delta s \tag{14}$$

Hence the convenience of Eq. 12 in determining the absorption process across the spectrum of interest. For each frequency (or wave number) there is a value of T which can be established if one knows the three quantities k, γ and d. These are tabulated in the Handbook [2] mentioned earlier, for those molecules best suited by the random band model (e.g., CO_2, H_2O).

Observe that when a $\to \infty$ Eq. 12 reduces to

$$\bar{T} = \exp\,(-\,kN') \tag{15}$$

which is the monochromatic transmittance $T_v = e^{-\tau_v}$. This corresponds to the non-physical case where the line thickness becomes much larger than the line spacing d. Clearly, at the limit, the "gappiness" of the molecular band would be smoothed out and Eq. 15 would be a good representation of T across the small interval $\Delta\nu$. Fig. 3 gives an illustration of k for the 2.7μ band of CO_2.

Clearly this cannot be physically accomplished and accuracy demands that one use the denominator of Eq. 12 and therefore that both k and a be known with accuracy. Those values have been determined experimentally by measuring T and N' in a variety of experimental conditions and plotting $\dfrac{\ln\bar{T}}{N'}$ in ordinate and N' in abscissa. Eq. 12 can be rewritten:

$$\left(\frac{\ln\bar{T}}{N'}\right)^{-2} = \frac{1}{k^2} + \frac{1}{4ak}\,N' \tag{16}$$

It is clear that at the limit N' $\to$ 0, the intercept of the curve on the vertical axis gives k^2, whereas the slope of the curve there is $\dfrac{1}{4ak}$. Hence the value of the "gappiness" a.

Most of the progress accomplished in recent years has been in the measurement of the gappiness a, which can be obtained only if one disposes of experiments for large values of N'. Notice on Fig. 4 the drastic increase in accuracy obtained [3] for the water vapor band at 2000° K, when a new 20 ft, burner (stars) was substituted to the smaller flame (solid line) for the measurement of a.

Finally, it is noticed that the line width γ, and therefore a, depend on temperature (see part II). It is therefore necessary, if one wishes to calculate the transmittance T (Eq. 12) through non-isothermal flames, to introduce an averaging procedure (such as the "Curtis Godson approximation") which is introduced in part II. Chapter 4 of the Handbook [2] gives a careful evaluation of this procedure.

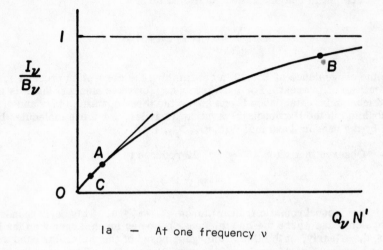

Ia   —   At one frequency ν

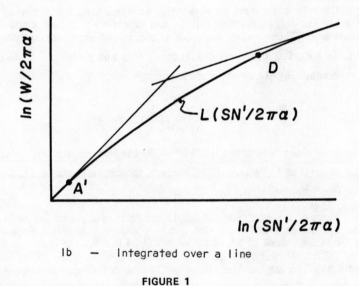

Ib   —   Integrated over a line

FIGURE 1

NOTE: Figure captions for this chapter are listed on page 498.

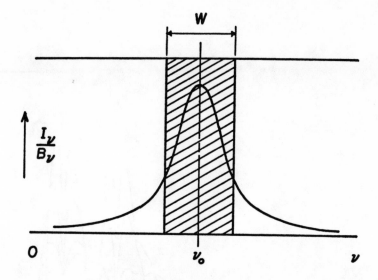

FIGURE 2a

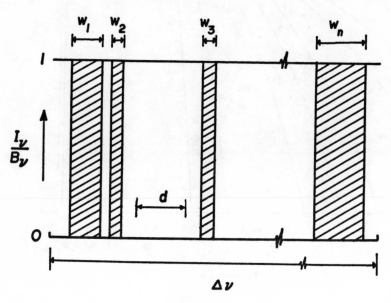

FIGURE 2b

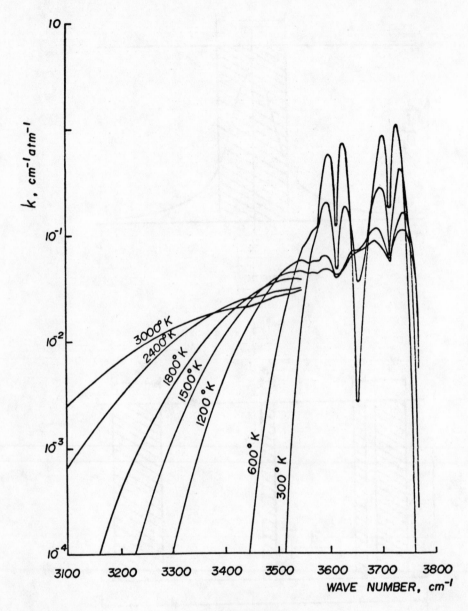

FIGURE 3

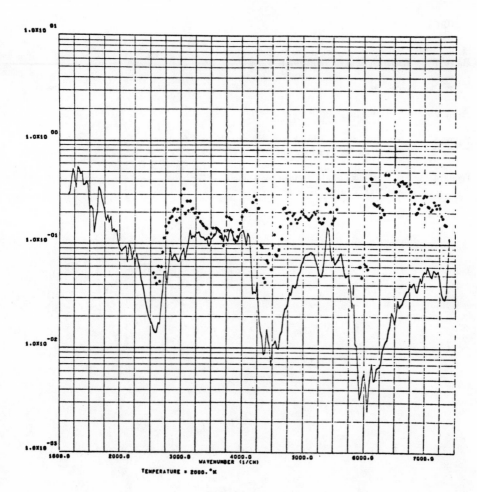

FIGURE 4

FIGURE CAPTIONS

Figure 1   Radiation from an Isothermal slab

Figure 2a   Single lines

Figure 2b   Group of lines

Figure 3   The absorption coefficient k (averaged over 25 cm^{-1} frequency
               intervals) for the 2.7 μ band of CO_2

Figure 4   The gappiness a as a function of wave number $\omega \equiv \dfrac{\nu}{c}$
               for H_2O at 2000°K

Chapter 2

INTRODUCTION TO THE USE OF
THE NASA HANDBOOK SP-3080[*]

R. GOULARD
Purdue University, USA

(This chapter is a reproduction of the illustrative example given in the General Appendix of the Handbook [2], pp. 377-383.)

Several radiation calculation methods have been discussed in the handbook. Some were very crude such as the thin gas or box models discussed in Section 3.1 and some were accurate but lengthy such as line by line calculations.

The main purpose of the handbook, however, is to provide a reasonably accurate representation of radiation fluxes without producing a computer overload. The two procedures outlined in Section 5.3 constitute, from the experience of the authors, an effective compromise. The multiple line group model (MLG) provides the possibility to consider n possible line groups, reflecting better the dependence of the band parameters on temperature. The single line group model (SLG) is usually satisfactory for relatively low temperature flames.

The calculation procedures for MLG and SLG (Section 5.3) make use of a number of properties and refer to this General Appendix for the corresponding

[*] This Handbook [2] provides analytical methods and necessary data to enable the engineer to make estimates of the spectral distribution of radiation from inhomogeneous multiconstituent gases. It includes a review of: (1) basic principles of gaseous radiation; (2) calculational techniques for homogeneous and inhomogeneous gas mixtures; (3) representations of the radiation of specific molecules; (4) accuracy of the calculational models; (5) experience with using the predictive techniques; and (6) tabulations of radiation data for several species. The band model prediction technique is emphasized and data are furnished based on the use of a random band model with exponential line strength distribution. The data are furnished in most cases for spectral intervals of 25 cm^{-1} at temperatures in the range of 300K to 3000K. Gases treated include H_2O, CO_2, CO, HF, HCl, CN, OH, and NO.

NOTE: All figures quoted in text are at the end of the chapter.

tables. These tables are given in Section A2. The rest of Section A1 will be devoted to a typical illustration of the procedures recommended in Section 5.3.

A1.1   Typical Calculation

Nonhomogeneous gases typical of flames have been studied experimentally by Simmons [4]. The example calculation will estimate the radiative flux emerging in the frequency range of the 2.7μ band from the left side of an inhomogeneous CO_2 gas sample at standard pressure conditions (p = 760 mm Hg) (see Fig. 5). The temperature distribution across the 60 cm path is given at 10 equidistant points (including both extremities) [4].

386, 528, 719, 953, 1130, 1160, 979, 737, 541, 387 in K

A1.1.1   Calculation Procedure

The emerging radiance* for a particular wavenumber ω can be written in general:

$$N_\omega = - \int_0^L N_\omega^0 (T) \frac{d}{d\ell} (\tau_\omega) d\ell \tag{A-1}$$

where the transmittance τ_ω is given by

$$\tau_\omega = \exp \left(- \sum_i X_{\omega,i}\right) \tag{A-2}$$

where i indicates the number of chemical components contributing to the flame radiation. Equation (A-1) is usually transformed into a numerical summation form such that

$$N_\omega = - \sum_{m=1}^{m=M} N_\omega (T_m) (\tau_{\omega,m} - \tau_{\omega,m-1}) \tag{A-3}$$

there the index m refers to the sections across the flame. If all sections are of equal width, Δl, then the total width L = M(Δl).

Whether one uses equation (A-3) or a more sophisticated integration procedure for equation (A-1), the two key parameters of the calculation are the Planck function $N_\omega^0(T)$ and the transmittance τ_ω. The Planck function is tabulated in Section A2.1 and the transmittance is given by equation (A-2).

*In the Handbook terminology (and in this part II), the radiance N and wavenumber ω replace systematically the intensity I and the frequency ν; N and I are identical, whereas the wave number ω is given in terms of ν by:

$$\omega \equiv \frac{\nu}{c} = \lambda^{-1}$$

where c is the speed of light in vacuo.
Also the symbol τ is used to designate the transmittance T (Eq. A-2).

The case under consideration involves a single chemical component (CO_2) so $i = 1$. Also, the temperature ratio across the flame is relatively low, so the single line group (SLG) model will be used. For the SLG model, the optical depth through each position extending from the observation side ($u = 0$, $m = 1$) to a given pathlength u (or section M) inside the flame[1] is

$$X = f(X^*, \bar{a}_c, \bar{a}_D) \tag{A-4}$$

where

$$X^* = \int_0^u k\, d' \quad \text{or} \quad X^* = \sum_{m'=1}^{m'=m} k_{m'}\, \Delta u_{m'} \tag{A-5}$$

$$\bar{a}_c = \frac{1}{X^*} \int_0^u \frac{\gamma_c}{d} k\, du' \quad \text{or} \quad \bar{a}_c = \frac{1}{X^*} \sum_{m'=1}^{m'=m} \left(\frac{\gamma_c}{d} k\right)_{m'} \Delta u_{m'} \tag{A-6}$$

$$\bar{a}_D = \frac{1}{X^*} \int_0^u \frac{\gamma_D}{d} k\, du' \quad \text{or} \quad \bar{a}_D = \frac{1}{X^*} \sum_{m'=1}^{m'=m} \left(\frac{\gamma_D}{d} k\right)_{m'} \Delta u_{m'} \tag{A-7}$$

The values of X^*, a_c and a_D will be necessary at each position ($0 < u' < u; 1 < m < < M$) in order to calculate N_ω (equations (A-1) or (A-3)). The functional form of f, the "curve of growth," is given by equation (5-25):

$$f(X^*, a_c, a_D) = X^* \left(1 - y^{-1/2}\right)^{1/2} \tag{A-8}$$

where

$$y = \left[1 - \left(\frac{X_c}{X^*}\right)^2\right]^{-2} + \left[1 - \left(\frac{X_D}{X^*}\right)^2\right]^{-2} - 1 \tag{A-9}$$

Finally, the values of X_c and X_D are given by:

$$X_c = X \left(1 + \frac{X^*}{4\bar{a}_c}\right)^{-1/2} \tag{A-10}$$

$$X_D = 1.7 a_D \left\{\ln\left[1 + \left(0.589\, \frac{X^*}{\bar{a}_D}\right)^2\right]\right\}^{1/2} \tag{A-11}$$

[1]The absorption coefficient tabulated in Table A-2 corresponds to standard density conditions. Therefore, the equivalent path length (Δu) over an assumed isothermal interval is obtained by multiplying the physical pathlength Δl by $\left(\frac{273}{T, K}\right) \left(\frac{P}{1\text{ atm}}\right)$.

The information needed for the solution of equations (A-4) through (A-11) includes the temperatures, pressures, and the mole fractions of the radiating species in addition to the four band model parameters:

absorption coefficient, k
line density, 1/d,
collision half-width, γ_C,
Doppler half-width, γ_D.

The absorption coefficient k and the line density 1/d can be found in tables of Section A2. A short index, at the beginning of the tables, points out the appropriate page numbers for each substance. The collision and Doppler half-widths vary with ambient conditions as expressed by:

$$\gamma_{c_i} = \left[\sum_j (\gamma_{ij})_{273} \, P_j \left(\frac{273}{T}\right)^{\eta_{ij}} \right] + (\gamma_{ii}^*)_{273} \, P_i \left(\frac{273}{T}\right)^{\eta_{ii}^*} \tag{A-12}$$

and

$$\gamma_D = (5.94 \times 10^{-6}) \frac{\omega}{M^{1/2}} \left(\frac{T}{273}\right)^{1/2}. \tag{A-13}$$

where i refers to the radiating component under study and j to the other components of the sample; p is the partial pressure; M is the molecular weight; and the values of $(\gamma_{ij}^*)_{273}$, $(\gamma_{ii}^*)_{273}$, η_{ij} and η_{ii}^* are given in Table 5-19 of Reference 2.

A1.1.2 Example Calculation

The formulas listed in equations (A-1) through (A-13) can now be applied to the inhomogeneous gas sample described at the beginning of this section (A1.1). For purposes of illustration, the sample calculation will be made using 10 equal increments of 6 cm along the path through the gas. A linear temperature variation will be assumed between the measured temperature points, and the temperature for each 6 cm section will be taken as the temperature at the center of the section. The resulting pathlength increment adjusted to STP conditions is

$$\Delta u = 6 \, cm \left(\frac{273}{T}\right) (1 \, atm)$$

Temperatures and Δu increments are listed in Table A1-1 along with radiation calculations for 4 wavenumbers: 3400, 3450, 3500, and 3600 cm^{-1}. Values of k (STP) and 1/d were obtained from the tables of Section A2 of the Handbook [2], using linear interpolation on temperature.

Since the gas sample is pure CO_2, the collision broadened half-width is evaulated using coefficients from Table 5-19 for CO_2 self broadening only: $\gamma_{ij} = 0.09$, $\eta_{ij} = 0.5$, $\gamma_{ii}^* = 0.1$, and $\eta_{ii}^* = 1.0$. Using these coefficients and one atmosphere partial pressure, equation (A-12) reduced to

$$\gamma_c = 0.09 \, (273/T)^{1/2} + 0.01 \, (273/T).$$

The Doppler half-width for CO_2 is obtained from equation (A-13) using the molecular weight of 44

$$\gamma_D = 0.895 \ (10^{-6}) \omega \ (T/273)^{1/2}$$

Values of γ_C and γ_D for each section of the gas are given in Table A1-1.

The fine structure parameters are averaged using equations (A-6) and (A-7) to give effective values of a_C and a_D from the edge of the gas through each successive increment. These averaged values are then used in equations (A-10) and (A-11) to estimate the optical depth at the end of each increment in the gas. These results are presented in Table A1-1.

Examination of the values of a_C and a_D indicates that a_C is generally an order of magnitude larger than a_D. As a result, when the optical depths become less than the linear limit (X*), the optical depth due to Doppler broadening becomes much less than that for collision broadening. In cases such as this, Doppler broadening can be neglected and the calculation procedure significantly shortened. However, to illustrate the procedure equations (A-8) and (A-9) were used in the sample problem to compute the combined optical depth X. Using this combined optical depth, the transmissivity was computed using equation (A-2) (i = 1) and the radiance was calculated by equation (A-3) for m = 1 to 10. The results are listed in Table A1-1 and the overall radiance (m = 10) agrees very well with the measured values of Figure 6.

The excellent agreement between the measured data and the sample problem is better than should be expected considering the simplification made in the solution. In a high pressure gas such as the example, the transmissivity changes too much in some of the 6 cm (Δu) increments, so a smaller increment should be used if the summations are to be a reasonable representation of the integrals. In addition, the interpolation of k and 1/d would probably be better represented by a curve fit or interpolation based on the logarithm of k or 1/d.

TABLE AI-I - Part I

Incre-ment	ω (cm⁻¹)	T (K)	Δu(STP) (cm-atm)	k(STP) (cm⁻¹-atm⁻¹)	1/d (cm)	γ_{C-1} (cm⁻¹)	γ_{D-1} (cm⁻¹)	X*	$\bar{a}_c$	$\bar{a}_D$
1	3400	440	3.723	4.305E-7	5.81	0.0771	0.00386	1.603E-6	0.4479	0.02243
	3450			1.695E-4	7.76		0.00392	6.310E-4	0.5983	0.008171
	3500			0.008402	9.38		0.00398	0.03128	0.7231	0.01003
	3600			0.5368	2.27		0.00409	1.999	0.1750	0.002911
2	3400	595	2.753	9.071E-7	10.5	0.0656	0.00449	4.100E-6	0.5946	0.03749
	3450			3.571E-4	14.3		0.00456	0.001614	0.8055	0.4004
	3500			0.1769	17.1		0.00462	0.07998	0.9659	0.05203
	3600			0.4985	3.61		0.00476	3.371	0.2002	0.008493
3	3400	777	2.108	0.006207	72.1	0.0569	0.00513	0.01308	4.102	0.3699
	3450			0.02216	51.0		0.00521	0.04832	2.831	0.2703
	3500			0.06746	33.0		0.00528	0.2222	1.550	0.1302
	3600			0.3832	15.3		0.00543	4.179	0.3297	0.02291
4	3400	979	1.673	0.01329	142.0	0.0503	0.00576	0.03531	6.018	0.6519
	3450			0.04703	92.7		0.00584	0.1270	3.966	0.4382
	3500			0.1239	50.8		0.00593	0.4295	2.035	0.2128
	3600			0.3349	28.7		0.00610	4.739	0.4615	0.04092
5	3400	1131	1.448	0.1862	195.	0.0466	0.00619	0.06227	7.347	0.8922
	3450			0.06574	124.		0.00628	0.2222	4.743	0.5842
	3500			0.1664	64.2		0.00637	0.6704	2.379	0.2833
	3600			0.2986	38.7		0.00656	5.172	0.5737	0.05872
6	3400	1159	1.413	0.01960	205.	0.0460	0.00627	0.08996	7.988	1.013
	3450			0.06919	130.		0.00636	0.3200	5.119	0.6581
	3500			0.1742	66.7		0.00645	0.9165	2.564	0.3226
	3600			0.2919	40.6		0.00664	5.584	0.6692	0.07430
7	3400	1006	1.628	0.01424	152.	0.0496	0.00584	0.1131	7.899	0.9876
	3450			0.05035	98.3		0.00593	0.4019	5.071	0.6429
	3500			0.1315	53.2		0.00601	1.131	2.577	0.3219
	3600			0.3285	30.5		0.00618	6.119	0.7429	0.08428
8	3400	798	2.053	0.006944	79.4	0.0561	0.00520	0.1274	7.511	0.9231
	3450			0.02474	55.3		0.00528	0.4527	4.850	0.6035
	3500			0.07333	34.8		0.00536	1.281	2.504	0.3062
	3600			0.3782	16.7		0.00551	6.895	0.7649	0.08515
9	3400	610	2.685	3.516E-4	14.2	0.0647	0.00455	0.1283	7.465	0.9213
	3450			0.001594	16.6		0.00461	0.4570	4.814	0.5987
	3500			0.02079	18.2		0.00468	1.337	2.449	0.2969
	3600			0.4231	4.31		0.00481	8.031	0.6962	0.07603
10	3400	456	3.592	4.797E-7	6.30	0.0756	0.00393	0.1283	7.465	0.9213
	3450			1.889E-4	8.43		0.00399	0.4575	4.811	0.5980
	3500			0.009361	10.2		0.00405	1.362	2.419	0.2922
	3600			0.5257	2.41		0.00416	9.443	0.6193	0.06617

TABLE AI-I   -   Part II

Incre-ment	ω (cm^{-1})	X_c	X_D	X	τ	$-\Delta\tau$	$N^o_\omega \times 10^6$ (watts/cm-st)	$\Sigma N^o_\omega (-\Delta\tau)$ (watts/cm-st)
1	3400	1.603E-6	1.603E-6	1.603E-6	1.0000	0	-	0.00
	3450	6.310E-4	6.310E-4	6.310E-4	0.9994	0.0006	0.64	0.00
	3500	0.03111	0.0207	0.03111	0.9694	0.0306	0.55	0.02
	3600	1.018	0.0172	1.018	0.2765	0.7235	0.43	0.31
2	3400	4.100E-6	4.100E-6	4.100E-6	1.0000	0	-	0.00
	3450	0.001614	0.001614	0.001614	0.9984	0.0010	11.65	0.01
	3500	0.07916	0.06834	0.07916	0.9249	0.0445	10.78	0.50
	3600	1.477	0.04204	1.481	0.2270	0.0495	9.21	0.77
3	3400	0.01307	0.01307	0.01307	0.9870	0.0130	86.45	1.12
	3450	0.04822	0.04819	0.04824	0.9529	0.0455	82.32	3.76
	3500	0.2183	0.1849	0.2183	0.8075	0.1174	78.34	9.69
	3600	2.047	0.1034	2.048	0.1285	0.0985	70.82	7.74
4	3400	0.03528	0.03531	0.03531	0.9653	0.0217	318.48	8.04
	3450	0.1265	0.1261	0.1266	0.8810	0.0719	309.02	25.98
	3500	0.4186	0.3391	0.4187	0.6580	0.1495	299.65	54.49
	3600	2.509	0.2022	2.512	0.0812	0.0473	281.29	21.05
5	3400	0.06220	0.06226	0.06226	0.9396	0.0257	627.42	24.16
	3450	0.2209	0.2198	0.2211	0.8017	0.0793	614.62	74.72
	3500	0.6480	0.5000	0.6482	0.5235	0.1345	601.71	135.42
	3600	2.867	0.2806	2.872	0.0568	0.0244	575.74	35.10
6	3400	0.08983	0.08996	0.08996	0.9139	0.0257	697.57	42.08
	3450	0.3175	0.3137	0.3177	0.7280	0.0737	684.32	125.15
	3500	0.8781	0.6330	0.8784	0.4160	0.1075	670.95	207.55
	3600	3.179	0.3478	3.185	0.0412	0.0156	634.87	45.00
7	3400	0.1129	0.1131	0.1131	0.8930	0.0209	364.53	49.70
	3450	0.3980	0.3894	0.3982	0.6715	0.0565	354.38	145.17
	3500	1.074	0.7053	1.074	0.3415	0.0745	344.31	233.20
	3600	3.499	0.3926	3.506	0.0300	0.0112	334.22	48.74
8	3400	0.1271	0.1271	0.1272	0.8808	0.0122	102.07	50.95
	3450	0.4475	0.4320	0.4477	0.6390	0.0325	97.42	148.34
	3500	1.206	0.7270	1.207	0.3000	0.0415	92.93	237.06
	3600	3.823	0.4052	3.830	0.0216	0.0084	84.42	49.45
9	3400	0.1280	0.1281	0.1281	0.8797	0.0011	15.40	50.97
	3450	0.4517	0.4361	0.4519	0.6360	0.0030	13.74	148.38
	3500	1.254	0.7280	1.255	0.2850	0.0150	13.27	237.26
	3600	4.075	0.3715	4.082	0.0170	0.0046	11.41	49.50
10	3400	0.1280	0.1281	0.1281	0.8797	0	-	50.97
	3450	0.4522	0.4372	0.4523	0.6359	0.0001	0.93	148.38
	3500	1.275	0.7271	1.276	0.2795	0.0055	0.82	237.26
	3600	4.305	0.3348	4.311	0.0134	0.0036	0.65	49.51

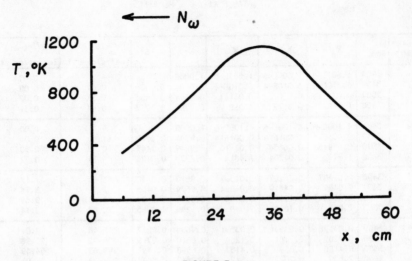

FIGURE 5

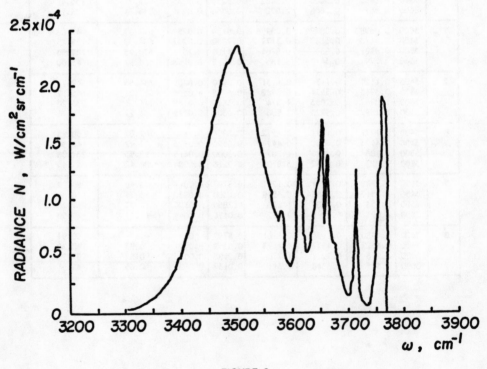

FIGURE 6

NOTE: Figure captions for this chapter are listed on page 507.

FIGURE CAPTIONS

Figure 5   Measured temperature profile of a CO_2 flame (Simmons – Ref. 4)

Figure 6   Measured outgoing radiance N_ω of the CO_2 flame shown on Fig. 5

Chapter 3

CARBON PARTICLE RADIATION

R. GOULARD
Purdue University, USA

Finally it should be stressed that the methods developed in the Handbook were conceived primarily in an aerospace context, where gaseous combustion products are predominant. Thus the problem of carbon particle radiation, although considered in the handbook (notably in paragraphs 5.4 and 8.2), is limited to the small size particles ($D \leq 0.4\mu$) usually found in rocket exhausts. In such cases, there exists a stronger dependence on λ (see Fig. 7 for instance) than if the average particle diameter was closer to the radiation wavelength (Mie theory). This latter situation is more characteristic [5] of furnace flames and oil fires which are likely to contain much larger soot particles. Therefore, the relationships proposed in the handbook for carbon particle radiation should be used with the utmost caution. The gaseous band models, of course, are reliable in all cases.

Granted this distinction, the rocket exhaust calculations shown on Figs. 7 and 8 exhibit some features which are relevant to fire combustion also. Figure 7 illustrates the radiance of a hydrocarbon-oxygen plume (H-1 Engine) at low altitude. Fuel-rich exhaust (about 1 ii mass fraction of carbon) mix with the low altitude (dense) airstream and continue to burn in a hot "after burning" layer around the jet. The spectral radiance shown on Fig. 7 (Fig. 8.3 in the Handbook) includes the continuum carbon particle radiation from the core of the plume with some absorption from the cooler gases surrounding the core (CO_2 4.3μ band). In this case the sophistication of the random band models — "gappiness" parameter a included — is clearly unnecessary: Eq. 15 of these notes would be adequate.

Figure 8 illustrates the radiance from the same engine (H-1) at high altitude, where the surrounding cold air is not dense enough to support an after burning layer around the jet nor to absorb in the CO_2 bands as it did on Fig. 7. In this case, gaseous radiation from the hot core contributes a significant addition to the continuum carbon radiation. Note, moreover, that a calculation based on a

NOTE: All figures quoted in text are at the end of the chapter.

single parameter k (Eq. 15), would lead to an average over prediction of 20 ii. Hence, in this case, the need to use the more sophisticated model proposed in the Handbook.

In summary, carbon particle content tends to dominate flame radiation whenever it is present. In the case of clean flames, or if exact spectral information needed, the information provided on gaseous radiators in the Handbook should yield accurate predictions.

REFERENCES

1. R.M. Goody, "Atmospheric Radiation — Theoretical Basis," Oxford Press, 1964.
2. C.B. Ludwig, W. Malkmus, J.E. Reardon and J.A.L. Thomson, "Handbook of Infra Red Radiation from Combustion Gases," NASA SP-3080, to appear ca. August, 1973.
3. C.B. Ludwig and W. Malkmus, "Band Model Representations for High Temperature Water Vapor and Carbon Dioxide," from the Proceedings of the Specialists' Conference on Molecular Radiation (R. Goulard, Ed.), NASA TM X-53711, October 1967.
4. F.S. Simmons, H.Y. Yamada, C.B. Arnold — NASA CR-72491, April, 1969.
5. H.C. Hottel and A.F. Sarofim, "Radiative Transfer," McGraw Hill, 1967.

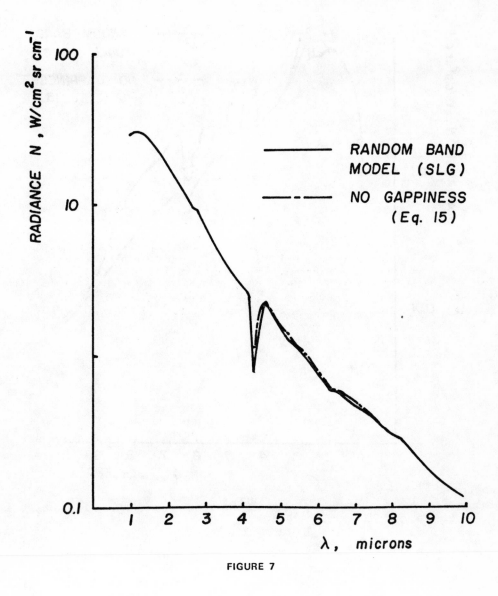

NOTE: Figure captions for this chapter are listed on page 513.

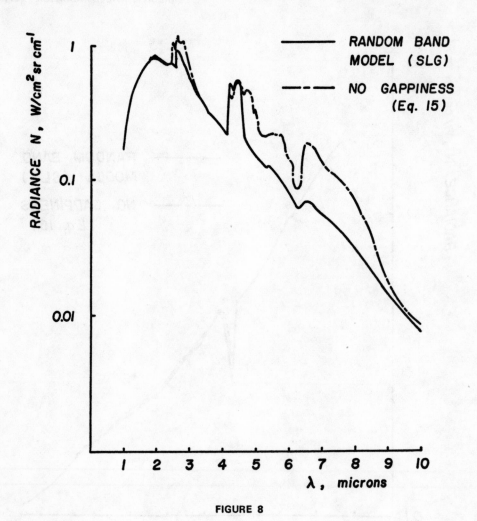

FIGURE 8

FIGURE CAPTIONS

Figure 7   H-1 Engine low altitude radiance N as a function of wavelength λ

Figure 8   H-1 Engine high altitude radiance N as a function of wavelength λ

INDEX